Lecture Notes in Physics

For information about Vols. 1–131, please contact your bookseller or Springer-Verlag.

Vol. 132: Systems Far from Equilibrium. Proceedings, 1980. Edited by L. Garrido. XV, 403 pages. 1980.

Vol. 133: Narrow Gap Semiconductors Physics and Applications. Proceedings, 1979. Edited by W. Zawadzki. X, 572 pages. 1980.

Vol. 134: γγ Collisions. Proceedings, 1980. Edited by G. Cochard and P. Kessler. XIII, 400 pages. 1980.

Vol. 135: Group Theoretical Methods in Physics. Proceedings, 1980. Edited by K. B. Wolf. XXVI, 629 pages. 1980.

Vol. 136: The Role of Coherent Structures in Modelling Turbulence and Mixing. Proceedings 1980. Edited by J. Jimenez. XIII, 393 pages. 1981.

Vol. 137: From Collective States to Quarks in Nuclei. Edited by H. Arenhövel and A. M. Saruis. VII, 414 pages. 1981.

Vol. 138: The Many-Body Problem. Proceedings 1980. Edited by R. Guardiola and J. Ros. V, 374 pages. 1981.

Vol. 139: H. D. Doebner, Differential Geometric Methods in Mathematical Physics. Proceedings 1981. VII, 329 pages. 1981.

Vol. 140: P. Kramer, M. Saraceno, Geometry of the Time-Dependent Variational Principle in Quantum Mechanics. IV, 98 pages. 1981.

Vol. 141: Seventh International Conference on Numerical Methods in Fluid Dynamics. Proceedings. Edited by W. C. Reynolds and R. W. MacCormack. VIII, 485 pages. 1981.

Vol. 142: Recent Progress in Many-Body Theories. Proceedings. Edited by J. G. Zabolitzky, M. de Llano, M. Fortes and J. W. Clark. VIII, 479 pages. 1981.

Vol. 143: Present Status and Aims of Quantum Electrodynamics. Proceedings, 1980. Edited by G. Gräff, E. Klempt and G. Werth. VI, 302 pages. 1981.

Vol. 144: Topics in Nuclear Physics I. A Comprehensive Review of Recent Developments. Edited by T.T.S. Kuo and S.S.M. Wong. XX, 567 pages. 1981.

Vol. 145: Topics in Nuclear Physics II. A Comprehensive Review of Recent Developments. Proceedings 1980/81. Edited by T. T. S. Kuo and S. S. M. Wong. VIII, 571-1.082 pages. 1981.

Vol. 146: B. J. West, On the Simpler Aspects of Nonlinear Fluctuating. Deep Gravity Waves. VI, 341 pages. 1981.

Vol. 147: J. Messer, Temperature Dependent Thomas-Fermi Theory. IX, 131 pages. 1981.

Vol. 148: Advances in Fluid Mechanics. Proceedings, 1980. Edited by E. Krause. VII, 361 pages. 1981.

Vol. 149: Disordered Systems and Localization. Proceedings, 1981. Edited by C. Castellani, C. Castro, and L. Peliti. XII, 308 pages. 1981.

Vol. 150: N. Straumann, Allgemeine Relativitätstheorie und relativistische Astrophysik. VII, 418 Seiten. 1981.

Vol. 151: Integrable Quantum Field Theory. Proceedings, 1981. Edited by J. Hietarinta and C. Montonen. V, 251 pages. 1982.

Vol. 152: Physics of Narrow Gap Semiconductors. Proceedings, 1981. Edited by E. Gornik, H. Heinrich and L. Palmetshofer. XIII, 485 pages. 1982.

Vol. 153: Mathematical Problems in Theoretical Physics. Proceedings, 1981. Edited by R. Schrader, R. Seiler, and D.A. Uhlenbrock. XII, 429 pages. 1982.

Vol. 154: Macroscopic Properties of Disordered Media. Proceedings, 1981. Edited by R. Burridge, S. Childress, and G. Papanicolaou. VII, 307 pages. 1982.

Vol. 155: Quantum Optics. Proceedings, 1981. Edited by C.A. Engelbrecht. VIII, 329 pages. 1982.

Vol. 156: Resonances in Heavy Ion Reactions. Proceedings, 1981. Edited by K.A. Eberhard. XII, 448 pages. 1982.

Vol. 157: P. Niyogi, Integral Equation Method in Transonic Flow. XI, 189 pages. 1982.

Vol. 158: Dynamics of Nuclear Fission and Related Collective Phenomena. Proceedings, 1981. Edited by P. David, T. Mayer-Kuckuk, and A. van der Woude. X, 462 pages. 1982.

Vol. 159: E. Seiler, Gauge Theories as a Problem of Constructive Quantum Field Theory and Statistical Mechanics. V, 192 pages. 1982.

Vol. 160: Unified Theories of Elementary Particles. Critical Assessment and Prospects. Proceedings, 1981. Edited by P. Breitenlohner and H.P. Dürr. VI, 217 pages. 1982.

Vol. 161: Interacting Bosons in Nuclei. Proceedings, 1981. Edited by J.S. Dehesa, J.M.G. Gomez, and J. Ros. V, 209 pages. 1982.

Vol. 162: Relativistic Action at a Distance: Classical and Quantum Aspects. Proceedings, 1981. Edited by J. Llosa. X, 263 pages. 1982.

Vol. 163: J. S. Darrozes, C. Francois, Mécanique des Fluides Incompressibles. XIX, 459 pages. 1982.

Vol. 164: Stability of Thermodynamic Systems. Proceedings, 1981. Edited by J. Casas-Vázquez and G. Lebon. VII, 321 pages. 1982.

Vol. 165: N. Mukunda, H. van Dam, L.C. Biedenharn, Relativistic Models of Extended Hadrons Obeying a Mass-Spin Trajectory Constraint. Edited by A. Böhm and J.D. Dollard. VI, 163 pages. 1982.

Vol. 166: Computer Simulation of Solids. Edited by C.R.A. Catlow and W.C. Mackrodt. XII, 320 pages. 1982.

Vol. 167: G. Fieck, Symmetry of Polycentric Systems. VI, 137 pages, 1982.

Vol. 168: Heavy-Ion Collisions. Proceedings, 1982. Edited by G. Madurga and M. Lozano. VI, 429 pages. 1982.

Vol. 169: K. Sundermeyer, Constrained Dynamics. IV, 318 pages. 1982.

Vol. 170: Eighth International Conference on Numerical Methods in Fluid Dynamics. Proceedings, 1982. Edited by E. Krause. X, 569 pages. 1982.

Vol. 171: Time-Dependent Hartree-Fock and Beyond. Proceedings, 1982. Edited by K. Goeke and P.-G. Reinhard. VIII, 426 pages. 1982.

Vol. 172: Ionic Liquids, Molten Salts and Polyelectrolytes. Proceedings, 1982. Edited by K.-H. Bennemann, F. Brouers, and D. Quitmann. VII, 253 pages. 1982.

Lecture Notes in Physics

206

Static Critical Phenomena in Inhomogeneous Systems

Proceedings of the XX Karpacz Winter School
of Theoretical Physics, February 20–March 3, 1984
Karpacz, Poland

Edited by A. Pękalski and J. Sznajd

Springer-Verlag
Berlin Heidelberg GmbH 1984

Editors

Andrzej Pękalski
Institute of Theoretical Physics, University of Wrocław
Cybulskiego 36, Wrocław, Poland

Józef Sznajd
Institute of Low Temperatures and Structural Research
pl. Katedralny 1
Wrocław, Poland

The XX Karpacz Winter School of Theoretical Physics
was organized by the Institute of Theoretical Physics
of the University of Wrocław

ISBN 978-3-540-13369-8 ISBN 978-3-540-38925-5 (eBook)
DOI 10.1007/978-3-540-38925-5

Originally published by Springer-Verlag Berlin Heidelberg New York Tokyo in 1984

2153/3140-543210

Lecture Notes in Physics

206

Static Critical Phenomena in Inhomogeneous Systems

Proceedings of the XX Karpacz Winter School
of Theoretical Physics, February 20–March 3, 1984
Karpacz, Poland

Edited by A. Pękalski and J. Sznajd

Springer-Verlag
Berlin Heidelberg GmbH 1984

Editors

Andrzej Pękalski
Institute of Theoretical Physics, University of Wrocław
Cybulskiego 36, Wrocław, Poland

Józef Sznajd
Institute of Low Temperatures and Structural Research
pl. Katedralny 1
Wrocław, Poland

The XX Karpacz Winter School of Theoretical Physics
was organized by the Institute of Theoretical Physics
of the University of Wrocław

ISBN 978-3-540-13369-8 ISBN 978-3-540-38925-5 (eBook)
DOI 10.1007/978-3-540-38925-5

Originally published by Springer-Verlag Berlin Heidelberg New York Tokyo in 1984

2153/3140-543210

PREFACE

The 1984 Karpacz Winter School of Theoretical Physics was the twentieth organised by the Institute of Theoretical Physics of the Wroclaw University. Traditionally, every third School is devoted to problems in statistical physics or solid state theory, while the other two deal with high energy or mathematical physics. This reflects the scientific interests at our Institute. The present School was devoted to static critical phenomena in inhomogeneous systems, with the main emphasis on magnetic models. It had the character of a course rather than a conference. Hence the approach in most of the lectures was tutorial in order to introduce the audience, consisting mostly of young research workers, to particular topics, which can be roughly divided into six groups. In the first one, general foundations for theoretical description of disordered systems close to critical points were given (lectures by M. Droz, R.B. Stinchcombe, J.M.J. Van Leeuwen). Spin glasses and frustration were reviewed by M. Droz, J. Zittartz, W. Kinzel, and A. Sütö. Critical behaviour of different random systems was discussed in the lectures by L. Sasvari, G. Sobotta, A. Sütö, and A. Pękalski. The presence of a surface brings inhomogeneity into the model, and such systems have been described by T. Burkhardt, T. Nattermann, W. Selke, and M. Ausloos. C. di Castro and W. Weller spoke about localisation problems and metal-insulator transitions. Finally, J. Sivardière and D.A. Huckaby showed us how useful the concept of the lattice gas can be.

We hope that the introductory character of most of the lectures will make this book a useful reference for all those wishing to enter into the fascinating and fast-growing field of phase transitions in disordered or inhomogeneous systems.

Regretfully, because of personal problems some of the invited lecturers from Poland and abroad were unable to come. This resulted in the omission of certain topics, like percolation.

In addition to the invited talks we decided, following the suggestion of Prof. J. Zittartz, to include also the contributed talks in the proceedings; in them more up-to-date and specific problems have been presented. However, we have decided not to include the discussions following the lectures, since this would delay the publishing of the proceedings.

We express our deep gratitude to the lecturers who came to Karpacz despite different obstacles and the usual lack of time. We hope that their presence and readiness to discuss physics, sometimes during long hours, helped to create a warm and informal atmosphere at the School. Even beautiful weather and good skiing conditions were not able to divert participants from attending the lectures.

The duties of organising the School have been distributed among our friends and colleagues from the Institute of Theoretical Physics of the University of Wroclaw: A. Borowiec, M. Dudek, Cz. Oleksy, D. Skrajna, J. Stelmach, and Z. Strycharski. Their help and eagerness in solving numerous organisational problems is gratefully acknowledged. We are also thankful to M. Ausloos for his active aid in contacting other lecturers.

Finally, it should be mentioned that it would have been impossible to organise the School without help and financial support from the University of Wroclaw.

Wroclaw, May 1984

A. Pękalski

J. Sznajd

C O N T E N T S

IV. SURFACE EFFECTS

V. LOCALISATION

VI. LATTICE GAS

CONTRIBUTED PAPERS

*) This paper arrived when the volume was already in press and due to technical reasons could not be included in the appropriate Section IV "Surface Effects".

*) This paper arrived when the volume was already in press and due to technical reasons could not be included in the appropriate Section IV "Surface Effects".

I. FOUNDATIONS

INTRODUCTION TO THE PHYSICS OF DISORDERED SYSTEMS

Michel Droz
Département de Physique Théorique
Université de Genève
1211 Genève 4, Switzerland

The aim of this introduction is to offer to the newcomer in this field a general overview of the aspects of the physics of disordered systems which will be discussed during this Winter School. This is not an exhaustive coverage of the subject but rather a random walk in the physics of disordered systems, reviewing the key ideas and concepts which will be developped in detail in the following lectures. This introduction is organized as follows. In chapter 1, we consider the case without disorder. The most commonly used models and qualitative features of these models are discussed. In Chapter 2, we add the disorder into the preceding models. We discuss the concepts of quenched and annealed disorder. Four different classes of disordered systems are then considered : diluted spin systems, random field systems, spin-glasses and Anderson's localization problem. Some of the key concepts useful to describe the properties of these systems are introduced.

1. SYSTEMS WITHOUT DISORDER

Before considering the case of disordered systems, let us look at these systems without disorder. As we shall see, many qualitative aspects of the "pure" systems (i.e. without disorder) will remain valid for the disordered systems; many questions asked in the pure case will be asked in the disordered case. We consider here two classes of systems : the spin systems and the electron systems.

1.1. The spin systems

One usually considers regular d-dimensional lattices Λ (Fig. 1a). Each site $\vec{i}$ of Λ is occupied by a spin variable $\vec{s}_{\vec{i}}$. The spin variables are characterized by their number of components n (Ising spin for $n = 1$, XY-spin for $n = 2$, Heisenberg-spin for $n = 3,\ldots$). One can consider quantum spins or classical spins (limit $s \to \infty$ of the quantum case). The spins at different sites $\vec{i},\vec{j}$ interact via the interaction $J_{\vec{i}\vec{j}}$. These interactions can be short-range or long-range interactions, having an unique sign (ferromagnetic or antiferromagnetic) or an oscillatory character (e.g. RKKY interaction, see next lecture). The system is described by a Hamiltonian H. Typically one has

$$H = -\sum_{\vec{i},\vec{j}} J_{\vec{i},\vec{j}} \vec{s}_{\vec{i}} \cdot \vec{s}_{\vec{j}} - \sum_{\vec{i}} \vec{h}_{\vec{i}} \cdot \vec{s}_{\vec{i}} \tag{1.1}$$

where the second term in (1.1) describes the coupling of the spins with an external field. However, more complicated Hamiltonians are also considered. Note also that the lattice is not always translationally invariant (as in Fig. 1a). For technical reasons it may be very useful to consider hierarchical lattices [1] or fractal lattices [2]. Examples are given respectively in Figs 1b and 1c.

The model being defined, let us look at the questions one is asking. Two types of properties can be studied : the static and the dynamic properties.

A. Static properties

The first question to answer is what is the phase diagram for a given model. What are the different phases as a function of the parameters describing the system and what are the orders of the transitions between those different phases [first order on second order (continuous) transition]. One may answer these questions qualitatively i.e. find a phase diagram topologically correct but incorrect as far as the details are concerned. Such a prediction can be usually obtained in the framework of mean-field like approximations. However, the mean-field like approximations can lead to qualitatively wrong predictions at low dimensions.

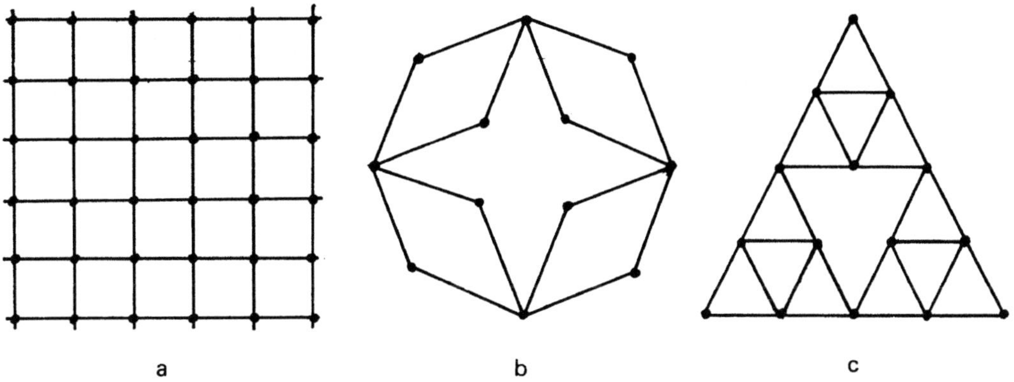

Fig. 1 : a) translationally invariant 2-dim regular lattice
b) hierarchical lattice
c) fractal lattice (Sierpinski gasket)

An important concept is the lower critical dimensionality d_L. If a given model has a phase transition at finite temperature $T_c(d)$ at high dimensions d, then $T_c(d)$ goes to zero at $d = d_L$ and for $d < d_L$ the system has no transition. It is thus important to know what is d_L in order to decide if the system can have or not a phase transition. The usual mean-field like approximations do not give d_L correctly and one has to go beyond mean-field to get more correct results. Once the phase diagram is known, one can ask more precise questions. Two temperature domains play a particular role :

i) Ground state and low temperatures

- What is the nature of the ground state ? its degeneracy ?
- What are the elementary excitations ?
- What is the temperature dependence of the interesting physical quantities at low T (specific heat, susceptibility,...) ?

ii) Vicinity of a phase transition

Very often a phase transition is associated with a broken symmetry. It is then suitable to describe the transition in terms of an order parameter ψ which is zero in the phase with the higher symmetry and non zero in the phase of lower symmetry. If the order parameter is discontinuous at the transition, the transition is said to be of first order. If the order parameter is continuous across the critical temperature T_c, the transition is said to be of second order. In this case, the physical quantities are singular in the reduced temperature $t = (T - T_c)/T_c$. The singularities are expressed by the standard critical exponents $\alpha, \beta, \gamma, \delta, \nu, \eta$ defined as [3]

$$
\begin{aligned}
& c \sim |t|^{-\alpha} \\
& \psi \sim |t|^{\beta} \qquad T < T_c \\
& \chi \sim |t|^{-\gamma} \\
& \psi(T_c) \sim h^{1/\delta} \\
& G(\vec{r}) = \langle \psi(\vec{o})\psi(\vec{r}) \rangle \sim \frac{1}{r^{d-2+\eta}} f(r/\xi) \\
& \xi \sim |t|^{-\nu}
\end{aligned}
\tag{1.2}
$$

where c is the specific heat, h the field conjugated to the order parameter, χ the susceptibility, ξ the correlation length and $G(\vec{r})$ the two-point correlation function of the order parameter. The natural questions to ask, given a model Hamiltonian are then :

- What is the order parameter ?
- What are the critical exponents ?

It turns out that the critical exponents depend upon a very restricted number of parameters such as the number of components of the order parameter, the dimensionality of the system and the range of the interactions. This allows to classify the second order transitions in a limited number of universality classes. All the models belonging to a same universality class have the same critical exponents. Physically, this means that a large portion of the information contained in the original Hamiltonian is irrelevant as far as the critical behaviour is concerned. This irrelevance is naturally explained in the framework of the renormalization group theory [4]. Each Hamiltonian belongs to the domain of attraction of a fixed point Hamiltonian which governs the critical behaviour. It is thus important to determine to which universality class a system belongs to.

Another useful concept is the one of upper critical dimensionality d_u. d_u is defined as the dimension above which the critical exponents given by a mean-field theory are correct. This means physically that above the upper critical dimensionality d_u, the fluctuations do not play a crucial role and can be often neglected [5]. Note that we have only considered here a simple critical point. The situation is more complicated if more than two phases are present. One has then multicritical points (see lecture of Dr. Savardière).

B. Dynamics

In this case we consider the non equilibrium aspects of the problem. The first step is to write the equations of motion for the quantities of interest. The model may not have an intrinsic dynamic and one has to write some phenomenological equations of motion (Glauber or Kawasaki dynamics [7]). Then, one studies the excitations of the system. The main concept is that, the system can be characterized by some typical frequencies $\omega_c(\vec{q})$, where $\vec{q}$ is the wave-vector. The dynamical scaling hypothesis asserts that [6]

$$\omega_c(q) = q^z\varphi(q\xi) = \xi^{-z}g(q\xi) \qquad (1.3)$$

where z is the dynamical exponent.

It follows for example that the characteristic frequency describing the relaxation of the order parameter goes to zero as ξ^{-z} when the correlation length ξ diverges. This slow relaxation of the order parameter close to T_c is called critical slowing down. The dynamics is much more complicated than the statics [7] and little will be said on dynamics in this School (see however Dr. Stinchcombe's lectures).

1.2. Electron systems

One considers the problem of independent electrons in a perfect crystal, i.e. in a periodic potential. One has to solve the Schrödinger equation. Working in the tight binding approximation, the one electron Hamiltonian is [8] :

$$H = \sum_{\vec{x}} V_o |\vec{x}><\vec{x}| + \sum_{<\vec{x},\vec{x}'>} t_o |\vec{x}><\vec{x}'| \qquad (1.4)$$

V_o is a local potential and t_o a hopping term between nearest-neighbour sites on the lattice. The relevant questions are then :

i) What are the eigenvalues and the density of states ?

ii) What are the properties of the associated eigenstates ?

It is relatively easy to show that :

a) There is a finite band of electronic states between $E_o(V_o,t_o)$ and $E_1(V_o,t_o)$.

b) All the states have spatial wave functions that extend throughout the crystal. The eigenstates are said to be extended.

2. DISORDERED SYSTEMS

We add some disorder in the pure systems considered above. In Section 2.1 we classify the different types of disorder for the spin systems. In Section 2.2 we discuss the concepts of quenched and annealed disorder. In Sections 2.3, 2.4, 2.5 we consider in more details the problems of diluted spin systems, random fields and spin-glasses. Finally in Section 2.6 we look at the Anderson's localization problem.

2.1. Disorder in spin systems

Three types of disordered spin systems are generally studied.

A. Diluted systems

Three situations may arise

1) Some bonds are missing; this is the bond disorder. The probability distribution $P(\{J_{ij}\})$ characterizes the disorder.
2) Some sites are empty; this is the site disorder. The probability distribution of the occupation variables $P(\{\varepsilon_i\})$ characterizes the disorder.
3) Both bond and site disorder can be present simultaneously.

B. Random field problem

The spins $\vec{S}_{\vec{i}}$ at each site $\vec{i}$ are submitted to random external fields $\vec{h}_i$, characterized by the probability distribution $P(\{\vec{h}_i\})$.

C. Spin-glasses

The coupling constants J_{ij} can be ferromagnetic like or antiferromagnetic like (with different amplitudes) leading to a competition between bonds. Moreover, the bonds J_{ij} are randomly distributed and characterized by the probability distribution $P(\{J_{ij}\})$.

Note that one can also consider systems in which the bonds are randomly distributed with different amplitudes but do not compete (see Dr. Zittartz's lecture). It is also interesting to consider the non random limit of such systems (see Dr. Burkhardt's lecture). To go further it is useful to say more on the types of disorder considered.

2.2. Quenched and annealed disorder

Two types of disorder can be present in disordered spin systems.

Let us illustrate this problem on a simple example, a site diluted Ising model. The Hamiltonian is :

$$H = -J \sum_{<\vec{i}\vec{j}>} (s_{\vec{i}}\varepsilon_{\vec{i}})(s_{\vec{j}}\varepsilon_{\vec{j}}) \qquad (2.1)$$

$\{\varepsilon_i\}$ are the site disorder variables, which can take the values 0 or 1. One asks that the configurational average of the ε_i , i.e. the probability that the site $\vec{i}$ is occupied is fixed : $<\varepsilon_i>_\varepsilon = x$. Two cases have to be distinguished.

A. Quenched disorder

This is the case where the configurational averages, given by a probability distribution $P(\{\varepsilon_i\})$, are independent of the thermal average (canonical average for example). Thus for any observable A, the quantity to compute is the so-called quenched average :

$$<<A>_T>_\varepsilon = \int \prod_i d\varepsilon_i P(\{\varepsilon_i\}) <A(\{\varepsilon_i\})>_T \qquad (2.2)$$

$$\text{where} \quad <A(\{\varepsilon_i\})>_T = \mathop{Tr}_{(s)} [A(s_i,\{\varepsilon_i\}) \exp{-\beta H}] / \mathop{Tr}_{(s)} [\exp{-\beta H}] \ . \qquad (2.3)$$

It is often (but not always) the case that physical quantities are self-averaging. This means that for a very large system (or more precisely in the thermodynamic limit $N \to \infty$), the value $<A(\varepsilon)>_T$ for a given configuration of disorder is equal to the quenched average "almost surely" (with probability one).

Quenched disorder is the usual situation in real magnets.

B. Annealed disorder

This is the case where the disorder degrees of freedom are in thermal equilibrium with the spin variables. Thus, the averages over the spin variables and disorder variables are carried out with the same weight factor (the Gibbs factor for example)

$$\rho = \text{const.} \exp[-\beta H(s,\{\varepsilon_i\}) - \mu \sum_i \varepsilon_i] \qquad (2.4)$$

where the chemical potential μ is fixed by the constrain $<\varepsilon_i>_\varepsilon = x$.

2.3. More on diluted systems and percolation

Let us consider some typical examples of models of diluted systems [9]. We limit ourselves to Ising spins, the extension to more general spins is evident. We consider quenched disorder.

- Typical site diluted model

$$H = -J \sum_{<ij>} (s_i \varepsilon_i)(s_j \varepsilon_j) \tag{2.5}$$

ε_i = independent random variables with probability distribution

$$P(\varepsilon_i) = (1 - x)\delta(\varepsilon_i) + x\delta(1 - \varepsilon_i) \ . \tag{2.6}$$

- Typical bond diluted model

$$H = -J \sum_{<ij>} \varepsilon_{ij} s_i s_j \tag{2.7}$$

ε_{ij} = independent random variables with probability distribution

$$P(\varepsilon_{ij}) = (1 - x)\delta(\varepsilon_{ij}) + x\delta(\varepsilon_{ij} - 1) \ . \tag{2.8}$$

The qualitative phase diagram for such a model is sketched on Fig. 2. Three domains have to be distinguished :

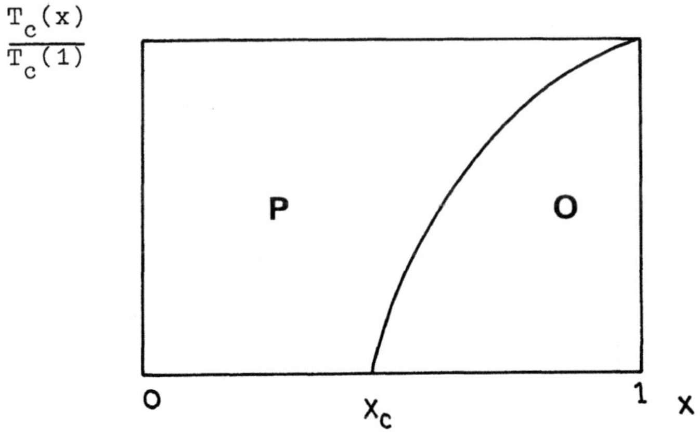

Fig. 2 : Qualitative phase diagram for a diluted spin system. The paramagnetic (P) and ordered phases (O) are separated by a line of transition.

a) The pure limit $x = 1$: one is in the pure case, the critical temperature is $T_c(1)$ and the critical indices $\{z_p\}$.

b) The diluted finite temperature regime $x_c < x < 1$: several questions should be answered in this regime. Is the transition still sharp ? If it is sharp, do the critical indices depend on x ? If the transition is sharp, then renormalization group arguments impose that $\{z(x)\} = \{z_d\}$ for all $x : x_c < x < 1$. The next natural question is then : is $\{z_d\} = \{z_p\}$ or not ? This question is answered by the Harris criterion [10]. According to this criterion established for small dilution, $\{z_d\} = \{z_p\}$ if the specific heat exponent α_p of the pure system is negative. If α_p is positive, then $\{z_d\} \neq \{z_p\}$ and in the renormalization group language a new "disordered fixed point" governs the critical behaviour. The Harris criterion can be understood heuristically as follows.

Small dilution generates a random perturbation of the local interactions which can be interpreted (via the Gibbs factor) as a random perturbation of local temperature. But the linear response to a perturbation of temperature is the specific heat. If $\alpha_p > 0$, the response diverges as $T \to T_c$, leading to an instability in the system.

c) Limit $T = 0$. Percolation problem : if one removes enough bonds (or sites), there will be no linkages path extending across the whole system. Thus there exists a critical value of x called the percolation threshold x_c, below which there is no infinite cluster in the system (see Fig. 3). Note that site dilution is more effective then bond dilution, thus $x_c^{bond} < x_c^{site}$.

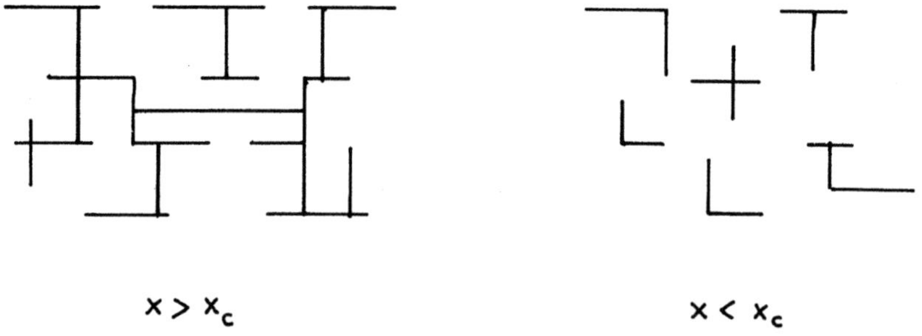

Fig. 3 : Bond percolation for $x > x_c$ there is an infinite cluster but for $x < x_c$ there is no infinite cluster in the system.

Although the percolation problem is a purely geometrical problem it has some similarities with the problem of second order phase transitions. Several quantities are singular in $x - x_c$ and one introduces critical exponents $\{z_{pc}\}$ to describe these singularities. For example : $P_\infty(x)$ the fraction of the sites (or bonds) belonging to the percolating cluster

$$P_\infty(x) \sim (x - x_c)^{\beta_{pc}} \qquad x > x_c$$

$S(x)$, the mean size of the isolated finite cluster $\sim |(x - x_c)|^{-\gamma_{pc}}$

$$\xi(x), \text{ the correlation length} \sim (x - x_c)^{-\nu_{pc}} . \tag{2.9}$$

Those critical exponents satisfy scaling laws similar to the ones for usual phase transitions [11].

The relevant questions to ask in this problem are :

i) What is the percolation threshold x_c (for a given lattice in d-dimension) ?

ii) What are the exponents $\{z_{pc}\}$?

iii) What is the conductivity of a random resistor network ? (each bond is replaced by a resistance).

Another related problem of importance is the so-called directed percolation, where there are some restrictions on the direction of flow through "open" or occupied sites or bonds [12].

2.4. More on the random field problem

The typical random field models are defined by the Hamiltonian

$$H = -J \sum_{\langle ij \rangle} \vec{s}_i \cdot \vec{s}_j - \sum_i \vec{h}_i \cdot \vec{s}_i \tag{2.10}$$

where the $\vec{h}_i$ are independent random variables characterized by the probability distribution $P(\vec{h}_i)$. Two cases are generally studied.

i) The two peaks distribution

$$P(\vec{h}_i) = \tfrac{1}{2}[\delta(\vec{h}_i - \vec{h}_o) + \delta(\vec{h}_i + \vec{h}_o)] . \tag{2.11}$$

ii) The gaussian distribution

$$P(h_i) = [(2\pi)^{1/2}\sigma]^{-1}\exp-(h_i^2/2\sigma) \qquad (2.12)$$

Several experimental realizations of random field systems can be obtained for example by a site disordered antiferromagnet in an uniform magnetic field. The relevant questions to ask in this problem are :

- How the random fields modify the phase diagram of the pure case ?
- What is the role played by the probability distribution $P(h_i)$? (universality classes ?)

For systems with continuous spin symmetry (i.e. with a number of components $n \geq 2$) it has been shown [13] that the critical exponents associated with the ferromagnetic transition in the presence of a random field are equal to those of the zero field transition in $(d - 2)$ dimensions and the lower critical dimension is $d_L = 4$. For Ising spins the situation is not so clear. There is presently a controversy (theoretically as well as experimentally) about the value of the lower critical dimension in the Ising case. Is $d_L = 2$ or 3 ? [14]. This controversy stimulated interesting works on both static and dynamic properties of interfaces. However it may well turn out that this problem bares some resemblance with the spin-glass problem, in the sense that it may be difficult to distinguish experimentally between a true equilibrium transition and an apparent transition related to metastable effects.

2.5. More on spin-glasses

I shall only discuss here certain models currently used in order to illustrate several important concepts in the physics of spin-glasses (frustration, replica trick, breaking of ergodicity). More realistic spin-glass models will be discussed by Drs Droz and Kinzel in this School.

The simplest models consider spins on a regular lattice interacting via random independent interactions J_{ij}. Thus the Hamiltonian is

$$H = -\sum_{ij} J_{ij} s_i s_j - h\sum_i s_i \qquad (2.13)$$

and the independent random variables J_{ij} are characterized by the probability distribution $P(J_{ij})$. We restrict ourselves to Ising spins for simplicity, although real spin-glasses are more likely to have Heisenberg spins. Two types of probability distributions are usually considered.

i) <u>The gaussian model</u>

$$P(J_{ij}) = [2\pi \bar{J}^2]^{-1/2} \exp-[(J_{ij} - J_o)/2\bar{J}^2] \ . \tag{2.14}$$

ii) <u>The two-peaks model</u>

$$P(J_{ij}) = \tfrac{1}{2}[\delta(J_{ij} - J_o) + \delta(J_{ij} + J_o)] \ . \tag{2.15}$$

<u>The frustration</u>

As shown by Toulouse [15], the frustration is a very important ingredient for a spin-glass. This concept allows to distinguish between real and apparent disorder. The simplest illustration is obtained by considering the Mattis model [16], defined by the Hamiltonian

$$H = -\sum J_{ij} s_i s_j \ , \qquad J_{ij} = J\varepsilon_i \varepsilon_j \tag{2.16}$$

where the ε_i are independent random variables with probability distribution $P(\varepsilon_i) = \frac{1}{2}[\delta(\varepsilon_i - 1) + \delta(\varepsilon_i + 1)]$.

Let us consider an elementary square, called plaquette, on a 2 dimensional square lattice (see Fig. 4a). For a given realization of the ε_i , the spins will choose, in the ground state, an orientation minimizing the energy. The spin configuration 4a looks disordered. However by a <u>gauge transformation</u> (under which the zero field Hamiltonian (2.16) is invariant),

$$s_i \to -s_i \qquad J_{ij} \to -J_{ij} \qquad j \text{ nearest neighbour of } i \tag{2.17}$$

one recovers an ordered configuration (Fig. 4b). Moreover, one sees from (2.16) that for every plaquette $\square$, one has

$$J^{-4} \prod_{\square} J_{ij} = +1 \ . \tag{2.18}$$

A configuration not allowed in the Mattis model, is the one drawn on Fig. 4c. Here the spin s_1 tells to s_2 to point upwards while the spin s_3 tells to s_2 to point downwards. s_2 receives contradictory orders and is said to be frustrated. Note that this disorder cannot be removed by a gauge transformation, this is a real disorder. Moreover for the plaquette of Fig. 4c, one has :

$$J^{-4} \prod_{\square} J_{ij} = -1 \ . \tag{2.19}$$

The plaquette is frustrated.

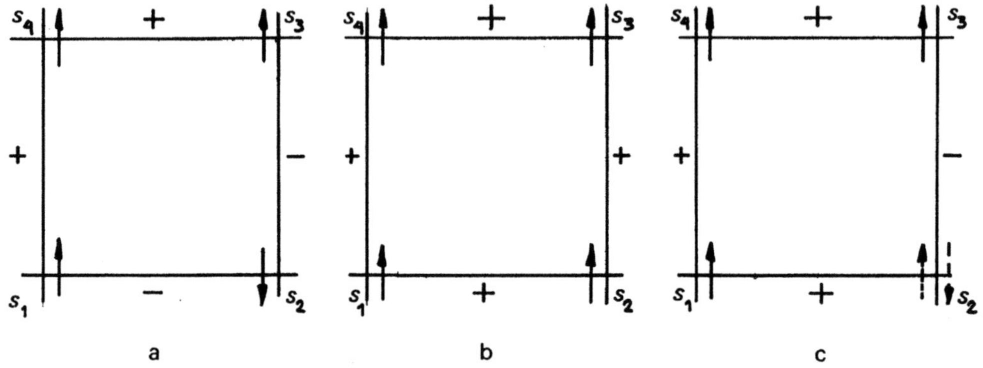

Fig. 4 : a) Apparent disorder in the Mattis model;
b) "a" after a gauge transformation;
c) Frustrated configuration not allowed in the Mattis model.

Although the zero field Mattis model is not frustrated and thus the thermodynamic quantities insensible to the disorder, several interesting observations can be made by looking at the response to an uniform field h (for $h \to 0$). One finds the following facts :

- The Edwards-Anderson order-parameter

$$q = \langle\langle s_i \rangle_T^2 \rangle_\varepsilon = N^{-1} \sum_i \langle s_i \rangle_T^2 \tag{2.20}$$

behaves as $(T_c - T)^\beta$ for $T < T_c$.

- The magnetic susceptibility χ, computed by linear response theory, satisfies the so-called Fischer's law

$$\chi(T) = (k_B T)^{-1}(1 - q) \tag{2.21}$$

showing a cusp at T_c and corresponding to a Curie law above T_c.

- The susceptibility associated to the E.A. order parameter (or the non linear magnetic susceptibility) diverges at T_c :

$$\chi_q(T) = (NT^2)^{-1} \sum_{i,j} [\langle s_i s_j \rangle_T - \langle s_i \rangle_T \langle s_j \rangle_T]^2 =$$

$$= \partial^2 q/\partial h^2 \sim (T - T_c)^{-\gamma} . \qquad (2.22)$$

Note that the frustration is not proper to disordered systems but plays also an important role in non disordered systems (see lectures of Drs Sutö and Zittartz). The presence of frustration leads generally to

- A high degeneracy of the ground state.
- A lowering of the critical temperature.

In conclusion a good spin-glass model should have both disorder and frustration.

Mean field theory

Given a microscopic model, one generally tries to solve it in a simple approximation, for example a mean field like approximation, before trying a more complete solution. The formulation of a good mean field theory for spin-glasses is a long and tortuous story [17)] and we shall only consider here one aspect of this story, the so-called Sherrington-Krikpatrick (SK) model [18)]. For an non disordered Ising model, one gets the mean field approximation by working with a weak infinite range interaction. Sherrington and Kirkpatrick applied the same idea to the gaussian model (2.13), (2.14). In order to ensure the extensivity of the free energy, the parameters J_o and $\bar{J}$ in (2.14) are rescaled according to $J_o \to J_o/N$ and $\bar{J}^2 \to \bar{J}^2/N$, where N is the number of spins. One then computes the quenched free energy

$$F = -k_B T \langle \ell n Z \rangle_J \qquad (2.23)$$

where Z is the partition function for a given bonds configuration. It is very difficult to compute $\langle \ell n Z \rangle_J$ directly. In order to overcome the difficulty, one uses the so-called replica trick. This trick uses the simple mathematical identity :

$$\ell n Z = \lim_{n \to 0} n^{-1} [Z^n - 1] . \qquad (2.24)$$

Thus one introduces n similar replicas of the original system (indexed by $\alpha = 1$ to n). One computes the partition function of these n replicas, i.e. Z^n, and one does the analytic continuation $n \to 0$ in (2.24). Note that this replica trick has be used successfully for other disordered systems than spin-glasses [19].

Without going into the details, let us summarize the main steps of the SK solution and the associated difficulties.

- The computation of $\langle Z^n \rangle_J$ leads to gaussian integrals which can be performed exactly [18]. The remaining trace is computed by a saddle point method where the variational parameters are $q_n^{\alpha\beta}$ and m_n. By symmetry of the original problem under replica permutation, SK looked for a symmetrical saddle point, i.e., $q_n^{\alpha\beta} = q_n$, $(\alpha \neq \beta)$. The analytical continuation for $n \to 0$ leads to the following predictions :
  - i) Two order parameters, coupled by self-consistent non linear equations, characterize the problem
    - a) The averaged magnetization

$$m = \lim_{n\to 0} m_n = \langle\langle s_i \rangle_T \rangle_J \ . \tag{2.25}$$

    - b) The Edwards-Anderson order parameter

$$q = \lim_{n\to 0} q_n = \langle\langle s_i \rangle_T^2 \rangle_J \ . \tag{2.26}$$

  - ii) For J_o not too large, there is a sharp transition at $k_B T_f = \bar{J}$ from a paramagnetic phase to a spin-glass phase, and below T_f, $q \sim (T_f - T)^\beta$.
  - iii) There is a cusp in the magnetic susceptibility $\chi_o(T)$ and in the specific heat at T_f.
  - iv) There is a singularity in the non linear susceptibility at T_f :

$$m(T,h) = \chi_o h - c\chi_o^3 (T - T_f)^{-1} h^3 + O(h^5) \quad \text{for} \quad T > T_f \ . \tag{2.27}$$

  - v) For $T \to 0$, the entropy $S(T)$ tends towards a negative value

$$S(T) \to S_o < 0 \quad \text{as} \quad T \to 0 \ . \tag{2.28}$$

This entropy catastrophe is not acceptable and thus the SK solution is not correct for $T < T_f$. What goes wrong ? Going back to the saddle point method, one sees that in the limit $n \to 0$ the free energy is maximum, instead of minimum for finite values of n. The SK solution is not stable below a line in the plane (T,h) known as the de Almeida-Thouless instability line. The only way to find a stable solution is to look for a saddle point which is non symmetric under the permutation of the replicas. This is the breaking of the replica symmetry below the de Almeida-Thouless instability line. How to break this symmetry is a very difficult question, not completely settled yet. However, there is more and more evidence that the hierarchical scheme proposed by Parisi [20] is the good one. The main output of Parisi's theory is the following.

i) The Edwards-Anderson order parameter q has to be replaced by a functional order parameter $q(x) : 0 \leq x \leq 1$.

ii) The susceptibility is given by

$$\chi(T) = T^{-1}[1 - \int_0^1 q(x)dx] \ . \tag{2.29}$$

For $T < T_f$ it turns out that $\chi(T)$ is constant (plateau) in contradistinction with Fischer's law.

iii) The phase diagram [21] (for Ising spins) is sketched on Fig. 5. In the lower domain delimited by the lines (1), (2) and (5), there is breaking of the replica symmetry. In the plane $h = 0$, the line (4) separates the spin-glass phase from the mixed phase in which both $q \neq 0$ and $m \neq 0$.

iv) For Heisenberg spins, one has two mixed phases [22]. Both are associated with the freezing of the transverse components of the spins. In the high temperature mixed phase the replica symmetry is not broken, but in the low temperature mixed phase this symmetry is broken.

It remains to give a physical interpretation to $q(x)$. It turns out that the quantity which is physically interpretable is the probability distribution $P(q) = dx/dq$ [23]. One can show that in the SK model, the states in the phase space are grouped in many valleys, separated by free energy barriers which heights diverge with the number of spins in the system. There is breaking of ergodicity [24]. $P(q)$ can then be interpreted in terms of the overlap of the different valleys [23]. Moreover, it was recently shown [26] that $P(q)$ is not self-averaging, and thus P(q) must itself be characterized by a probability distribution ! This model shows another remarkable property in the spin-glass phase namely the ultrametric topology of the space of states [25].

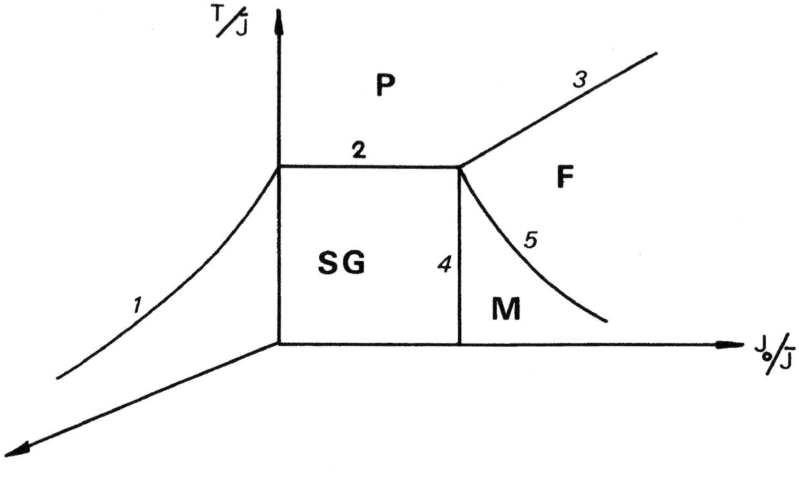

Fig. 5 : Phase diagram for the Ising case. Phases P, SG, M and F are respectively the paramagnetic, spin-glass, mixed and ferromagnetic phases.

Thus one sees that even the mean field theory is not an easy problem and still many questions are open. The next step is obviously to go beyond mean field and look at the role played by the fluctuations.

2.6 The Anderson's localization problem

We return to the tight binding model (1.4) and consider the case where V_o or t_o are no longer constant but vary randomly throughout the crystal. Thus V_o or t_o are characterized by a probability distribution. One asks similar questions than in the pure case, as a function of the amount of disorder in a given dimension. It turns out that the disorder induces localized states. Such localized states cannot conduct electricity at zero temperature. An important concept is the mobility edge, i.e. the energy E_c separating localized from extended states. The transition from extended to localized states is called the (Anderson's) localization transition [26]. One is also interested to study the conductivity σ as a function of the disorder.

Note that the model considered here is a simple, one electron model. Such a model cannot explain the experimental results (see Dr. Di Castro's lecture) and it is crucial to take into account the Coulomb interaction among the electrons and thus to solve a much more complicated many electron problem.

In conclusion, I hope that this random walk in the physics of disordered systems has convinced you that the basic questions one is asking and the basic concepts are very similar in the different classes of disordered problems which will be discussed in great details in the following lectures.

REFERENCES

1. Berker A.N. and Ostlund S., J. Phys. C12, 4961 (1975).

2. Mandelbrot B.B., 1982. The Fractal Geometry of Nature (Freeman Ed., San Francisco).

3. Stanley H.E., 1971. Introduction to Phase Transitions and Critical Phenomena (Oxford University Press, Oxford).

4. Ma S.K., 1976. Modern Theory of Critical Phenomena (Benjamin Reading, Massachusetts).

5. Ginzburg V.L., Soviet Phys. Sol. State 2, 1824 (1960).

6. Halperin B.I. and Hohenberg P.C., Phys. Rev. Lett. 19, 700 (1967).

7. Dynamic Critical Phenomena and Related Topics, edited by C.P. Enz, Lecture Notes in Physics, Vol. 104 (Springer Verlag, Berlin 1975).

8. Ashcroft N.W. and Mermin N.D., Solid State Physics (Ed. Holt, Rinehart and Winston, New York 1976).

9. Stinchcombe R.B. in Phase Transitions and Critical Phenomena, Vol. 7 (ed. by C. Domb and J.L. Lebowitz, Academic Press, London 1983).

10. Harris A.B., Jour. Phys. C7, 1671 (1974).

11. Percolation Structures and Processes, Ann. Israel Phys. Soc. 5 (1983). G. Deutscher ed.

12. Kinzel W. in Ref. [11].

13. Imry Y. and Ma S.K., Phys. Rev. Lett. 35, 1399 (1975).

14. Imry Y. in Proceedings of Statphys. 15 (to be published).

15. Toulouse G., Comm. on Physics 2, 115 (1977).

16. Mattis D.C., Phys. Rev. Lett. 20, 1356 (1971).

17. Toulouse G., Congrès de la Société Française de Physique, Clermont-Ferrand 1981 (Les éditions de Physique, 1982).

18. Sherrington D. and Kirkpatrick S., Phys. Rev. Lett. 35, 1792 (1975), Phys. Rev. B17, 4384 (1978).

19. Emery V.J., Phys. Rev. B11, 239 (1975); Grinstein G. and Luther A.H., Phys. Rev. B13, 1329 (1976); Stephen M.J. and Grest G.S., Phys. Rev. Lett. 38, 567 (1977).

20. Parisi G., Phys. Rev. Lett. 43, 1754 (1979), Jour. Phys. A13, L115, 1101, 1887 (1980).

21. Toulouse G., Jour. de Physique Lettres 41, L447 (1980).

22. Toulouse G. and Gabay M., Jour. de Physique Lettres 42, L103 (1981).

23. Parisi G., Phys. Rev. Lett. 50, 1946 (1983).

24. Palmer R.G., Adv. Phys. 31, 669 (1982).

25. Mézard M., Parisi G., Sourlas N., Toulouse G. and Virasoro M., preprint (1983).

26. Anderson P.W., Phys. Rev. 109, 1492 (1958).

SCALING AND CRITICALITY IN DILUTE SYSTEMS

R.B. Stinchcombe
Department of Theoretical Physics
1 Keble Road, Oxford OX1 3NP

1. Introduction

These lectures describe the treatment of static critical properties of diluted systems by simple scaling methods.

This is an area in which geometrical concepts, as well as the notions of criticality, self-similarity (scale invariance) and scaling all play an important role.

The basic phenomenon contained in substitionally diluted lattice-based systems is the percolation transition, which is the simplest geometrical transition and, of all transitions, the one most readily described by scaling methods. When the percolation behaviour is combined with a thermal phase transition, as for example in dilute magnets, extremely rich behaviour can result, as many recent experiments on phase boundaries, critical thermodynamic properties, correlation lengths and correlation functions clearly exhibit.The basic theoretical questions raised by such recent experiments and the theoretical problems naturally arising in the development of a basic theoretical description of the area have caused it to be the subject of extensive study in recent years, and our discussion here is necessarily selective.

To provide a reasonably coherent viewpoint we concentrate here on the simple diluted systems which show percolative critical behaviour, by itself or in combination with thermal critical behaviour, treating specifically percolation, percolative conductivity, dilute and mixed Ising and Heisenberg magnets, and diluted quantum systems. We do not attempt to include any discussion of more complicated problems such as the spin glass or random field systems. Some parts of the present account overlap slightly with lectures given at the Varenna Summer School "Highlights of Condensed Matter Physics" [1]; other parts are developed more fully in a review on "Dilute Magnetism" [2].

We begin by describing the percolation transition, and using it to illustrate the idea of criticality at a continuous phase transition and the underlying concept of self similarity or scale invariance (diverging correlation length). It is this property which is exploited in the scaling method employed in later sections.

2. Percolation; criticality and scale invariance

The percolation transition [3] occurs in e.g. a randomly bond-diluted lattice, where removal of bonds disconnects the lattice when the bond concentration p reaches a critical value p_c. The transition is purely geometric, and is from a phase with infinitely extended linkage paths (i.e. containing an infinite cluster of connected bonds)

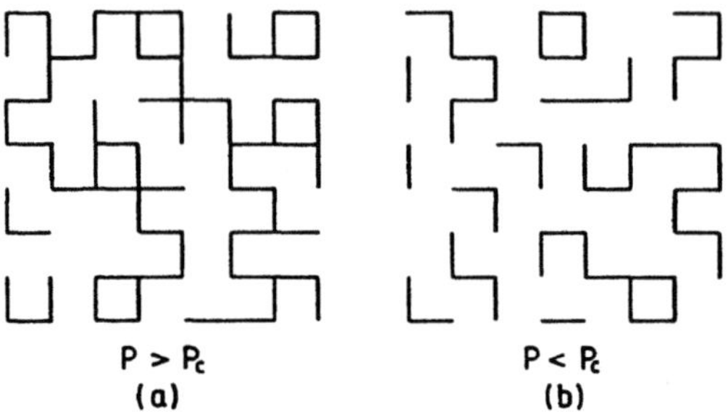

Fig.1. Bond dilution: (a) above, (b) below the percolation threshold

as shown in Figure 1(a), to a phase with only finite clusters (Figure 1(b) At p_c a characteristic "correlation length, ξ, diverges like

$$\xi/a = C|p-p_c|^{-\nu_p} \qquad (1)$$

where a is the lattice spacing and ν_p is a critical exponent. The critical behaviour of ξ is analogous to that occurring in thermal continuous phase transitions in pure systems, where the divergence occurs as the temperature approaches its critical value T_c. In such systems the diverging correlation length causes singularities in the thermodynamic properties (e.g. a divergence of the susceptibility or compressibility in a magnet or gas, and a singular specific heat). In general, critical behaviour at a continuous phase transition is induced by a diverging correlation length.

The divergence of ξ is equivalent to the statement that the system is self similar at the transition (looks the same at all magnifications insufficient to resolve the lattice spacing). This scale invariance can be seen for percolation in Figure 2, which shows part (the backbone of the infinite cluster) of the diluted lattice at the percolation threshold.

The basic idea of renormalisation group theory [4] is to exploit the divergence of ξ by a scaling approach. The transformation ("renormalisation") of parameters under a length scaling of the system is carried out in such a way as to preserve the absolute range of correlations. The original, and still the most controlled way of achieving

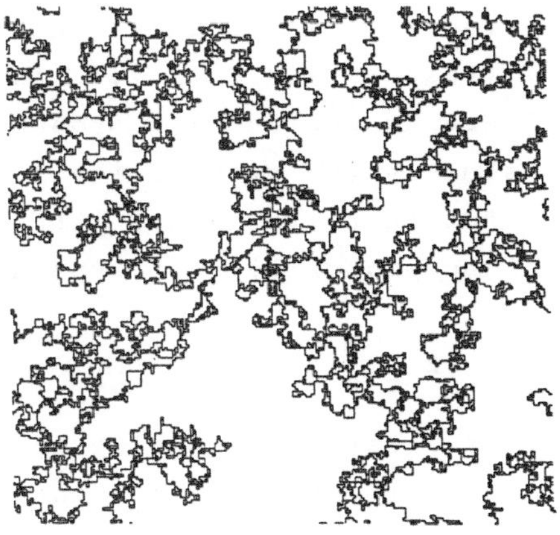

Fig.2. Scale-invariant structure of the infinite cluster backbone at the percolation threshold

this is to use field-theoretic methods in momentum space. A simpler and more direct method, sufficient for our purposes, is to work in position space. This procedure will be illustrated for the percolation problem in the next section, where it will be shown how the resulting transformation yields the critical concentration p_c and the exponent ν_p

3. Scaling treatment of percolation

We consider [5] a diluted square lattice with bond concentration p and lattice spacing a as shown in the left hand side of figure 3. Removal of the open-circle sites of the lattice produces the square lattice shown on the right of the figure, which has bond concentration p' and lattice spacing a'. Clearly $a' = ba$, where $b = \sqrt{2}$ is the scale factor of the dilatation of the original lattice. It is convenient for

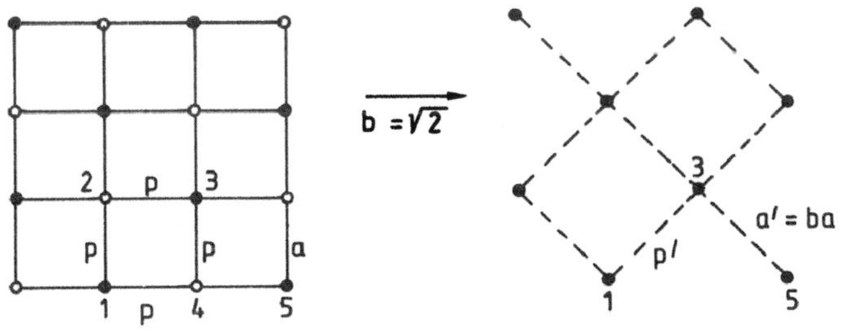

Fig.3. Dilation of a square lattice by scale factor $\sqrt{2}$ by decimating open-circle sites

us to consider dilatation processes which maintain the absolute connectivity properties of those sites, e.g. 1,3, that remain. In the new lattice, sites 1 and 3 are connected with probability p' (since the bond concentration is the same as the probability a given nearest neighbour linkage is present). But in the original lattice the probability that 1 and 3 are linked is the probability of a path of bonds from 1 to 3 via 2 and or 4. Thus

$$p' = 2p^2 - p^4 \equiv R(p) \qquad (2)$$

(where the right hand side is obtained by adding the probabilities $p^2(1-p^2)$, $p^2(1-p^2)$, p^4 of the following independent events: connection via 2 and not 4, via 4 and not 2, and via 2 and 4). (2) is the renormalisation group transformation of the concentration parameter p for dilatation by $b = \sqrt{2}$. The transformation can be exploited as follows.

(i) <u>At the transition</u> ($p = p_c$) the system is self similar (the only length which which the lattice spacing a can be compared is the correlation length, and that is infinite). So the system does not notice the dilatation, i.e. no change of parameters is needed to maintain the connectivity properties. Thus p_c is one of the special ("fixed point") values, normally denoted by p^*, which transform into themselves:

$$p^* = R(p^*). \qquad (3)$$

This gives $p_c = 0.618$ for the critical concentration (and two other fixed point values $p^* = 0,1$ corresponding to the empty and pure lattices).

(ii) <u>Near the transition</u> ($p \sim p_c$) the correlation length in the original lattice diverges according to (1), and that in the new lattice satisfies a similar equation

$$\xi'/a' = C|p' - p_c|^{-\nu_p}. \qquad (4)$$

However, since the absolute range of correlations was preserved, ξ' and ξ are equal, while $a' = ba$, $b = \sqrt{2}$ and p' is related to p by (2). Near p_c, equation (2) becomes

$$p' - p_c = \lambda(p - p_c) \qquad (5)$$

where

$$\lambda = (dR/dp)_{p = p_c} \qquad (6)$$

is the eigenvalue of the transformation (2) linearised at the fixed point p_c. Using (5), the ratio of (4) to (1) gives

$$\frac{1}{b} = \lambda^{-\nu_p} \qquad (7)$$

from which the critical exponent ν_p is found to be 0.818, by inserting the values of b and $\lambda(= 1.528)$.

(iii) Away from the transition the relationship between correlation length and concentration can in principle be found for all p by solving a functional equation: if $\xi/a = f(p)$, then (since $\xi'/a' = \xi/(ab)$, and p' and p are related by (2)),

$$f(R(p)) = \frac{1}{b} f(p). \tag{8}$$

This is similar to the procedure in (ii), but without the linearisation. In the case of one dimension, where it is easy to show that $R(p) = p^2$, equation (8) is easily solved, with the usual result $f(p) \propto [\ell n(1/p)]^{-1}$ for the exact correlation length.

The above illustration exhibits the following features of the renormalisation group scaling method:

(a) the transformation of parameters accompanying a length scaling of the system is constructed;

(b) fixed points of such transformations give critical parameters;

(c) eigenvalues of the linearised transformation give exponents.

The above illustration does not give an exact treatment of the two dimensional percolation problem, as can be seen by comparing the critical parameter with the result ($p_c = 1/2$) from duality. This is because our discussion is over-simplified in neglecting further-neighbour linkages and correlations which are generated on the transformed lattice. This means that the original system scales into a more complicated system and it becomes necessary to consider a generalised model with more parameters than just the nearest neighbour bond concentration. This extension of parameter space can be truncated by considering only parameters generated on a small part of the lattice, e.g. a square cluster of nine sites. Such approaches are capable of giving ν_p to about one per cent accuracy [5]. The accuracy of the "decimation" method just described is, however, eventually limited by the omission of a step analogous to the scaling of the spin magnitude in the full renormalisation group treatment of thermal problems [4]. It nevertheless provides a very simple way of implementing the scaling technique, and its directness allows one to keep contact with intuitive physical ideas and to readily treat systems (e.g. those combining geometric with thermal features) whose treatment by more sophisticated methods becomes very involved. The method is not, of course, limited to two dimensions.

The method has been applied to many non-thermal problems, such as percolation cluster problems, percolative conductivity, random and interacting walks, aggregation problems, and localization. Its application to thermal problems, such as magnetic systems is well known, and it has helped to elucidate the essential physical features of many random thermal systems, such as dilute and mixed magnets, and the spin glass. References to all these applications are given in [6,2].

4. Percolative conductivity: scaling a distribution

In this section we discuss percolative conductivity [7]. This is one of the simplest problems which combine percolative and other features and is of interest in its own right and because of its relation to static and dynamic critical properties of dilute Heisenberg magnets near the percolation threshold [8,9,10]. Moreover it raises in a simple way an important point which occurs in the scaling treatment of other random problems, such as dilute magnets and spin glasses [11], namely the scaling of a probability distribution function.

The essential features of the percolative conductivity problem can be understood by considering a diluted lattice, such as that in Figure 1, where each nearest neighbour bond, present with probability p, has conductance σ. It is clear that below the percolation threshold p_c the absence of extended linkage paths makes the bulk conductivity $\sigma(p)$ zero for a large diluted resistor network of this type. For $p > p_c$, $\sigma(p)$ is

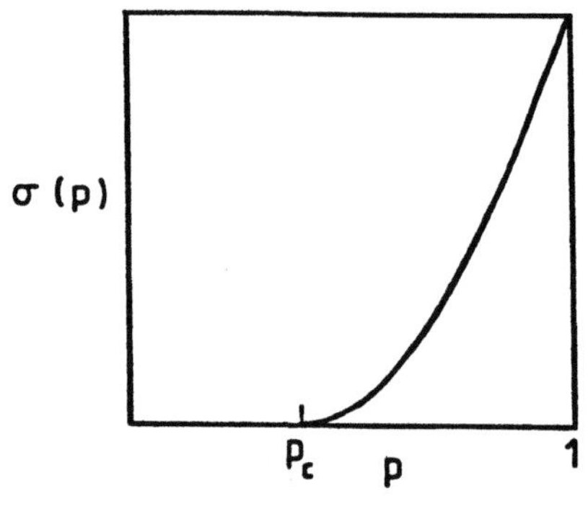

Fig.4. Percolative conductivity of a diluted resistor network

expected to have the form shown in Figure 4, that is (a) non-zero, (b) increasing with concentration p, (c) small near p_c because of the loosely connected nature of the infinite cluster backbone (Figure 2) on which the conduction takes place. The way in which $\sigma(p)$ vanishes as p approaches p_c from above can be characterised by a critical exponent t:

$$\sigma(p) = \gamma\sigma\,(p-p_c)^t \qquad (9)$$

where γ is a numerical constant, and σ is the single-bond conductance parameter.

The exponent t can be shown [8] to be related to the dynamic exponent in diluted Heisenberg magnets near the percolation threshold: for wavelengths long, but short compared to ξ, the spin wave excitations in such a system have frequency

$$\omega = D(p)\, q^2, \tag{10}$$

where the spin wave stiffness D(p) is given by the ratio of $\sigma(p)$ to the percolation order parameter (which vanishes like $(p-p_c)^{\beta_p}$, where β_p is a familiar static exponent for percolation). Thus

$$D(p) \propto (p-p_c)^{t-\beta_p}. \tag{11}$$

It will be shown in section 6 that t is also directly related to an exponent characterising static critical properties in diluted Heisenberg magnets.

The scaling treatment of percolation conductivity [7] is, as in the case of percolation, based on the transformation of the system under a dilatation. However, instead of scaling a parameter, the full distribution function for the random conductance variable σ_{ij} has to be scaled. Initially, σ_{ij} is distributed according to

$$P(\sigma_{ij}) = (1-p)\,\delta(\sigma_{ij}) + p\delta(\sigma_{ij}-\sigma) \tag{12}$$

(conductance σ present with probability p, absent with probability 1-p).

If the scaling shown in Figure 3 is now carried out for the two-dimensional resistor network, the bond 13 of the scaled lattice has a random conductance

$$\sigma_i' = \frac{\sigma_{12}\sigma_{23}}{\sigma_{12}+\sigma_{23}} + \frac{\sigma_{14}\sigma_{43}}{\sigma_{14}+\sigma_{43}} \equiv f\{\sigma_{ij}\} \tag{13}$$

where σ_{ij}, ij = 12 etc, are the random conductances of the bonds ij of the original network which have been combined in series and parallel to give the result (13). This relation implies that the probability distribution $P(\sigma)$ transforms as follows:

$$P(\sigma_{ij}) \to P(\sigma_i') = \int d\sigma_{12} \ldots d\sigma_{43}\, P(\sigma_{12}) \ldots P(\sigma_{43})\,\delta(\sigma_i' - f\{\sigma_{ij}\}). \tag{14}$$

The original binary form (12) is not retained under the transformation. It is nevertheless possible to follow the scaling of the distribution, and to characterise its transformation by its approach to special invariant forms [7] associated with fixed point values of the initial parameters (p,σ). The scaling of the distribution near such fixed points is described by eigenvalues which determine the critical exponents ν_p,t. A simple approach maintaining the basic features of the full treatment and illustrating more easily the essential points is to approximate $P(\sigma_i')$ by a distribution of the same form as (12) but with scaled parameters p', σ' chosen to reproduce the weight of the $\delta(\sigma_i')$ part of $P(\sigma_i')$ (as given by (14)) and its first moment. That results in

$$p' = 2p^2 - p^4 \equiv R(p) \qquad (15)$$

$$p'\sigma' = p^2\sigma \, . \qquad (16)$$

As we could have expected, this transformation contains the percolation scaling. We may now carry out steps analogous to (i), (ii), (iii) in section 2:

(i) the (doubly unstable) fixed point of (15), (16) controlling the criticality is at $(p^*,\sigma^*) = (p_c, 0)$, where p_c is as previously obtained in section 2.

(ii) Near the transition located by this fixed point, (9) applies together with a corresponding form involving transformed variables for the scaled lattice. Since $\sigma(p)$ has been preserved under the scaling, the ratio of the two equations gives

$$1 = \frac{\sigma'}{\sigma}\left(\frac{p'-p_c}{p-p_c}\right)^t = p_c \, \lambda^t \, . \qquad (17)$$

The last expression arises from the use of the linearised form (5) of (15), and a corresponding linearisation of (16) about the unstable fixed point, and so contains the eigenvalues λ, p_c of the linearised transformations. (17) results in the value $t = 1.34$ for the critical exponent, fortuitously close to the most accurate recent estimate [12].

(iii) Away from the critical region the dimensionless conductance $\sigma(p)/\sigma$ becomes a more general function, $g(p)$ say, than that shown in (9). From (15), (16), the functional equation determining $g(p)$ is

$$p^2 g(R(p)) = R(p)g(p) \qquad (18)$$

The (numerical) solution of this equation gives a result very like that shown in Figure 4.

More precise treatments of the diluted resistor network problem, and of the associated dynamic and thermal properties of dilute Heisenberg magnets can be obtained by performing the scaling on larger clusters, and this is especially necessary in the treatment of higher dimensional cases.

5. Dilute magnets: qualitative description

A real site-diluted magnet, such as $Co_pZn_{1-p}Cs_3Cl_5$, or the bond diluted magnetic models which will for simplicity be discussed here, exhibits percolation effects. As the concentration p of e.g. magnetic bonds is reduced through p_c, the critical temperature $T_c(p)$ falls to zero, as shown in Figure 5. This is because an ordered state can only exist where a macroscopically large number of spins are coupled by the

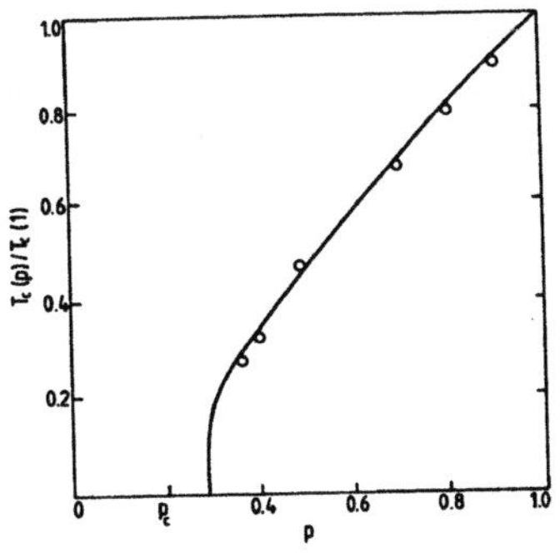

Fig.5. Phase diagram of diluted Ising magnet, from theory [13] (full line) and experiment [14](points)

magnetic bonds (exchange interactions J), which is only the case for the spins on the infinite cluster existing for $p > p_c$.

In a ferromagnet at T = 0 all spins in the same cluster are parallel, so the correlation length is that for the percolation problem and so diverges at p_c. As has been seen in neutron scattering experiments, the correlation length diverges all along the boundary separating ordered and paramagnetic phases, if the transition is continuous, so a scaling approach is suited to locating the boundary and treating the critical behaviour near it.

Such comments apply equally to Ising or Heisenberg magnets, to ferro or antiferromagnets, and to site or bond diluted cases (though only the former actually occurs). However, details of the phase boundary and critical behaviour differ between Ising, Heisenberg and other models, so we first present the scaling treatment of the diluted Ising model, and then give brief discussions of other diluted magnetic systems.

6. Dilute Ising magnets

In random magnets, as in the percolative conductivity problem, a randomly distributed variable occurs, and its distribution has to be scaled. In the bond diluted case this variable is the exchange interaction J_{ij} for a particular bond ij or, more conveniently in the Ising case,

$$t_{ij} \equiv \tanh J_{ij}/K_B T. \tag{19}$$

This variable is described by a distribution of the same form as in (12), but with σ_i, σ replaced by t_i, $t \equiv \tanh J/K_B T$.

For the bond diluted square lattice Ising model, the transformation of paramaters p, t can be obtained by again scaling the square cluster

1234 of Figure 3 to a single bond 1,3, first arriving at a distribution function scaling analogous to (14). The new feature is that the problem now combines thermal aspects with dilution. Thermal scaling can be obtained by carrying out a partial trace over decimated spins, and preserving Boltzmann probabilities between the undecimated spins (which preserves the absolute correlation length in a thermal problem). This has the consequence that, in place of (13), the scaled random variable for the bond 13 is, in the Ising case [11]

$$t_i' = \frac{t_{12}t_{23}+t_{14}t_{43}}{1+t_{12}t_{23}t_{14}t_{43}} \equiv g\{t_{ij}\}, \tag{20}$$

where the notation should be obvious. $g\{t\}$ is the function replacing $f\{\sigma\}$ in equation (14) to obtain the analogous equation for the scaling of the distribution in the Ising case.

By procedures analogous to those described in section 4, the transformation of parameters can be extracted:

$$p \to p' = R(p) \tag{21}$$

$$t \to t' = T(p,t)\ . \tag{22}$$

The non-trivial fixed points of these equations are

$$(p^*,t^*) = (p_c,1),\ (1,t_c) \tag{23}$$

where p_c is the percolation threshold given previously, and t_c is the critical value of $t \equiv \tanh J/K_BT$ for the pure Ising model. The two fixed points, denoted hereafter by A, B respectively, are the zero-temperature percolation fixed point and the pure Ising fixed point. These points are joined by a flow line AB (a line which transforms into itself under (21), (22)), which is the critical "surface" dividing ordered and disordered phases and along which ξ diverges, The phase boundary shown in Figure 5 was obtained by converting to the temperature versus concentration space the critical flow line derived from an analogous scaling treatment of a three dimensional Ising model [13]. The infinite slope of such Ising critical curves arises from the conversion from the $\tanh J/K_BT$ variable, which results in the exponential behaviour

$$\exp[-2J/K_BT_c(p)] \propto (p-p_c) \tag{24}$$

near the percolation fixed point. The critical curves obtained by such methods have been shown to be remarkably accurate, by comparing with limiting slopes, etc, known exactly for dilute two dimensional Ising models [2]. And they agree well with experiment, as is suggested by reference to Figure 5, which includes with the theoretical curve experimental results on a real diluted Ising magnet, $Co_pZn_{1-p}Cs_3C\ell_5$ [14].

Critical indices are obtained, as in previous sections, by linearising the transformation equations (in this case (21), (22)) about the fixed points ((23)). At the percolation point A, the eigenvalue arising from (21) is the percolation eigenvalue λ found previously, so in the dilute Ising model at T=0 the percolation behaviour (1) is recovered. The second eigenvalue at A, related to the low temperature behaviour near the threshold, turns out to be again λ. This has the consequence that in this neighbourhood the change from percolation to thermal critical behaviour is controlled by a "crossover" exponent with value unity [11]. This is an exact result for the dilute Ising model [15,10], and is related to the unit power of $(p-p_c)$ on the right hand side of (24). One of the eigenvalues at point A yields the pure Ising correlation length divergence at p=1; the other, which is less than 1, implies that the pure Ising critical behaviour is not modified by the randomness. This cannot be regarded as a reliable result, since a stability criterion [16] for the pure Ising fixed point is not always satisfied by the simple scaling approach, and the possible existence of a random Ising fixed point in two dimensions is a very delicate question.

The simple scaling method is nevertheless a very useful approach to dilute Ising magnets and it has been used extensively in a very wide range of situations [2]. An example is its application to thermodynamic properties of dilute magnets, including magnetisation, susceptibility and other free energy derivatives in addition to the quantities (correlation length and phase boundary) discussed above. Figure 6 shows specific heat curves obtained using the method [17]. These can be compared with the experimental results [18] (actually on a different type of system) shown in Figure 7. Finally, we remark that many variations of the model or randomness have been treated. From this area we mention only mixtures of Ising magnets, such as $A_p B_{1-p}$ where A and B are both magnetic [19]: critical curves have been obtained for various ratios J_B/J_A; in the limit of zero ratio the dilute system is recovered.

7. Dilute Heisenberg magnets

We now briefly discuss how the scaling method can be applied to static critical phenomena in Heisenberg systems. Here the scaling is difficult because of commutation problems. These problems can be avoided by considering the classical (infinite spin) version of the model [9]. Alternatively, cluster approximations neglecting commutation with spins outside of the cluster can be used for, say, the spin-½ model [13].

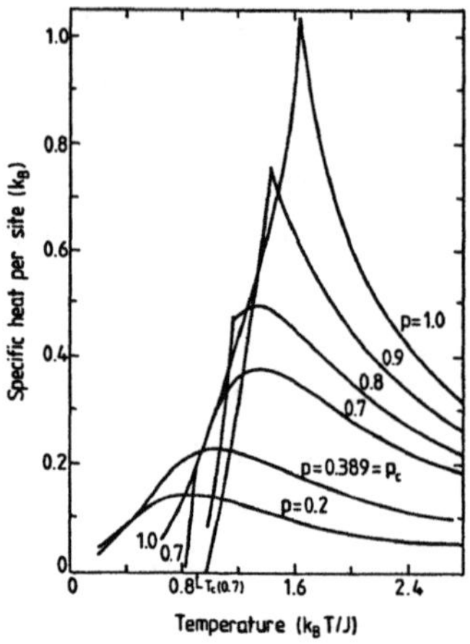

Fig.6. Specific heat of the bond-diluted two-dimensional Ising model, as determined by a position space renormalisation group method [17].

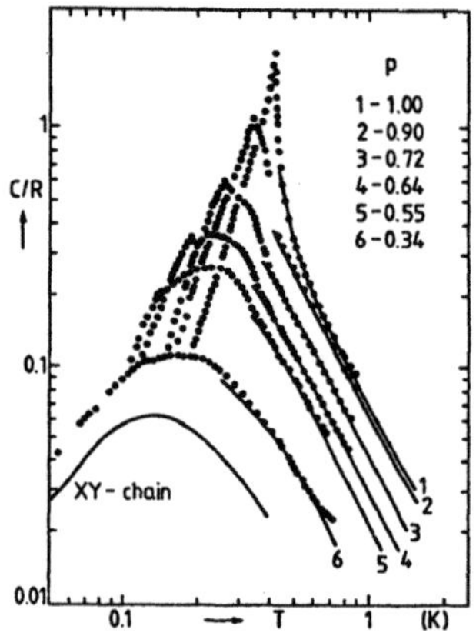

Fig.7. Experimental results [18] for specific heat of diluted XY magnet

In both cases it can be shown that for the random-bond Heisenberg model at low temperatures the scaling of $K \equiv J/K_BT$ on (for example) the cluster 1234 of figure 3 is

$$K' = \frac{K_{12}K_{23}}{K_{12}+K_{23}} + \frac{K_{14}K_{43}}{K_{14}+K_{43}} \equiv f\{K_{ij}\} \ . \qquad (25)$$

This function may then be used in the scaling equations for the distribution of K_{ij} (c.f. equation (14)) to obtain the transformation of parameters K, p and hence by a route analogous to that described in the previous section, to find the low temperature critical properties of the dilute Heisenberg model.

The approximate cluster calculation for the spin-½ model is not limited to low temperatures, so full phase diagrams and critical behaviour can be obtained [13]. Because of the different thermal scaling, the thermal critical indices are different from those of the corresponding diluted Ising model, but the percolation threshold and exponent again occur. The critical curve shows an obvious difference from the Ising case: in place of (24), the transition temperature of the diluted Heisenberg model approaches zero as $p \to p_c$ in the following power law manner:

$$T_c(p) \propto (p-p_c)^{1/\phi}. \tag{26}$$

ϕ is here the Heisenberg crossover exponent, which can be calculated from the eigenvalues at the percolation fixed point. The exponent ϕ is exactly related to the percolation conductivity exponent t, for the following reason [9,10]. The relationship (25) is the same as (13) for the conductivity problem; that is, for this particular example, the variable K_{ij} has combined at low temperatures in the same way as the conductance variable σ_{ij}. This result is actually quite general. That means that the thermal exponent for the dilute Heisenberg scaling in the neighbourhood of the (zero temperature) percolation fixed point is exactly related to the corresponding exponent in the dilute resistor network. Moreover, both problems have the percolation eigenvalue λ as their other eigenvalue at the percolation fixed point. It then follows exactly that

$$1/\phi = t. \tag{27}$$

The low temperature properties of mixed Heisenberg magnets can also be treated using scalings based on (25). The extension of the treatment to general temperatures has also recently been carried out [20] using a generalisation of the approximate cluster calculation for the dilute spin-½ Heisenberg model.

8. Dilute quantum systems, Potts models, anisotropic magnets, etc.

The Ising model in a transverse field Γ exhibits "quantum" critical phenomena at a zero temperature phase transition at a critical value Γ_c of the field [21]. The quantum critical behaviour in d-dimensions is related to the usual critical properties of the zero field model in d+1 dimensions [22]. Reasons have been given [23] to believe that the zero temperature diluted "transverse Ising model" (TIM) should have novel features, including a discontinuity in the dependence of critical field on concentration at the percolation threshold. This has been verified by position space scaling methods [24,25]. The simplest way of proceeding [24] is to overcome the quantum difficulties by deducing from the related two-dimensional Ising model the exact scaling for the one-dimensional TIM, and to build from it an approximate scaling for the two-dimensional system. In place of (20) one then finds

$$\lambda' = \frac{1}{2}(\lambda_{12}\lambda_{23} + \lambda_{14}\lambda_{43}) = h\{\lambda_{ij}\} \tag{28}$$

where $\lambda_{ij}=J_{ij}/\Gamma$. Procedures like those described previously can then be used to obtain exact and approximate descriptions of the one and two-dimensional transverse Ising models, respectively. Alternatively, a direct position space approach can be developed [25] not exploiting the contact with the higher dimensional Ising model. This approach can also be used to obtain the generalisation to non-zero temperatures, including the crossover from quantum to classical critical behaviour, as well as the percolation-thermal crossover.

Diluted q-state Potts models have also beeen treated by the position space approach. A scheme like that discussed above for the diluted Ising model can be followed [26], but a new thermal variable u replaces t, and its scaling is slightly more complicated than (20). Critical curves ($u_c(p)$ versus p) have been obtained for various q. That for q=1 corresponds to a doubly diluted model, and the simple scaling method gives it exactly [26]. For a more precise treatment of other features, more sophisticated position space scaling methods have to be applied [27].

Two types of anisotropy can be distinguished in spin systems, and diluted models involving each type have been treated by scaling methods. In the first type, the anisotropy is with respect to lattice direction. An example is the layer magnet, in which the nearest neighbour exchange coupling between spins in different layers is smaller by a factor $\gamma \ll 1$ than that between spins in the same layer. The resulting critical curves exhibit a "dimensional" crossover between the percolation concentrations characteristic of the uncoupled and coupled layer systems if γ is very small [28]. The second type of anisotropy is with respect to spin direction, and it is this anisotropy which distinguishes XY or Ising-like magnets from (isotropic) Heisenberg ones. In the diluted magnet, as in the pure case, this type of anisotropy is relevant, in the sense that it drives the critical behaviour of a weakly anisotropic uniaxial or planar model to that of the diluted Ising or XY model respectively, as well as giving rise to many additional effects [29].

In all the models discussed (conductivity, Ising, Heisenberg, transverse Ising, Potts) and in other dilute cases, the scalings give some emphasis to one-dimensional features near the percolation threshold (because the infinite cluster at p_c is of lower dimensionality than the pure lattice). This is especially evident in the application of the Migdal method to dilute magnets [13,17]. In such scaling treatments the thermal variable at the percolation fixed point is closely connected to the inverse critical correlation length K_1 of the corresponding pure one-dimensional problem [2]. Any special features K_1 may have then appear in the dilute higher dimensional magnet at the percolation threshold. This is the reason for the discontinuity in Γ_c at p_c in the diluted transverse Ising model. It also has the consequence that in the diluted weakly unaxially anisotropic Heisenberg model a crossover is seen (experimentally and in the scaling treatment [2]) between Heisenberg power law behaviour (26) and Ising exponential behaviour (24) as the critical temperature decreases as p approaches p_c: this effect is related to the anisotropy-induced crossover in the temperature dependence of K_1 for the pure weakly anisotropic Heisenberg chain.

9. Conclusion

We have tried to demonstrate that an adequate treatment of criticality requires the use of a scaling method, and that local randomness makes it convenient to carry out the scaling in position space. For these reasons, and because of the intrinsic simplicity of the method, it is especially suited to the treatment of random critical systems. We have mainly discussed here the special case of random dilution, where intuitive viewpoints (e.g. about percolative behaviour) can be useful and illuminating and where, it seems to us, the position space method has the advantage of not obscuring the intuitive picture. Clearly, for more sophisticated and subtle problems it has to be used in a much more developed form than indicated here, or possibly replaced by a more controlled scaling approach.

No attempt has been made here to give a complete review, and since only the basic approach to each problem has been discussed, and space is limited, only pioneering work on basic examples has been referred to. Other important work, e.g. that of Stanley and coworkers and many others on percolation and other geometric problems, of Bernasconi and Straley on percolation conductivity, and of a multitude of authors on thermal percolative problems, has not been referred to A more complete survey, particularly of the last mentioned areas, can be found in [1,2,6,8].

This account has discussed static phenomena. Dynamic phenomena in pure and disordered systems can also be treated by a position space scaling approach [1], as will be discussed elsewhere in these proceedings [30].

References

1. R.B. Stinchcombe, "Static and Dynamic Critical Phenomena in Dilute Systems", in "Highlights of Condensed Matter Physics" ed. F. Bassani, F. Fumi and M.P. Tosi, Proceedings of the Varenna Summer School, 1983 (North-Holland).
2. R.B. Stinchcombe, "Dilute Magnetism", in "Phase Transitions and Critical Phenomena", ed. C. Domb and J.L. Lebowitz, (Academic Press, 1983).
3. J.W. Essam, Rept. Prog. Phys. 43, 833 (1980).
4. See for example S-K Ma "Modern Theory of Critical Phenomena" (Benjamin, 1976).
5. A.P. Young and R.B. Stinchcombe, J. Phys. C8, L535 (1975); C9, L643
6. T.W. Burkhardt and J.M.J. Van Leeuwen "Real Space Renormalisation" (Springer - Topics in current physics, Vol. 30, 1982).
7. R.B. Stinchcombe and B.P. Watson, J. Phys. C9, 3221 (1976).
8. S. Kirkpatrick, "Models of Disordered Materials", in "Ill-Condensed Matter", ed. R. Balian, R. Maynard and G. Toulouse (North-Holland 1979).
9. R.B. Stinchcombe, J. Phys. C12, 2625 (1979).
10. A. Coniglio, Phys. Rev. Lett. 46, 250 (1981).
11. A.P. Young and R.B. Stinchcombe, J. Phys. C9, 4419 (1976).
12. B. Derrida and J. Vannimenus, J. Phys. A15, L557 (1982).
13. R.B. Stinchcombe, J. Phys. C12, 4533 (1979).
14. E. Lagendijk and W.J. Huiskamp, Physica 62, 444 (1972).
15. D.J. Wallace and A.P. Young, Phys. Rev. B17, 2384 (1978).
16. A.B. Harris, J. Phys. C7, 1671 (1974).
17. C. Jayaprakash, E.F. Riedel, and M. Wortis, Phys. Rev. B18, 2244 (1978).
18. H.A. Algra, L.J. de Jongh, W.J. Huiskamp, and J. Reedijk, Physica 86-88B, 737 (1977).
19. J.M. Yeomans and R.B. Stinchcombe, J. Phys. C12, L169 (1979).
20. C. Tsallis, R.B. Stinchcombe and B. Buck, J. Phys. C17, 301 (1984).
21. R.B. Stinchcombe, J. Phys. C6, 2459 (1973).
22. A.P. Young, J. Phys. C8, L309 (1975).
23. A.B. Harris, J. Phys. C7, 3082 (1974).
24. R.B. Stinchcombe, J. Phys. C14, L263 (1981).
25. R.R. dos Santos, J. Phys. C15, 3141 (1982).
26. J.M. Yeomans and R.B. Stinchcombe, J. Phys. C13, L239 (1980).
27. W. Kinzel and E. Domany, Phys. Rev. B23, 3421 (1981).
28. A. Mujeeb and R.B. Stinchcombe, J. Phys. C15, 3163 (1982).
29. R.B. Stinchcombe, J. Phys. C14, 397 (1981); see also A.O. Caride, C. Tsallis and S.I. Zanette, Phys. Rev. Lett. 51, 145 (1983).
30. R.B. Stinchcombe, "Dynamic Scaling in Pure and Dilute Systems" Lecture notes in this volume.

DYNAMIC SCALING IN PURE AND DILUTE SYSTEMS

R.B. Stinchcombe
Department of Theoretical Physics
1 Keble Road, Oxford OX1 3NP

1. Introduction; dynamic scaling

This account is concerned with the use of scaling methods to treat dynamic properties of pure and random systems (c.f. [1] for static properties). Such methods are designed particularly to treat critical dynamics [2,3], where they exploit the divergence of the controlling (correlation)length. But, in the form developed here, the scaling methods can be applied equally well to ordinary non-critical dynamic properties: they for example provide a new and very efficient approach for the numerical generation of densities of states, etc. [4-9]. This feature is particularly valuable in dealing with random systems.

The dynamic method here presented is carried out in position space. This has the advantage that strong local randomness (e.g. dilution) can be easily treated with the method, which makes it ideally suited to the discussion of random critical dynamic behaviour [2,3,10,11], e.g. that occurring near the percolation threshold in a dilute system [12,13,14]. The position space method can also treat very directly spatially inhomogeneous systems, such as fractals [15], for which momentum space methods are unsuitable [3,16,17,18,19,20,21]. As an illustration we later derive some exact results for dynamic critical properties of fractals.

To deal with dynamics, a discrete scaling method of the decimation type is here applied directly to the equation of motion which in all our examples is linear [22,2,4,5]. One or more dynamic parameters, e.g. a dimensionless frequency, are transformed to new values so as to preserve certain properties of the system under a discrete length dilatation. The transformation of the frequency is sufficient to determine the frequency-wave vector relationship, density of states, etc, and the dynamic exponents associated with such quantities [2]. Even in the very simplest cases (e.g. the pure chain) the frequency transformation is non-linear, and has novel properties, not normally found in recursion relations for static scaling, related to properties of the iterative maps of theories of chaos [23,24].

As well as characteristic frequencies, response functions are of importance for the specification of dynamic properties. These can also be obtained from the scaling method if the equation of motion employed is that for the appropriate Green function, and the dimensionless frequency and Green function are both scaled [3,6]. Dynamic scaling forms for such quantities were specified long ago [25], and since we make contact with these forms we give them now. The dynamic response function χ depends on frequency ω, wave vector k and parameters such as temperature or, in the case of dilution-induced transitions [1], the concentration p which can approach critical values (p_c) at which the correlation length ξ diverges. According to the dynamic scaling hypothesis, the dependence on the three variable reduces, in the critical region, to

$$\chi(k,\omega,p) = k^{-(2-\eta+z)} G(k\xi,\omega/k^z) \quad \omega,k,1/\xi \to 0 \tag{1}$$

where η and z are static and dynamic exponents, G is a scaling function of two variables only, and the dependence on p is all contained in $\xi(p)$. If the response χ has a well defined peak position or width, a characteristic frequency can be determined and that must, from (1), also have a scaling form [25]:

$$\omega = k^z F(k\xi). \qquad k,\ 1/\xi \to 0. \tag{2}$$

The forms (1), (2) which are used very extensively in the analysis of experiments on critical dynamics, can be deduced using the scaling methods to be discussed subsequently [2,3].

The basic scaling method is illustrated in the next section, using the example of linear dynamics in a pure chain. The results are, of course, the same as those available using standard elementary methods. But the application brings out the essential features of the scaling method, including the derivation of the transformation of frequency (iterative map)etc and the extraction of critical dynamics, of frequency-wave vector relationships and densities of states etc. The example also indicates the largely straightforward generalisations to cases not tractable by standard methods. Subsequent sections deal with the treatment of such cases: dynamics of non-random systems in higher dimensions including fractals (§3);scaling for disordered cases (§4); the dilute chain, including the relationship of the crossover between pure and finite cluster behaviour to that between chaotic and stable orbit behaviour in the iterative map (§5): dilute systems near the percolation threshold (§6). The account continues with a section (§7) on scaling and crossover relationships, between dynamic, static and density of states exponents,and concludes with a discussion section (§8).

2. Scaling for pure chain: iterative map, critical dynamics

As in the position space methods described for static problems in

[1], we study the transformation of parameters under a dilatation of the lattice in which the lattice spacing a increases by a discrete factor b. The dilatation is achieved by the decimation method, elimination of (dynamic) variables on a sublattice [22,2].

For lattice-based systems, with dynamic variables $\{u_i\}$ where i is a site label, the simplest dynamical problem involves linear equations of motion of the type

$$\Omega u_i = \Sigma M_{ij} u_j , \qquad \Omega \propto \omega^n \tag{3}$$

where M_{ij} is a real symmetric dynamical matrix, and Ω (which can be real or imaginary) arises from an nth derivative with respect to time (e.g. n = 1 for ferromagnetic spin wave systems, electron tight-binding models, Ising Glauber dynamics,hopping diffusion etc, n = 2 for phonon problems, etc). In the commonest such systems M_{ij} is vanishing except when i, j are nearest neighbours (n.n.) and it is simplest for us to consider that case.

Two categories of problem have to be distinguished:the 'Goldstone' system (e.g. spin waves, Glauber dynamics, diffusion and phonons) where

$$M_{ii} = \sum_j M_{ij} \tag{4}$$

and systems without this relation (e.g. tight binding models). These have different critical dynamics since, in the Goldstone systems,(4) makes $\Sigma_i u_i$ a constant of the motion. The scaling method applies equally well to both types, but to be specific we consider here the Goldstone type. This is because its critical dynamics is more interesting, and because it covers many common systems, including spin waves, Ising

Fig.1. Decimation of a linear chain, used in length scaling approach to dynamics

dynamics, and diffusion problems where dilution-induced criticality can be an additional effect of real as well as theoretical interest [12,13,26].

The standard equation of motion (3), with (4), for such a Goldstone system with n.n. coupling becomes in the case of the one-dimensional system shown in the left hand side of Figure 1:

$$(2 - \Omega)\, u_a = u_{b_1} + u_{b_2}. \qquad (5)$$

Here the strength of the n.n. coupling (i.e. of the n.n. matrix element M) has been incorporated into Ω, which is now the nth power of a dimensionless frequency, and we have taken the specific case where i is the site a of the figure.

Similar equations relate u_{b_1}, u_{b_2} to their nearest neighbours (u_a, u_{c_1} and u_a, u_{c_2}). Elimination of u_{b_1}, u_{b_2} is the equivalent of decimation as used in [1], and results in

$$[(2 - \Omega)^2 - 2]u_a = u_{c_1} + u_{c_2}. \qquad (6)$$

This is the equation relating u_a to its neighbours on the scaled lattice shown in the right hand side of Figure 1, and it is of a form identical to (5), but with a transformed factor $(2 - \Omega) \rightarrow (2 - \Omega')$ where [2]

$$\Omega' = 4\Omega - \Omega^2. \qquad (7)$$

This equation gives the transformation of the frequency parameter Ω which must accompany the dilatation illustrated in Figure 1, in which the lattice spacing doubles ($a \rightarrow a' = ba$, where $b = 2$), if the relationship between u_a and u_{c_1}, u_{c_2} is to be kept the same as in the original lattice. In the case where a periodic wave exists on the lattice, this last condition is equivalent to the preservation of its absolute wave vector. A more useful quantity for us is the (dimensionless) "relative" wave vector k (absolute wave vector multiplied by lattice spacing), and this clearly doubles in the transformation: $k \rightarrow k'$ where

$$k' = 2k. \qquad (8)$$

In cases where a wave vector is inappropriate, k can be replaced by an inverse dimensionless dynamical length, e.g. a localisation length.

The parameter transformations (7), (8) can be used to determine the frequency - wave vector relationship [2], dynamic exponent [2] and density of states [6,9,10] as follows.

(a) Dispersion relation

We suppose that the relationship between reduced frequency Ω and relative wave vector k is

$$\Omega = f(k). \qquad (9)$$

(This "dispersion relation" is more general than (2), in the sense that it applies for all k. However it is more special in corresponding only to the pure case, where ξ does not appear: more precisely, ξ is infinite and $F(k\xi)$ is a constant). Using also a corresponding relationship between Ω' and k' for the scaled system, (7) and (8) then imply

$$f(2k) = 4f(k) - f(k)^2. \tag{10}$$

This non-linear functional equation has the solution

$$f(k) = 2(1 - \cos\alpha k) \tag{11}$$

with α a constant. Inserted into (9), this gives the usual form for the spin wave frequency of a n.n. spin chain, or (using $\Omega \propto \omega^2$) the Born-Von Karman result for the phonon spectrum.

We point out that the transformation (7) has the form of a non-linear iterative map of the type much studied recently because of its remarkable properties [24]. The more general case of quadratic map corresponding to (7) with 4 replaced by a general parameter λ shows bifurcations, chaos, and special universal behaviour. Except when λ takes one or two special values the related functional equation of type (1) is insoluble. This situation will arise frequently in our later applications. Then one has to proceed numerically or adopt methods, as in (b) immediately below, specifically for special regions like the critical region where a linearisation of the iteration equations becomes adequate.

(b) Dynamic exponent

Equations (7),(8) have a fixed point

$$(\Omega^*, k^*) = (0,0). \tag{12}$$

The neighbourhood of this fixed point is the critical region where Ω and k are both small, and we suppose that there the frequency-wave vector relationship (9) reduces to

$$\Omega = ck^z \qquad k \sim 0 \tag{13}$$

where c is a constant (c.f.(2)). z is the dynamic exponent.

For small Ω, k, (7), (8) become

$$\Omega' = \lambda_\Omega \Omega \tag{14}$$

$$k' = bk \tag{15}$$

where $\lambda_\Omega = 4$, $b = 2$. The ratio of (13) to a corresponding equation for the scaled system gives, using (14), (15)

$$\lambda_\Omega = b^z. \tag{16}$$

Thus the eigenvalue λ_Ω of the linearised frequency scaling equation, for dilatation b, yields through (16) the dynamic exponent z. In the present case we obtain z = 2, which of course can be obtained alternatively from the long wavelength limit of (11). The procedure just used is exactly like that [1] for obtaining static critical exponents from renormalisation group transformations.

(c) Density of states

Another way of proceeding from (7), (8) will now be discussed. It is now well known that under successive iterations of (7) (or the more general equation of this type with 4 replaced by λ where λ exceeds a critical value $\lambda_c \sim 3.57$) the successive values into which Ω scales sample one or more continuous ranges in an erratic fashion. This is the so-called "chaotic" behaviour. If the asymptotic sampling density is $\rho(\Omega)$ (so $\rho(\Omega)\,d\Omega$ is a normalised measure of the number of visitations of the region $d\Omega$ after many iterations), $\rho(\Omega)$ is, in our case, the same as the density of states in energy, as we now show [10]. Because of (8), successive iterations double the relative wave vector k; so (using (9)) the nth iterate of $\Omega = f(k_o)$ is $f(2^n k_o)$. However, because of (11), f is an (even) periodic function so the nth iterate depends only on $2^n k_o$ modulo π/α. For large n this last quantity takes random values uniformly covering the range 0 to π/α (except when k_o takes special values $N\pi/2^M\alpha$, which have zero measure), so the iterative sampling of Ω is just like finding the energies for states uniformly distributed in k-space. The sampling density $\rho(\Omega)$ is thus the usual density of states in energy.

This result provides a very powerful numerical method [4-9] for obtaining densities of states, since it is a very simple matter to numerically generate the sampling density under a specified recursive procedure.

(d) Dynamic response functions

If, in place of (5) a Green function equation of motion is used, the scale of the Green function is set by an inhomogeneous term. The matching of this term in original and scaled equation (the appropriate generalisation of (6)) then determines a transformation of the Green function, in addition to the tranformations (7), (8) previously discussed. The combined transformations are sufficient to determine the dependence of the Green function (and hence the dynamic response function) on frequency and wave-vector [3,6]. This is typically done numerically, or by linearisation procedures like (b) for the critical regime.

The example treated in this section (and the obvious generalisations like vibrations of a diatomic chain or spin waves in the Heisenberg antiferromagnet, etc) is more easily discussed by the standard procedure of diagonalising the equations of motion in k-space, which leads to identical results for dispersion relation, dynamic exponent, density of states etc. However the standard method cannot be applied if the system is spatially inhomogeneous or random. The scaling method does not have such limitations, as is shown in the following sections.

3. Scaling for dynamics of non-random systems in higher dimensions, including fractals

When applied to hypercubic lattices of more than one dimension, the decimation scaling method described above leads to an extenstion of parameter space: a dynamical matrix coupling only nearest neighbour sites on the original lattice is transformed to one coupling all sites of the transformed lattice when the dynamical variables on a sublattice are eliminated. In some other cases (e.g. the honeycomb lattice) elimination of nearest neighbours of a given site generates a lattice with more nearest neighbour bonds than in the original lattice. In both situations a cluster or bond-moving prescription can be developed to avoid the proliferation of couplings or reduce the generated lattice to one of the original form, and it is straightforward to do this in such a way that the Goldstone relation (4) is maintained [27]. For example, for a $b = \sqrt{3}$ dilatation of a honeycomb lattice (removing sites which are next nearest neighbours of each other on the original lattice, and moving half the new nearest neighbour bonds) the resulting transformations of reduced "frequency" Ω and wave vector k are

$$\Omega' = 3 - \tfrac{1}{2}\left[(3-\Omega)^2 - 3\right] = 3\Omega - \tfrac{1}{2}\Omega^2 \qquad (17)$$

$$k' = \sqrt{3}\, k. \qquad (18)$$

(17) is not now exact, but it is sufficient to give the exact dynamic exponent: using the linearisation procedure (b) of section 2 we obtain $\lambda_\Omega = 3$, $b = \sqrt{3}$, which inserted into (16) yields the exact value $z = 2$.

A more interesting application of the scaling method is to spatially inhomogeneous systems, such as fractals, where the standard k-space diagonalisation of the equations of motion (sufficient to treat the honeycomb lattice example) cannot be applied. For many fractal lattices the scaling method gives an exact treatment of the dynamics [3,16,17,20]. The results are of real significance, since fractals can be used to represent real systems at the percolation threshold [21,28] (see section 6).

As an illustration we consider the dynamics of the triangular Sierpinski gasket fractal shown in Figure 2. This fractal is constructed

Fig.2. Non-random fractal sometimes used to represent infinite cluster backbone in two dimensions. Dynamics on such fractals can be discussed exactly by scaling method given in text.

by dividing the outer triangle into four equal ones, removing the centre one of the four and dividing the remaining three triangles as in the first step, and continuing the process. In this example as everywhere else in this presentation, the dynamics is of the simplest linear Goldstone type (specified by equations (3), (4) with M_{ij} coupling only nearest neighbour sites). We consider a triangle made up of four of the smallest triangles in the gasket. This triangle contains altogether six sites: three vertices a, c_1, c_2 and three sites b_1, b_2, b_{12} respectively midway between a and c_1, a and c_2, c_1 and c_2. The adjacent triangle of the same size sharing the vertex a has the following sites (in cyclic order around the triangle) a b_3 c_3 b_{34} $c_4 b_4$ (labelled in the same way as before). The equation (c.f. (5)) for the vertex site a is

$$(4 - \Omega)\, u_a = \sum_{i=1}^{4} u_{b_i} . \tag{19}$$

By writing corresponding equations for sites $b_1, b_2, b_3, b_4, b_{12}, b_{34}$ these six sites can be eliminated to yield an equation of the same form as (19) but relating u_a to u_{c_i} $(i = 1, \ldots, 4)$ and containing, in place of Ω,

$$\Omega' = 5\Omega - \Omega^2 . \tag{20}$$

The new equation corresponds to a Sierpinski gasket larger by $b = 2$ than the original one, since u_{c_i} now plays the role of the original nearest neighbour u_{b_i}. So (20) is the (exact) transformation of the frequency variable for dilatation by $b = 2$. The treatment can be generalised to d-dimensional hypertetrahedral Sierpinski gaskets, in which case the coefficient 5 in (20) is replaced by $d + 3$. So in these cases we again arrive at a quadratic iterative map, now with control parameter $\lambda=d+3$. The functional equation of the form (10) for the dispersion relation is now insoluble (except for d=1, where (10) is recovered with coefficient $\lambda=4$). The linearisation procedure of section 2(b) can however be applied to treat the critical dynamics: equation (16) yields for the d-dimensional hypertetrahedral Sierpinski gasket the dynamic exponent [3,16]

$$z = \ln (d+3)/\ln 2 \ . \tag{21}$$

The recursive sampling method of section 2(c) for the density of states yields a bizarre structure, which has recently been discussed in a related context elsewhere [29].

The low energy density of states in energy can be characterized by an exponent which can be obtained in the following way [16,20,30]. In the dilatation by b = 2 the number of bonds in the gasket is reduced by a factor d+1. This can be used to define an effective fractal dimensionality, d_f, as follows [19,21]:

$$d + 1 = b^{d_f} \ , \qquad d_f = \ln(d+1)/\ln 2. \tag{22}$$

At the same time according to (20), or its generalisation for arbitrary dimensions, the frequency scales by a factor d + 3, in the critical region (Ω small). The number of states up to Ω therefore changes by a factor d + 1 when Ω changes by d + 3; the number of states up to Ω is therefore proportional to $\Omega^{1-\alpha}$ where

$$\alpha = 1 - \frac{\ln(d+1)}{\ln(d+3)} = 1 - \frac{d_f}{z} \equiv 1-d_s \tag{23}$$

is an effective density of states exponent. The relationship (23) of dynamic and density of states exponents to the fractal and "spectral" dimensions d_f and d_s was first given in [16].

The dynamics of other fractals can also be discussed simularly. As a final example we refer to the Berker lattice obtained by replacing a single bond by n pairs in parallel and repeating the process indefinitely. A scaling by b = 2 can be achieved by decimating the sites inserted in the last step of the process. This yields the scaling equation, for b = 2,

$$\Omega' = \frac{2(n+1)}{n}\,\Omega - \Omega^2 \tag{24}$$

again of the quadratic map type [27,31]. The exact dynamic exponent is thus

$$z = \ln (2(n+1)/n)/\ln 2. \tag{25}$$

It is easy to arrive also at the fractal dimensionality ($d_f = \ln 2n/\ln 2$) and hence, using the relationship given in (23), the density of states exponent α.

In this section we have illustrated the application of the scaling method to non-random uniform or spatially inhomogeneous systems, including fractals; as will be discussed later, some of the results obtained for fractals are relevant to dynamics at the percolation threshold.

4. Scaling for disordered cases:introduction

The basic dynamical system described by (3) (lattice vibrations, spin waves, and so on) may show strongly modified behaviour in disordered lattices. Examples are the localisation of states caused by disorder in one dimensional system, and the dilution-induced critical dynamics at the percolation threshold. In the remainder of this paper we indicate how the scaling method can treat (usually approximately) such dynamic behaviour.

It is usual to distinguish site and bond randomness, where the randomness is associated with site or bond variables respectively [26]. Both are possible in some systems, e.g. vibrating random lattices where the mass may vary from site to site and the spring constant may be a bond-random variable. In this discussion we consider only bond randomness, where the (nearest neighbour) element M_{ij} of the dynamical matrix of equation (3) depends on the particular n.n. bond label ij.

We shall be primarily concerned with dilution, when the distribution of M_{ij} is binary, corresponding to non-zero or zero outcomes for M_{ij} with respective probabilities p, 1-p where p is the bond concentration. As in treatments of static phenomena in diluted systems [1], the scaling method transforms the probability distribution.

The position space scaling method is convenient for treating the strong local disorder associated with dilution, and is particularly suited to the discussion of critical properties. However, its iterative nature also results in other advantages already illustrated for pure dynamics, e.g. the possibility of efficient numerical generation of densities of states and other quantities by accumulating the number of visitations of a given frequency range by the iterated frequency variable. All the main ideas already encountered in the preceding treatments of pure dynamics have a straightforward analogue in the disordered case, e.g. frequency scaling (though it is now derived from the scaling of a distribution), eigenvalue which determine critical exponents, and so on.

The first application considered is the bond diluted chain [10].

5. Dilute chain: dynamic scaling, chaos

For the bond-diluted chain it is convenient to rewrite the equation of motion (3) as

$$u_i = \Sigma K_{ij} u_j \tag{26}$$

where $K_{ij} \equiv M_{ij}/\Omega$ is a bond-random dimensionless inverse "frequency" variable taking the values $1/\Omega$, 0 with respective probabilities p, 1-p. As in §2, we construct the decimation scaling shown in Figure 1. However in place of (5) we now begin with an equation for u_a involving the bond variables K_{ab_i}, i = 1,2. This equation is transformed, by elimination of u_{b_1}, u_{b_2} using their equations, to one of similar form but now involving u_a, u_{c_1}, u_{c_2} and, in place of K_{ab_i}, scaled bond variables K_{ab_i} corresponding to the nearest neighbour bonds joining a to c_i in the scaled lattice (on the right hand side of Figure 1). These scaled variables are given in terms of K_{ab_i}, and $K_{b_ic_i}$, i = 1,2, and so take various values (including zero). The non-zero outcomes for K'_{ab_i} have total weight p' and mean value $1/\Omega'$ where [10]

$$p' = p^2 \tag{27}$$

$$\Omega' = (1-p)^2(3\Omega-\Omega^2)+p(1-p)\Omega(8-9\Omega+2\Omega^2)/(1-\Omega)+p^2(4\Omega-\Omega^2) \equiv \Delta(\Omega). \tag{28}$$

(28) is the required frequency transformation and, as in static scalings [1] for dilute systems, it depends on the concentration parameter p, which also scales (according to the exact equation (27)). (28) is approximate except at p=1. Since the scaling was achieved with a dilatation by b=2, we have an additional scaling equation (c.f. (8)):

$$k' = 2k . \tag{29}$$

The procedures introduced in §3(b) can now be used to find a dynamic scaling form for the diluted chain [10]. Near the fixed point

$$(p^*, \Omega^*, k^*) = (1,0,0) \tag{30}$$

the scaling equations take the linearised form

$$\Omega'=\lambda_\Omega\Omega, \quad \delta p' = \lambda_p \delta p, \quad k' = bk \tag{31}$$

where $\lambda_\Omega = 4$, $\lambda_p = 2$, $b = 2$ and $\delta p \equiv p - p^*$. (31) implies the following functional form for the frequency:

$$\Omega = k^z f(k\delta p^{-\nu}) \tag{32}$$

where

$$z = \ell n\, \lambda_\Omega/\ell n b = 2 \tag{33}$$

$$\nu = \ell n\, b/\ell n\, \lambda_p = 1 . \tag{34}$$

(32) is the dynamic scaling result (2), with explicit values (33), (34) for the dynamic exponent z and the exponent ν describing the static correlation length $\xi(p)$ (c.f.[1]). The regime just treated, near the fixed point (30), is the dynamic critical region in which Ω, k and $1/\xi \propto (\delta p^{\nu})$ are all small. Since we are considering a chain, the critical concentration is $p^* = 1$ (pure system).

There is another fixed point, at which $p^* = 0$, corresponding to the empty chain. At the two fixed point values of p ($p^* = 0,1$), the frequency scaling (28) reduces to

$$\Omega' = \lambda\Omega - \Omega^2 \tag{35}$$

with $\lambda = 3,4$ respectively. (35) is the general quadratic map, which exhibits stable cycle behaviour of the point Ω under iteration if λ is moderately small. Increase of λ introduces successive period-doubling bifurcations until at $\lambda \sim 3.57$ a limit point of such bifurcations occurs, after which the iterated point moves chaotically, ergodically sampling a continuous range of frequencies. This latter behaviour occurs in particular for the case $\lambda = 4$, which results from $p = 1$, i.e. the pure limit of our dilute chain: it was already discussed for the pure chain in §3(a), (c). The nearly empty chain ($p \sim 0$ where $\lambda \sim 3$, corresponds to a stable orbit behaviour under iteration. The chaotic and stable orbit behaviours of the frequency variable in these two limits for the dilute chain are associated respectively with pure (extended state) dynamics, and (localised) dynamics on the finite clusters into which the very dilute chain decomposes. And the chaotic or obit sampling generates continuous or discrete densities of states associated respectively with extended or localised behaviour[10].

6. Dilute systems near the percolation threshold: dynamic and density of states exponents

In dilute systems of dimensionality greater than one, the percolation threshold occurs at a concentration p_c between 0 and 1, and the associated divergence of the percolation correlation length ξ gives rise to long wavelength dynamics at the threshold of a different character from that in the pure system [12-14]. A scaling form of the type (2) typically applies, with a dynamic exponent z different from the usual (integer) value found in the pure system. The density of states at low frequency is also modified near the percolation threshold. These effects are observable in for example dilute Heisenberg or Ising magnets, though so far only limited experimental investigations have been carried out.

The scaling method applied to the dilute chain in the previous section can be used to investigate these effects in higher-dimensional dilute systems, and we first give an example of results from this approach. It can be combined with the bond moving technique, introduced at the beginning of section 3, to obtain the generalisation to higher dimensions [3]. The $b=\sqrt{3}$ dilatation of the honeycomb lattice (for example) then leads, in place of (17), to a scaled bond-random variable K' whose value depends on the random variables K of bonds in a corresponding region of the original lattice. The weight and mean value of the non-zero outcomes for K' can then be used, as in the previous section, to obtain the scalings of the bond concentration p and characteristic reduced frequency Ω; in addition, the reduced wave vector scales with the lattice constant:

$$p' = R(p) \equiv 2p^2 - p^4 , \tag{36}$$

$$\Omega' = S(\Omega, p), \tag{37}$$

$$k' = b\ k. \tag{38}$$

The neighbourhood of the fixed point $(p^*, \Omega^*, k^*)=(p_c,0,0)$ is again the critical region, and there the scaling equations reduce to a linearised form of the type (31), with eigenvalues $\lambda_\Omega = 4.56$, $\lambda_p = 1.53$ (for $b=\sqrt{3}$). These are sufficient (c.f. (32)-(34))to yield the dynamic scaling form (2) near the percolation threshold, with $z = 2.76$, $\nu = 1.30$ [3]. Thus the frequency-wave vector relationhip for long wavelength spin waves in a diluted layer Heisenberg ferromagnet at the percolation threshold should be $\omega \propto k^{2.76}$, and a crossover from this behaviour should occur as k is reduced if p is not precisely equal to p_c (see section 7). Work is presently in progress to obtain corresponding forms for antiferromagnets, and three dimensional systems, etc.

An alternative way of approaching critical dynamics at the percolation threshold makes use of the fractal results derived in section 3. Exactly at p_c, the infinite percolation cluster (and in particular its backbone) is self similar and so involves all length scales, essentially because ξ is infinite.This self-similarity is the principal feature of fractals. However, the infinite cluster backbone is strictly a random fractal. Nevertheless its representation by a non-random fractal can be useful if the fractal dimensionalities are close to those of the infinite cluster backbone, and this is the case for d-dimensional hypertetrahedral Sierpinski gaskets [21,28]. So the dynamic and density of states exponents of excitations on the infinite cluster backbone in d-dimensional dilute systems at the percolation threshold should be close to those derived in section 3 for the d-dimensional gasket fractals. For example, in two dimensions the resulting dynamic exponent is (using (21)) $z = 2.32$ (compare the value $z = 2.76$ from the scaling of the random system); the density of states exponent is (from (23)) $\alpha = 0.32$, while the random system scaling also gives 0.32. The density of states exponent for the infinite cluster at p_c has recently been worked out for the two dimensional dilute system by a numerical approach using the Lanczos algorithm [30]. The result, $\alpha = 0.33 \pm .03$, is the most accurate direct evaluation available, but indirect arguments to be given in the next section result in more precise values.

7. Scaling and crossover relationships

It has been known for some ten years [12,13] that the spin wave stiffness in dilute Heisenberg ferromagnets is the ratio of two static percolation quantities, the conductivity and the percolation probability P, which are respectively proportional to $(p-p_c)^t$, $(p-p_c)^\beta$ near the percolation threshold. Values of the two static exponents t, β have been obtained by various methods. Using $\xi \propto (p-p_c)^{-\nu}$ (which defines the correlation length exponent ν) the result for the spin wave stiffness implies

$$\omega \propto \xi^{(\beta-t)/\nu} k^2. \tag{39}$$

This form of spin wave frequency actually holds in the regime k, $1/\xi \to 0$, with $k\xi$ small, which is a special case of the situation described by (2). Matching (39) with (2) in the limit $k\xi \to 0$ thus yields [3,14]

$$z = 2 + \frac{t-\beta}{\nu} , \tag{40}$$

as well as the asymptotic behaviour of the scaling function F for small argument.

The general result (2) shows that as $k\xi$ varies from small to large values the behaviour (39) crosses over to the form $\omega \propto k^z$ which applies for any small k if p is sufficiently close to p_c to make $\xi \gg 1/k$. Alternatively, if p is held fixed close to p_c (as in a real experiment) reduction of k causes a crossover from the form $\omega \propto k^z$ to the form (39).

(40) is an important scaling relationship of dynamic exponent z to static exponents. If we adopt a recently speculated form for the conductivity exponent t [20] the expression (40) for z reduces to

$$z = \frac{3}{2} (d - \beta/\nu). \tag{41}$$

For two dimensions this reduces still further, since there $\beta/\nu = 5/48$, exactly [32].

Corresponding reductions are also possible for the density of states exponent. At the percolation threshold, the infinite cluster has fractal dimensionality [21]

$$d_f = d - \frac{\beta}{\nu} . \tag{42}$$

This is because the percolation probability, which is a measure of the density of bonds in the infinite cluster, satisfies $P \propto \xi^{-\beta/\nu}$, and so the number of bonds in the infinite cluster changes by a factor $b^{d-\frac{\beta}{\nu}}$ under a dilatation by b. Under such a dilation Ω changes by b^z (see(16)), so by arguments like those preceding (23) we again obtain for the density of states exponent α,

$$\alpha = 1 - \frac{d_f}{z} . \tag{43}$$

Using (40), (42), this becomes

$$\alpha = \frac{(2-d)\nu+t}{2\nu+t-\beta} . \tag{44}$$

If we use (41) in place of (40) we obtain instead

$$\alpha = 1/3 \tag{45}$$

(in any dimension). The small error in the current value of one or more static exponentsis sufficient to make (40) and (41) indistinguishable, and (44) and (45) the same, within the errors. The results from (41), (45) for d=2 ($z = 91/32$, $\alpha=1/3$) can be compared will those quoted for random scaling, fractal, and numerical approaches at the end of the previous section.

Finally, we comment on relationships between some apparently different problems. From the discussion of equation (3) in section 2, it must be obvious that ferromagnetic spin wave problems, and diffusion problems (both Goldstone problems with $\omega \propto \Omega$) will be essentially the same on the same disordered lattice. The only distinction is between real frequency (spin waves) and imaginary frequency (diffusion). This equivalence can be used to obtain (39) using the relation between diffusion constant D and conductivity. However, diffusion does not always involve a width Dk^2: as in the crossover discussion above, there are two distinct regimes, $k\xi$ greater or less than unity, in critical percolative dynamics, and a form for the diffusion width like (2) is necessary to describe both [33]. Discrete diffusion models, of the "ant in the labyrinth" type, are convenient systems for the application of computer techniques to discuss the dynamics on dilute systems in both crossover regions: a computation of root mean square diffusion distance R versus time t (number of diffusion steps) can yield extremely accurate values for the exponent γ defined by

$$R \propto t^{\gamma} . \qquad (t, \xi >> 1). \tag{46}$$

Comparison with the dilute magnet discussion shows that γ should take the two values 1/2, 1/z in the two regimes $R >> \xi$, $R << \xi$. Monte Carlo computations of γ for the ant in the labyrinth problem [34,35], combined with the above relationships, yield the most accurate results so far available for dynamic exponents such as z (e.g. z=3.12±.018 in two dimension:

8. Conclusion

This excursion, using scaling methods for dynamics, started from pure regular systems, where familiar results are obtained by a novel route encountering iterative and chaotic phenomena of considerable current interest. We then moved on to dynamics of fractals, whose self similarity requires the use of a position space scaling method. Critical dynamics near the percolation threshold was then approached, where the scaling method can handle both criticality and local randomness. Richer iterative behaviour and crossover occur there, and we have to some extent concentrated on those features available from a straightforward generalisation of the viewpoints used earlier on. A number of extensions have already been made which there was not space to include (including some comparisons with experiment [36,37,38]), and many other very obvious extensions have yet to be worked out. More important omissions from this very selective account are a more detailed discussion of higher dimensional systems [5] and some discussion of densities of states and correlations in random tight binding and other models [4,6-9], where iterative techniques have been shown to be very powerful.

References

1. R.B. Stinchcombe, "Scaling and Criticality in Dilute Systems", lectures notes in this volume.
2. R.B. Stinchcombe, in "Highlights of Condensed Matter Physics", ed. F. Bassani, F. Fumi and M.P. Tosi, Proceedings of the Varenna Summer School, 1983 (North-Holland).
3. C.K. Harris and R.B. Stinchcombe, Phys. Rev. Lett. 50, 1399 (1983).
4. C.E.T. Goncalves da Silva and B. Koiller, Solid St. Commun. 40, 215 (1981).
5. B.W. Southern, A.A. Kumar, P.D. Loly and A-M.S. Tremblay, Phys. Rev. B27, 1405 (1983).
6. A-M.S. Tremblay and B.W. Southern, to be published.
7. M.O. Robbins and B. Koiller, Phys. Rev. B27, 7703 (1983).
8. C. Wiecko and E. Roman, to be published.
9. E. Anda, to be published.
10. R.B. Stinchcombe, Phys. Rev. Lett. 50, 200 (1983).
11. R.B. Stinchcombe and C.K. Harris, J. Phys. A17 (1984) to be published
12. S. Kirkpatrick, in "Ill-Condensed Matter", ed. R. Balian, R. Maynard and G.Toulouse (North-Holland, Amsterdam, 1979).
13. I. Ya. Korenblit and E.F. Shender, Usp. Fiz. Nauk 126, 233 (1978) [Sov. Phys. Usp. 21, 832 (1978)].
14. T.A.L. Ziman, J. Phys. C12, 2645 (1979).
15. B.B. Mandelbrot, Fractals (Freeman, San Francisco, 1977).
16. R. Rammal and G. Toulouse, J. Phys. (Paris) Lett. 44, L13 (1983).
17. R. Rammal and G. Toulouse, Phys. Rev. Lett. 49, 1194 (1982).
18. Y. Gefen, A. Aharony, B.B. Mandelbrot and S. Kirkpatrick, Phys. Rev. Lett. 47, 1771 (1981).
19. Y. Gefen, A. Aharony and S. Alexander, Phys. Rev. Lett. 50, 77 (1983)
20. S. Alexander and R. Orbach, J. Phys. (Paris) Lett. 43, 625 (1982).

21. Y. Gefen, A. Aharony, B. Mandelbrot and S. Kirkpatrick, in "Disordered Systems and Localisation", ed. C. Castellani, C. di Castro and L. Peliti (Springer, Berlin 1981).
22. L.G. Marland, D.Phil. Thesis, University of Oxford, 1978 (unpublished).
23. R.M. May, Nature 261, 459 (1976).
24. M. Feigenbaum, J. Stat. Phys. 19, 25 (1978), and 21, 669 (1979).
25. B.I. Halperin and P.C. Hohenberg, Phys. Rev. 117, 952 (1969).
26. J.M. Ziman, "Models of Disorder" (C.U.P., 1979).
27. R.B. Stinchcombe, to be published.
28. M.J. Stephen, Phys. Lett. A87, 67 (1981).
29. E. Domany, S. Alexander, D. Bensimon and L.P. Kadanoff, to be published.
30. S.J. Lewis and R.B. Stinchcombe, to be published.
31. J-M Langlois, A-M.S. Tremblay and B.W. Southern, to be published.
32. M.P.M. den Nijs, J. Phys. A12, 1857 (1979).
33. A. Aharony, STATPHYS. Proceedings, J. Stat. Phys. 1984, to be published.
34. R.B. Pandey and D. Stauffer, Phys. Rev. Lett. 51, 527 (1983).
35. R.B. Pandey, D. Stauffer, A. Margolina and Z.J. Zabolitzky, to be published.
36. A.C. Maggs and R.B. Stinchcombe, J. Phys. A17 (1984) to be published.
37. J.P. Boucher, C. Dupas, W.J. Fitzgerald, K. Knorr and J.P. Renard, J. Phys. (Paris) 39, L86 (1978).
38. Y. Endoh, G. Shirane, R.J. Birgeneau and Y. Ajiro, Phys. Rev. B19, 1476 (1979).

REAL SPACE RENORMALIZATION FOR FLUIDS AND INTERFACES

J.M.J. van Leeuwen
Laboratorium voor Technische Natuurkunde
2600 GA Delft, The Netherlands

In these lectures we discuss the possibilities for real space renormalization of the gas-liquid phase transition and critical point and the renormalization of the interface between two coexisting phases separated by a weak external field. Before we discuss these topics, we make a few remarks on the position of real space renormalization.

§1. Real space renormalization

The difference between real space renormalization and the field theoretic renormalization based on the Landau-Ginzburg-Wilson (LGW) hamiltonian is not so much the difference between working in real space and in momentum space. The method of Wilson [1)] is based on the following two ideas:

- long wave length fluctuations in the order parameter are responsible for the (non-classical) critical singularities.
- the effect of the long wave length fluctuations can be calculated by using self-similarity properties of a critical system under scale transformations.

The question then is, in what sense is the system selfsimilar. The field theoretic approach considers the appropriate LGW hamiltonian for the long wave length fluctuations of the order parameter. Scaling amounts to the integration over the short wave length components and determining the effect on the long wave components. When this effect is vanishing the scale transformation has a fixed point. Such a fixed point determines the critical singularities, which are universal i.e. insensitive for details of the molecular interaction. To a large extent universality results from the LGW hamiltonian as starting point in which most of the details of the interaction have been smoothed out.

Real space renormalization is more ambitious as it wants to realize the self-similarity for a microscopic hamiltonian. Here scaling means that the variables of the system are combined to new similar variables. For lattice systems (Ising model) this seems possible [2)] at least for a limited region of the phase diagram including the critical point. Universality here shows up through the properties of the (selfsimilar or) fixed point hamiltonian. It is attractive in the irrelevant (or non-singular) directions and so a critical fixed point dictates the singular properties for a whole surface of critical hamiltonians to be the same. In general two problems appear in real space renormalization

- to give the rules relating the new variables to the old ones;
- to work out the properties of the relation between the hamiltonians for the new and old variables.

It may seem that the second problem is the most difficult one and in a technical sense this is true. In fact one has usually a more difficult problem in determining the properties of a renormalization transformation then in evaluating directly the partition function. The hope is however that approximations, which fail at the critical point for the partition function, may be carried out successfully for a renormalization transformation. This depends of course crucially on the choice of the renormalization transformation i.e. on the first part of the problem. In particular in how far the rules for the transformation leave the essential long wave length fluctuations in the system and merely sum out the irrelevant short wave length fluctuations. In the overwhelming rich variety of renormalizations this aspect is often obscured.

In the Ising model the problem can be formulated as follows. The hamiltonian of the spinsystem is written in the general form

$$\mathcal{H}(s) = \sum_{\alpha} K_\alpha s_\alpha \tag{1.1}$$

where s_α stands for a spinfunction of type α. E.g. the spinfunctions corresponding to the order parameter s_ω and energy s_ε are

$$s_\omega = \sum_i s_i \qquad\qquad s_\varepsilon = \sum_{(ij)} s_i s_j \tag{1.2}$$

with s_i the spin of site i (and (ij) nearest neighbors sites). The coupling constants K_α represent the hamiltonian: K_ω is the magnetic field and K_ε the nearest neighbor coupling for the standard Ising model.

The renormalization transformation is defined as

$$e^{\mathcal{H}'(s')} = \sum_{\{s\}} P(s',s)\, e^{\mathcal{H}(s')} \tag{1.3}$$

with P(s',s) the weight function satisfying

$$P(s',s) \geq 0\,, \qquad \sum_{\{s'\}} P(s',s) = 1 \tag{1.4}$$

(1.3) is the defining relation for the new hamiltonian. $\mathcal{H}'(s')$ can be written again in the form (1.1) defining new coupling constants K'_α.

The relation

$$K'_\alpha = K'_\alpha(K_\beta) \tag{1.5}$$

is equivalent with the formal definition (1.3).

One sees that the conditions (1.4) leave a wide variety of possible transformations. (1.4) only garantees that the K'_α are real and that the partition functions of $\mathcal{H}$ and $\mathcal{H}'$ are the same.

A special class of weight functions is projective i.e. for a given configuration of {s} only one configuration {s'} is connected by a non-zero weight. The majority rule [3] assigning a new spin as the sign of the majority of a block of spin, is such a projective weight. In particular for these weight functions a very powerful method (dealing with the second half of the problem) was developed by Swendsen [4]. He observed that a Monte Carlo simulation of the spinsystem governed by $\mathcal{H}$ (s) is at the same time a simulation of the renormalized system governed by $\mathcal{H}'$(s'). So explicit knowledge of $\mathcal{H}'$(s') is not necessary for a Monte Carlo simulation to compute the correlation functions of s'_α, as well as mixed correlation function between s'_α and s_β. These two types of correlations are sufficient for determining the derivative matrix

$$T_{\alpha\beta} = \partial K'_\alpha \, / \, \partial K_\beta \tag{1.6}$$

which in turn yields directly the critical exponents.

This shows that the problem of the analysis of a renormalization transformation presents surmountable difficulties when the transformation is well chosen such that a critical system renormalizes sufficiently fast to a fixed point. In a recent publication [5] Swendsen has improved his method by using the freedom in choosing P(s',s) such that the fixed point coincides with the critical unrenormalized hamiltonian. It demonstrates that the choice in P(s',s) is to a certain extent free with no influence on the critical exponents.

§2. Renormalization of fluids

The above mentioned problems of real space renormalization come into focus when one intends to design a real space renormalization scheme for a fluid. The fluid is believed to be in the same universality class as the Ising model since its order parameter, the density, has one component. Thus the critical exponents are known quite accurately. The fluid has also a long history of approximate theories which give the non-universal features, such as location of critical point and coexistence region, rather satisfactorily but which all lead to classical critical behavior.

A theory which gives a complete picture of universal and non-universal properties is hardly conceivable, but it would be desirable to have a description which is as accurate as the classical approximations further away from the critical point and comparably accurate to real space renormalization theories in the vicinity of the critical point.

Conceptually the problem is also whether a renormalization on a microscopic level is possible and whether a lattice structure is necessary for a successful renormalization. Schofield and Hubbard [6)] have constructed a map by which the fluid properties are associated with a (general) Ising model. In this map the (small) asymmetry of the fluid is obscured. M.S. Green [7)] studied the problem in the context of Mac-Millans solution theory and the approach we will sketch here (based on the Ph.D. thesis of F. van Dieren [8)]) is in spirit similar to Green's attempt.

The basic ingredients of the renormalization procedure are the reduction of the number of particles under a simultaneous reduction of the size of the system. Already in this respect a fundamental difference appears between the Ising model and the fluid. In the Ising model a combining of spins to a block spin immediately dictates the reduction of the size of the system as the ratio of block distance to site distance. In the fluid the two operations are independent and have to be combined in an intelligent way.

A second difference between lattice models and a fluid is that for a lattice a finite reduction of the system size is the most natural situation whereas for the fluid the elimination of an infinitesimal fraction of the particles is achieved most easily. In this respect the fluid is an advantage as it allows to make the transformation formulae more explicit. On the other hand the continuous potentials are much more difficult to handle than the discrete interaction constants for a lattice.

Before we discuss the renormalization we have to give a few definitions of the system.

§2.1. Definitions for the system and the renormalization transformation

We consider a system of particles in an external field $\mu(\vec{r})$ interacting through 2-body potentials $\varphi(\vec{r}_1,\vec{r}_2)$, 3-body potentials $\varphi(\vec{r}_1,\vec{r}_2,\vec{r}_3)$ etc. We are interested in the case where 3-body and higher potentials are absent, but for the general structure of the theory they are necessary. Also the external field $\mu(\vec{r})$, or thermodynamic potential, can be taken as constant as well as the potentials translational invariant. Here it is advantageous to consider μ as a 1-particle potential and keep the $\vec{r}$-dependence where it does not give a complication. As is customary in renormalization theory the factor $-1/k_BT$ is incorporated in the potentials (and $1/k_BT$ in μ).

The energy of ℓ particles is given by

$$H(\vec{r}_1 \ldots \vec{r}_\ell) = \sum_{j=1}^{\ell} \mu(\vec{r}_j) + \sum_{(ij)} \varphi(\vec{r}_i, \vec{r}_j) + \ldots + \varphi(\vec{r}_1 \ldots \vec{r}_\ell) \tag{2.1}$$

which leads to the probability

$$W(\vec{r}_1 \ldots \vec{r}_\ell) = \frac{1}{Z_{gr}} \frac{1}{\ell!} \exp H(\vec{r}_1 \ldots \vec{r}_\ell) \tag{2.2}$$

of finding ℓ particles at positions $\vec{r}_1 \ldots \vec{r}_\ell$. The normalizing factor Z_{gr} reads

$$Z_{gr} = \sum_{N=0} \frac{1}{N!} \int d\vec{r}_1 \ldots d\vec{r}_N \exp H(\vec{r}_1 \ldots \vec{r}\ ) \tag{2.3}$$

For a translational invariant system of volume V, Z_{gr} gives the pressure p as

$$p\, V = \log Z_{gr} \tag{2.4}$$

(where $1/k_BT$ is also included in p).

In analogy with the Ising model we define a renormalization transformation on the basis of a set of weight functions P with the property

$$\begin{aligned} &P(\vec{r}_1' \ldots \vec{r}_M'; \vec{r}_1 \ldots \vec{r}_N) \geq 0 \\ &\sum_M \int d\vec{r}_1' \ldots d\vec{r}_M' \, P(\vec{r}_1' \ldots \vec{r}_M'; \vec{r}_1 \ldots \vec{r}_N) = 1 \end{aligned} \tag{2.5}$$

We consider only $M \leq N$ i.e. we make only connections of new configurations $\vec{r}_1', \ldots \vec{r}_M'$ with equal or less particles than the old configurations $\vec{r}_1 \ldots \vec{r}_N$. From P follows the renormalization transformation

$$W'(r_1' \ldots r_M') = \sum_N \int d\vec{r}_1 \ldots d\vec{r}_N \, P(\vec{r}_1' \ldots \vec{r}_M'; \vec{r}_1 \ldots \vec{r}_N) \, W(\vec{r}_1 \ldots \vec{r}_N) \tag{2.6}$$

The first property (2.5) of P guarantees the W' to be positive and the second property (2.5) forces the normalization

$$\sum_N \int d\vec{r}_1 \ldots d\vec{r}_N \; W(\vec{r}_1 \ldots \vec{r}_N) = 1 \tag{2.7}$$

to hold also for the new W'.

From the probabilities W' follow the new energies $H'(\vec{r}_1 \ldots \vec{r}_\ell)$ and the new potentials $\mu'(\vec{r}_1')$, $\varphi'(\vec{r}_1', \vec{r}_2')\ldots$. Formally they can be obtained by considering (2.6) successively for M = 0 yielding the new Z'_{gr} (and therefore p'), for M = 1 yielding $\mu'(\vec{r}_1')$ etc. In order that the new potentials are acceptable they must satisfy cluster properties which will follow from cluster properties to be imposed on $P(\vec{r}_1' \ldots \vec{r}_M'; \vec{r}_1 \ldots \vec{r}_N)$. We view P as the conditional probability of a stochastic process by which configurations $\vec{r}_1 \ldots \vec{r}_N$ are transformed in $\vec{r}_1' \ldots \vec{r}_M'$. An example may help to visualize the working of P.

§2.2. Spatial rearrangement

A special of transformation is of the form

$$P(\vec{r}_1' \ldots \vec{r}_M'; \vec{r}_1 \ldots \vec{r}_N) = \delta_{MN} \, \delta(\vec{r}_1' - \vec{r}_1 - \vec{a}_1) \ldots \delta(\vec{r}_N' - \vec{r}_N - \vec{a}_N) \tag{2.8}$$

i.e. particle i is displaced over a distance $\vec{a}_i$. The conditions (2.5) on P are satisfied when the $\vec{a}_i$ do not depend on the $\vec{r}_i'$ (but they may depend on $\vec{r}_1 \ldots \vec{r}_N$). The transformation (2.6) can be worked out easily for the case of uniform rescaling generated by

$$\vec{a}_i = (b^{-1} - 1)\vec{r}_i \tag{2.9}$$

with the result

$$W'(\vec{r}_1' \ldots \vec{r}_N') = b^{dN} W(b\vec{r}_1', \ldots b\vec{r}_N') \tag{2.10}$$

i.e. the particle position $\vec{r}_i = b\vec{r}_i'$ are reduced to $\vec{r}_i'$. The new coordinates $\vec{r}_i'$ occupy a volume $V' = b^{-d}V$. For p and the potentials then follows

$$\begin{cases} p' = b^d p \\ \mu'(\vec{r}_1') = d \log b + \mu(b\vec{r}_1') \\ \varphi'(\vec{r}_1',\vec{r}_2') = \varphi(b\vec{r}_1',b\vec{r}_2') \ \ldots \end{cases} \tag{2.11}$$

We are most interested in the infinitesimal generator of the transformation which is obtained by putting

$$b = 1 + dt \tag{2.12}$$

$$p' - p = \dot{p}\,dt \qquad \mu'(\vec{r}') - \mu(\vec{r}') = \dot{\mu}(\vec{r})dt \quad \text{etc.}$$

and ignoring terms beyond the linear terms in dt:

$$\begin{cases} \dot{p} = pd \\ \dot{\mu}(\vec{r}) = d + \vec{r}.\vec{\nabla}\mu(\vec{r}) \\ \dot{\varphi}(\vec{r}_1,\vec{r}_2) = (\vec{r}_1.\vec{\nabla}_1 + \vec{r}_2.\vec{\nabla}_2)\ \varphi(\vec{r}_1,\vec{r}_2) \\ \quad \ldots \qquad \ldots \qquad \ldots \\ \dot{\varphi}(\vec{r}_1\ldots\vec{r}_\ell) = \sum_{i=1}^{\ell} (\vec{r}_i.\vec{\nabla}_i)\varphi(\vec{r}_1\ldots\vec{r}_\ell) \end{cases} \tag{2.13}$$

The transformation (2.11) or its infinitesimal generator (2.13) reduces the size of the system without changing the number of particles. The advantage of the infinitesimal form (2.13) is that it can be added simply to infinitesimal generators of other (decimation) transformations.

A generalization of (2.13) is obtained by writing

$$\vec{a}_i = \vec{v}_i(\vec{r}_1\ldots\vec{r}_N)dt \tag{2.14}$$

where the $\vec{v}_i$ are the "velocities" of the displacement $\vec{a}_i$. By doing the integrals or by simple considerations of conservation of probability one finds for (2.14)

$$\dot{W}(\vec{r}_1\ldots\vec{r}_N) = -\sum_i \vec{\nabla}_i.[\vec{v}_i(\vec{r}_1\ldots\vec{r}_N)W(\vec{r}_1\ldots\vec{r}_N)] \tag{2.15}$$

It can be converted in a flow for the energies as

$$\dot{H}(\vec{r}_1 \ldots \vec{r}_N) = -\sum_i \vec{v}_i(\vec{r}_1 \ldots \vec{r}_N)\, \vec{\nabla}_i H(\vec{r}_1 \ldots \vec{r}_N) - \sum_i \vec{\nabla}_i . \vec{v}_i(\vec{r}_1 \ldots \vec{r}_N) \qquad (2.16)$$

For $\vec{v}_i = -\vec{r}_i$ the transformation (2.13) is retrieved. In order that acceptable potentials result the velocity field must have certain cluster properties.
A way to realize this is to write

$$\vec{v}_i = -\vec{\nabla}_i \Psi(\vec{r}_1 \ldots \vec{r}_N) \qquad (2.17)$$

and expand the Ψ in single particle, 2-particle etc. contribution

$$\Psi = \sum_i \phi_1(\vec{r}_i) + \sum_{ij} \phi_2(\vec{r}_i, \vec{r}_j) + \ldots \qquad (2.18)$$

The choice $\phi_1(\vec{r}) = r^2/2$ realizes the uniform rescaling given by (2.13). The second term yield with

$$\phi(\vec{r}_i, \vec{r}_j) = \phi(\vec{r}_i - \vec{r}_j) \qquad (2.19)$$

in a spatially uniform case

$$\dot{\varphi}(r) = 2 \frac{d\phi(r)}{dr} \frac{d\ (r)}{dr} + 2\Delta_r \phi(r)$$

$$\dot{\varphi}(r_1, r_2, r_3) = \vec{\nabla}_1 \phi(r_{12}) \cdot \vec{\nabla}_1 \varphi(r_{13}) + \vec{\nabla}_1 \phi(r_{13}) \cdot \vec{\nabla}_1 \varphi(r_{12}) \qquad (2.20)$$

$$+ \ldots + \vec{\nabla}_3 \phi(r_{23}) \cdot \vec{\nabla}_3 \varphi(r_{13}) + \ldots$$

with no change in the higher order potentials nor in μ or p. In the flow for $\dot{\varphi}(\vec{r}_1 \vec{r}_2 \vec{r}_3)$ only the 2-particle terms are indicated.

§2.3. Weighted decimation

Without removing particles or replacing groups of particles by lesser particles renormalisation is impossible. In this section we consider the processes by which an N-particle configuration is decimated to an (N-1)-particle configuration.

Moreover we study again the infinitesimal generator of such a process by writing

$$P(\vec{r}'_1 \ldots \vec{r}'_M; \vec{r}_1 \ldots \vec{r}_N) = \delta_{MN}\delta(\vec{r}'_1-\vec{r}_1) \ldots \delta(\vec{r}'_N-\vec{r}_N) + T(\vec{r}'_1 \ldots \vec{r}'_M; \vec{r}_1 \ldots \vec{r}_N)dt \tag{2.21}$$

T is the transition rate of the process. Including only N → N and N → N-1 processes we have

$$T(\vec{r}'_1 \ldots \vec{r}'_{N-1}; \vec{r}_1 \ldots \vec{r}_N) = \sum_{j=1}^{N} S_j \, \delta(\vec{r}'_1-\vec{r}_1) \ldots \delta(\vec{r}'_{j-1}-\vec{r}_{j-1})\delta(\vec{r}'_j-\vec{r}_{j+1}) \ldots \delta(\vec{r}'_{N-1}-\vec{r}_N)$$

$$T(\vec{r}'_1 \ldots \vec{r}_N; \vec{r}_1 \ldots \vec{r}_N) = R\,\delta(\vec{r}'_1-\vec{r}_1) \ldots \delta(\vec{r}'_N-\vec{r}_N) \tag{2.22}$$

S_j is the decimation probability by which the j-th particle is removed out of 1...N. S_j will depend on the configuration $\vec{r}_1 \ldots \vec{r}_j \ldots \vec{r}_N$ (but there is no use to have it also dependent on $\vec{r}'_1 \ldots \vec{r}'_{N-1}$ in view of the δ-functions). The normalization of P as expressed by (2.5) implies that

$$R(\vec{r}_1 \ldots \vec{r}_N) = \sum_{j=1}^{N} S_j(\vec{r}_1 \ldots \vec{r}_N) \tag{2.23}$$

A value of S_j independent of the configuration will be called simple decimation. In general we make the expansion

$$S_j(\vec{r}_1 \ldots \vec{r}_N) = s + \sum_{\ell \neq j}^{N} s(r_{\ell j}) + \ldots + s_j(\vec{r}_1 \ldots \vec{r}_N) \tag{2.24}$$

When (2.21) is inserted into (2.6) and the expressions (2.22) - (2.24) are used for T we can derive explicit expressions for the flow of the potentials. The result is for the spatial homogeneous case [8)]

$$\begin{cases} \dot{\mu} = -s + sz\int d\vec{r}_2\ (f(r_{12}) + s(r_{12})W(r_{12})) \\ \\ \dot{\varphi}(r_{12}) = -\ 2s(r_{12}) + z\int d\vec{r}_3[\,sf(r_{13})f(r_{23}) + 2s(r_{13})W(r_{13})f(r_{23}) \\ \qquad\qquad + s_3(r_1\, r_2\, r_3)W(r_{13})W(r_{12})] + \ldots \\ \\ \dot{\varphi}(\vec{r}_1\,\vec{r}_2\,\vec{r}_3) = -\ s_1(\vec{r}_1\,\vec{r}_2\,\vec{r}_3) - s_2(\vec{r}_1\,\vec{r}_2\,\vec{r}_3) - s_3(\vec{r}_1\,\vec{r}_2\,r_3) \\ \qquad\qquad + sz\int d\vec{r}_4\, f(r_{14})\, f(r_{24})\, f(r_{34}) + \ldots \end{cases} \tag{2.25}$$

z and f stand for the abbreviation

$$z = e^{\mu}, \qquad f(r_{12}) = W(r_{12}) - 1 \tag{2.26}$$

In (2.25) terms involving $\varphi(\vec{r}_1\,\vec{r}_2\,\vec{r}_3)$ in the equation for $\varphi(r_{12})$ and $\varphi(\vec{r}_1\,\vec{r}_2\,\vec{r}_3)$ have been omitted as well as the 4-particle potential which occurs in the expression for $\varphi(\vec{r}_1\,\vec{r}_2\,\vec{r}_3)$. In general the flow of the n-particle potential is determined by the potentials up to order n+1. Such a structure of the hierarchy would have to be expected. The remarkable aspect is that the flow can be made explicit, in contrast to the Ising model where the transformation equation can only be given approximately. This is a fruit of studying the infinitesimal generator. A finite resolution of the number of particles would lead also to a hierarchy where in principle all the potentials occur in each of the transformation equations.

We have generalized the simple decimations for two reasons

- It is known [9)] that simple (or unweighted) decimation cannot lead to an acceptable fixed point in the flow. The reason is that the correlation functions do not change under simple decimation. (They do however change under the influence of the 2-particle and higher weights $s(r_{12})$, $s_1(\vec{r}_1\,\vec{r}_2\,\vec{r}_3)$ ...). As the system becomes more and more dilute the $W(\vec{r}_1 \ldots \vec{r}_N)$ must finally become equal to the original correlation functions (being invariant). An additional uniform spatial rescaling as (2.13) will not change this picture but only collapse all structure to a point.
- One observes that the 3-particle $s_1(\vec{r}_1\,\vec{r}_2\,\vec{r}_3)$ occurs in the flow of the 3-particle potential $\varphi(\vec{r}_1\,\vec{r}_2\,\vec{r}_3)$. This weight can be used to cancel terms arising from the simple decimation process. Thus the undesirable growth of the many particle potentials due to decimation from a situation where it is initially absent may be suppressed by well chosen many particle weights $s_1(\vec{r}_1\,\vec{r}_2\,\vec{r}_3)$ etc.

§2.4. Outlook

The analysis of the flow equation based on a combination of the flows of (2.13), (2.20) and (2.25) is difficult and delicate. Most promising is to make a definite choice for the functions $\phi(r)$, s and s(r) and perform a Monte Carlo simulation of a fluid system and analyze the data on the basis of the Monte Carlo Renormalization dictated by $\phi(r)$, s and s(r).

As an exploration for the success of such an attempt we have investigated schemes which contribute (2.13), (2.20) and (2.25) ignoring the influence of 3 and more particle potentials. These equations do show a critical fixed point but the eigenvalues are poor. So it means that the influence of higher potentials is important. This conclusion is surprising as the many spin interactions are rather unimportant in the real space renormalization of the Ising model. It means that these effects must be included through Monte Carlo Renormalization or by an appropriate choice of the higher weights. The best would be a variational principle by which the influence could be minimalized.

§3. Real space renormalization for the interface

At a phase transition two phases are in coexistence. With an external force the phases can be separated as e.g. gravity separates liquid from vapor. In the context of an Ising model the separating field should be a spatially varying magnetic field. In general it is the field conjugated to the order parameter. The separating field should vary slowly as function of positions because strongly varying fields will deform interfaces. The interesting cases are fields that hardly vary on a molecular basis (as the gravity potential in the laboratory). Before we enter into the problem from the renormalization side let us review the mean field treatment in the Ising model.

§3.1. Mean field theory

We change the Ising hamiltonian (1.1) to

$$\mathcal{H} = \sum_i H_i s_i + K \sum_{(ij)} s_i s_j \tag{3.1}$$

with the local field H_i depending on position as

$$H_i = -gz_i \tag{3.2}$$

z_i being the z-coordinate of the position of the i-th lattice point. We have chosen the proportionality as -g to stress the analogy with the gravity for the fluids.

In the mean field theory the magnetisation m_i at site i is given by

$$m_i = \text{tgh}\,(H_i + K \sum_\delta m_{i+\delta}) \tag{3.3}$$

where the sum order δ runs through the neighboring sites $i+\delta$ of site i. Inverting (3.3) yields

$$H_i = \frac{1}{2} \log \frac{1+m_i}{1-m_i} - K \sum_\delta m_{i+\delta} \tag{3.4}$$

Now first assume that $m_{i+\delta}$ equals m_i , then (3.4) becomes

$$H_i = \mu(m_i) = \frac{1}{2} \log \frac{1+m_i}{1-m_i} - K c\, m_i \tag{3.5}$$

with c the coordination number of the lattice. (In the fluid μ is the thermodynamic potential). (3.5) is a local relation between field and magnetisation. When the field H_i is prescribed as in (3.2) the magnetisation will follow.
In fig. 1 we have sketched the function $\mu(m)$ for a temperature above T_c or equivalently for a coupling $K < K_C = 1/C$ (dotted line).

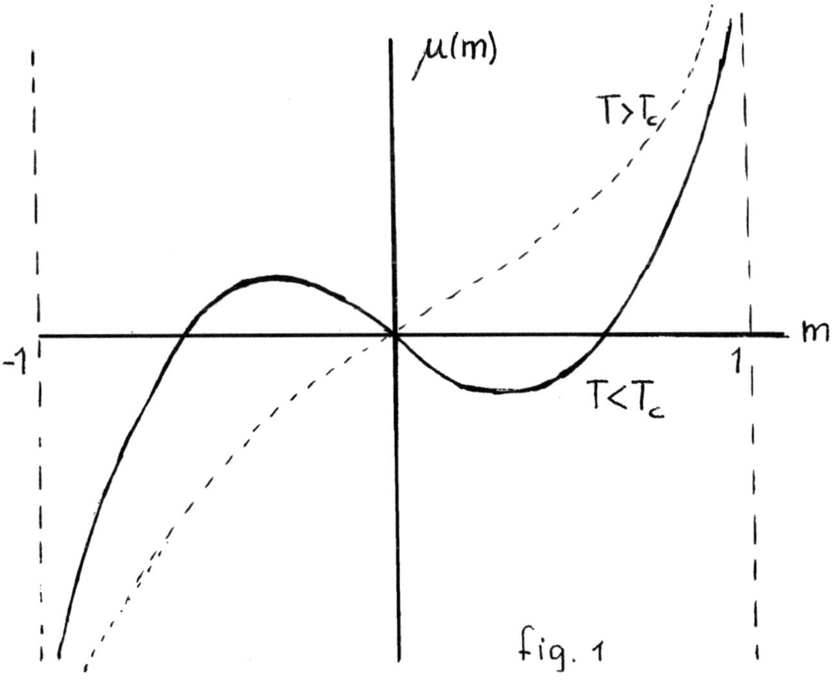

fig. 1

Turning the picture 90° gives the magnetisation as function of the μ or of the position z as μ (or H) varies linearly with z. Below T_C the curve $\mu(m)$ exhibits a Van der Waals loop and the magnetisation profile would jump from the ("liquid" or low z) branch to the ("vapor" or high z) branch with a sharp interface at $z = 0$ (or $\mu = 0$). This is obviously against the assumption $m_{i+\delta} \simeq m_i$ near the interface and we want to do better by writing

$$\sum_\delta m_{i+\delta} = \sum_\delta m(z_{i+\delta}) \simeq c\, m(z_i) + a^2 \frac{\partial^2 m}{\partial z^2} \tag{3.6}$$

Then (3.5) changes in the (continuous) form

$$H(z) = \mu(m(z)) - K a^2 \frac{\partial^2 m}{\partial z^2} \tag{3.7}$$

As H(z) is given $m(z)$ can be solved from this equation. Most interesting is the limit $g \to 0$ and $K \to K_C$ from above (i.e. $T \to T_c$ from below). Then m is small and we may write

$$\mu(m) \simeq (1 - K/K_c)m + \frac{m^3}{3} + \ldots \tag{3.8}$$

With (3.8) in (3.7) it can be solved for $g \to 0$

$$m(z) = -\sqrt{3t}\ \mathrm{tgh}\left(\sqrt{\frac{t}{2K}}\left(\frac{z - z_0}{a}\right)\right) \tag{3.9}$$

with $t = K/K_c - 1 = Kc - 1$. The important point is that the interface remains in absence of a separating field ($g \to 0$) apart from an arbitrary location z_0. Thus the interface is said to be intrinsic.

We want to make two comments on this result:

- The result has all the classical shortcomings of the mean field theory in its dependence on t (or $T - T_c$)(it should be t^β outside the tgh and t^ν inside the tgh (see (3.17)).
- The tgh-shape comes essential from the Van der Waals loop, which is a dubious feature for an equation of state in equilibrium statistical mechanics.

§3.2. Improvement of the mean field theory

One way to improve then the mean field theory is to replace in (3.7) the functions $\mu(m)$ by the true $H(m)$ relation as it applies in a fully homogeneous system.

For fluids this amounts to inserting the $\mu(\rho)$ relation for the fluid. Above T_c this is a unique procedure as the equation of state is known with all its singular (universal) behavior. The second derivative in (3.7) then yields the deviations from the density profile $\rho(z)$ as it would be calculated from local relation

$$-gz = \mu(\rho(z)) \tag{3.10}$$

which is nothing else than the barometric formulae for a non-ideal gas. Equation (3.7) with the true $\mu(\rho)$ function has been solved by Sengers and Van Leeuwen [10] for the scaling regime near and above T_c. The deviations from the local profile (3.10) are small and only appreciable in a narrow temperature regime near T_c (and densities $\rho(z)$ near ρ_c).

Below T_c, which is the more interesting regime, equation (3.7) with the true $\mu(\rho)$ has been solved by Fisk and Widom [11]. They inserted a guess for the Van der Waals loop which is consistent with the scaling properties for the equation of state above and below T_c outside the coexistence region. Although the resulting profiles are presumably quite satisfactory for fluids (detailed measurements are difficult) the procedure is theoretically unsatisfactory. The reason is that the Van der Waals loop is obtained from equilibrium statistical physics whereas one knows that a strict evaluation of the partition function always gives a flat piece in the equation of state.

§3.3. Simple renormalization

A preliminary idea about the behavior of the interface may be obtained from simple renormalization procedures. Let us introduce the variable $t = (K_c - K)/K_c \simeq (T - T_c)/T_c$. Near the (critical) fixed point the scaling equations are (in the homogeneous system)

$$\begin{cases} t' = b^{y_T} t \\ H' = b^{y_H} H \\ z' = b^{-1} z \end{cases} \quad \Rightarrow \quad m' = b^{y_M} m \quad (y_M = d - y_H) \tag{3.11}$$

We have added the trivial scaling relation for distances $z' = b^{-1}z$. Now assume [12] that for a slowly varying magnetic field $H = -gz$ the local relation (3.11) applies then one finds for g the scaling relation

$$g'z' = b^{y_H} gz \qquad \text{or} \qquad g' = b^{y_H+1} g \tag{3.12}$$

Consequently we find the scaling law for the magnetisation m as function of g, z and t

$$m(b^{y_H}g,\ b^{-1}z,\ b^{y_T}t) = b^{y_M} m(g, z, t) \tag{3.13}$$

This shows that m will, near the critical point, be a homogeneous function of g, z, t and not simply a function of the local combination gz as dictated by (3.10).

The solution of (3.13) can be written in the form

$$m(z, g, t) = g^{\beta\varphi}\, \tilde{m}(z\, g^{\nu\varphi},\ tg^{-\varphi}) \tag{3.14}$$

where we have translated the exponents $y_T = \nu^{-1}$ and $y_H = d\delta / (1+\delta)$ in the usual β, δ and ν and where φ stands for the combination

$$\varphi = 1 / (\beta\delta + \nu) \tag{3.15}$$

The homogeneous function m must be obtained from the renormalization trajectories and the boundary value.

The solution obtained by Sengers and Van Leeuwen [10] for the density profile of a fluid in a gravitational field for $T > T_c$ has indeed the form (3.14). For small g (or large second argument in $\tilde{m}$) the form (3.14) must become a function of the combination zg (as far as z and g are concerned) as in the solution of (3.10). It does indeed and as a consequence of (3.14) the profile obtains in the scaling regime (where (3.14) holds) the character

$$m = t^{\beta}\, \hat{m}\, (zg / t^{\beta\delta}) \tag{3.16}$$

For the fluid the functions $\tilde{m}$ and $\hat{m}$ are calculated for $T > T_c$ (or $t > 0$) [10] and by universality these functions also apply to the Ising model.

It is not possible to extract the behavior for $T < T_c$ out of the form for $T > T_c$. In fact the behavior for $T < T_c$ is much more interesting and shows an interface. In the Fisk-Widom theory [11] one obtains the limit of (3.14) for $g \to 0$. In that theory the limit exists and thus the profile is said to be intrinsic. (3.14) tells us that it should have the form

$$m = t^{\beta}\, \bar{m}\, (z\, t^{\nu}) \qquad (3.17)$$

The interfacial profile has a length scale $t^{-\nu} \sim \xi$ (the correlation length) and a magnitude t^{β} which is proportional to the spontaneous magnetisation.

Whether such an intrinsic profile exists has been a debate since the work of Buff, Lovett and Stillinger [13] on the capillary waves.

§3.4. Capillary waves

Imagine an interface with a finite thickness (e.g. (3.17)) and consider a surface wave of wave length long with respect to its thickness. In a gravitational field the energy of such a wave can be expressed as

$$E = \int d\vec{r}_{\perp}\, (g\, \Delta\rho\, h(\vec{r}_{\perp}) + \sigma_b\, |\, \vec{\nabla}_{\perp} h(\vec{r}_{\perp})\, |^2) \qquad (3.18)$$

Here $\vec{r}_{\perp}$ is a coordinate along the interface and $h(\vec{r}_{\perp})$ is the height at the point $\vec{r}_{\perp}$ (the field is perpendicular to the interface). $\Delta\rho$ is the density difference between the phases and σ_b a "bare" surface tension. In the Ising language σ_b would correspond to the combination Ka^2 in (3.7) and $\Delta\rho$ to the jump in spontaneous magnetisation. In terms of fourier components

$$h_{k_{\perp}} = \int d\vec{r}_{\perp}\, e^{i\vec{k}_{\perp}\cdot\vec{r}_{\perp}}\, h(\vec{r}_{\perp}) \qquad (3.19)$$

the energy reads

$$E = \sum_{\vec{k}_{\perp}} (g\, \Delta\rho + \sigma_b\, k_{\perp}^2)(h_{k_{\perp}})^2 \qquad (3.20)$$

Now it is a standard calculation in statistical mechanics to compute the free energy of such independent waves (the Gaussian model) with the result

$$\Delta\sigma = -\frac{k_B T}{2}\int \frac{dk_\perp}{(2\pi)^{d-1}} \log\left[\frac{2\pi k_B T}{g\Delta\rho + \sigma_b k_\perp^2}\right] \tag{3.21}$$

For an appropriate short wavelength (high $k_\perp$) cut-off this expression is convergent (also for $g \to 0$) and leads to an amount $\Delta\sigma$ to be added to the bare surface tension σ_b.

However the mean square displacement

$$\langle h^2\rangle = \frac{2}{\Delta\rho}\frac{\partial\Delta\sigma}{\partial\Delta\rho} = k_B T\int \frac{d\vec{k}_\perp}{(2\pi)^{d-1}} \frac{1}{g\Delta\rho + \sigma_b k_\perp^2} \tag{3.22}$$

diverges at the low $k_\perp$- side for $g \to 0$ for $d \leq 3$. For $d = 3$ the divergence is as log g and it is this catastrophe which causes the doubt on the existence of an intrinsic profile. For the long wave lengths the expression (3.18) is correct and the divergence is therefore not an artifact of the model. Although $\langle h^2\rangle$ for realistic values of g or realistic sizes of the container (which produce also a low $\vec{k}_\perp$ cut-off) remains of the order of microscopic distances the capillary waves are a serious conceptual difficulty in defining the interface. The general expression (3.14) is not in question but the problem is to deduce the small g behavior. It is not likely to be given by (3.17) in view of the capillary waves. The problem is intimately connected with the meaning of the Van der Waals loop in the equation of state, which allows the limit $g \to 0$ to be taken with a remaining intrinsic profile.

§3.5. Real space renormalization

Jasnow and Rudnick [14)] have studied the interfacial problem in the context of the ε-expansion and directly for $d = 3$ by field theoretic methods. Their conclusion is that for $d > 3$ an intrinsic profile results of the form (3.17) but that for $d = 3$ the capillary waves affect the interface such that even the universality of the profile is affected. (Their result can be cast in the form (3.14) but an extra non-universal cut-off parameter appears in the theory).

As real space renormalization is more suited to deal with spatially varying fields we now apply to the hamiltonian (3.1) a real space renormalization. For simplicity we disregard the proliferation of the coupling constants such that the renormalized hamiltonian is of the form

$$\mathcal{H}' = \sum_{i'} H'_{i'} s'_{i'} + K' \sum_{(i'j')} s_{i'} s_{j'} \tag{3.23}$$

In the appendix the Migdal transformation is described which indeed is restricted to a transformation of field variables H_i and a nearest neigbor coupling constant K. For our purpose it is sufficient to know that the transformation leaves the free energy invariant

$$F(\{H_i\},\ K) = F(\{H'_{i'}\},\ K') \tag{3.24}$$

and that the local response of the renormalized field $H'_{i'}$ at i' to a variation of H_i at i can be calculated

$$T_{i'i} = \delta H'_{i'} \ / \ \delta H_i \tag{3.25}$$

In fact (3.25) is a generalization of (1.6) for spatially varying coupling constants. In the appendix $T_{i'i}$ is computed for the Migdal transformation (see (A.9)).

From (3.24) and (3.25) we may deduce the transformation of the magnetisation profile

$$m_i = \frac{\partial F(\{H_i\},K)}{\partial H_i} = \frac{\partial F(\{H'_{i'}\},K')}{\partial H'_{i'}} \cdot \frac{\delta H'_{i'}}{\delta H_i} = \sum_{i'} m'_{i'}\ T_{i'i} \tag{3.26}$$

The essence of (3.26) is that $T_{i'i}$ relates the profile to the renormalized profile $m'_{i'}$. Repeated renormalization thus establishes an expression of the profile in terms of a limiting profile far away from the critical point. Normally one would expect that the limiting profile is the $T = 0$ interface and there is every reason to take this as a sharp interface. Not only all the excitations (e.g. capillary waves) are frozen out but also the field gradient (3.2) becomes stronger under renormalization as indicated in (3.12).

The consequence of such an argumentation is that with a sharp interface at $T = 0$ an intrinsic (and universal) interface results near T_c which indeed has the form (3.14) and for which the $g \to 0$ limit can be taken as in (3.17). The advantage of this renormalization derivation is that in no way the Van der Waals loop is used.

For $d \leq 3$ one must doubt whether such a result is valid as it misses the roughening transition which is known to exist for the $d = 3$ Ising model. For $T < T_R$ the interface is smooth and exists for $g \to 0$ but for $T > T_R$ the interface is rough and has to be located by a $g \neq 0$ or an equivalent constraint. Above the roughening transition the capillary waves play a role in determining the shape of the Ising interface. This is a problem in which the fluid and Ising model are likely not to be in the same universality class as the liquid surface is believed to be always in the rough phase (the reason is that in the Ising only discrete excitations exist whereas the liquid surface can always be perturbed by excitations with arbitrary small energy).

§3.6. Anisotropic renormalization

A possible explanation why a straightforward renormalization misses the roughening transition is that an equal reduction of scale in all directions washes out too fast the delicate details of the interfaces. A more subtile analysis can be made when the reduction of scale parallel and perpendicular to the gradient direction is taken different [16)]

$$b_{\parallel} = b_{\perp}^{a} \tag{3.27}$$

where a is a parameter to be adjusted.
Migdal's transformation allows such an anisotropic reduction. In the appendix (A.6) we find for $b = 1 + dt$

$$\begin{aligned} \dot{K}_{\perp} &= (d-2+a)K_{\perp} + B(K_{\perp}) \\ \dot{K}_{\parallel} &= (d-1)K_{\parallel} + a\,B(K_{\parallel}) \end{aligned} \tag{3.28}$$

with B(K) given by (A.5)

$$B(K) = (\sin hK)(\cos h K)\log \mathrm{tgh} K \tag{3.29}$$

As a consequence of the anisotropy (3.27) in the renormalization the nearest neighbor coupling parallel ($K_{\parallel}$) and perpendicular to the field gradient ($K_{\perp}$) become different (even when started in an isotropic $K_{\parallel} = K_{\perp}$ case).
We anticipate to have $a < 1$ as we want to reduce the scale in the field direction (perpendicular to the interface) slower to keep the interface details better.
Then one sees from (3.28) and (3.29) that $K_{\parallel}$ grows more rapid than $K_{\perp}$ and the Ising model develops below T_c into the SOS model, which can be considered as the extreme anisotropic Ising model.

The parameter a can now be chosen such that the capillary waves are invariant. The mean square displacement scales as $b_{\parallel}^{-2} = b_{\perp}^{-2a}$ and the field gradient scales as

$$g' = b_{\perp}^{y_H + a}\, g \tag{3.30}$$

with y_H given by (A.12)

$$y_H = -2\{(d-1)B(K_\perp) + aB(K_{/\!/})\} \tag{3.31}$$

For $1 < d < 3$ the quantity $\langle h^2\rangle$ from (3.22) behaves for $g \to 0$ as

$$\langle h^2\rangle \sim g^{\frac{3-d}{2}} \tag{3.32}$$

In order that this proportionality remains true after renormalization we must require for a

$$a = \frac{(d-1)(3-d)(-2B(K_\perp))}{d+1+2(3-d)B(K_{/\!/}))} \tag{3.33}$$

With this value of a we may ask whether (3.28) has another (interface) fixed point for $K^*_{/\!/} = \infty$. Indeed this is the case for a^* and K^* given by

$$\begin{aligned} a^* &= -(3-d)B(K^*_\perp) \\ [d-2-(3-d)B(K^*_\perp)]K^*_\perp + B(K^*_\perp) &= 0 \end{aligned} \tag{3.34}$$

This system has a solution for $1 < d \leq 3$ and K^* is plotted in fig. 2 as function of d.

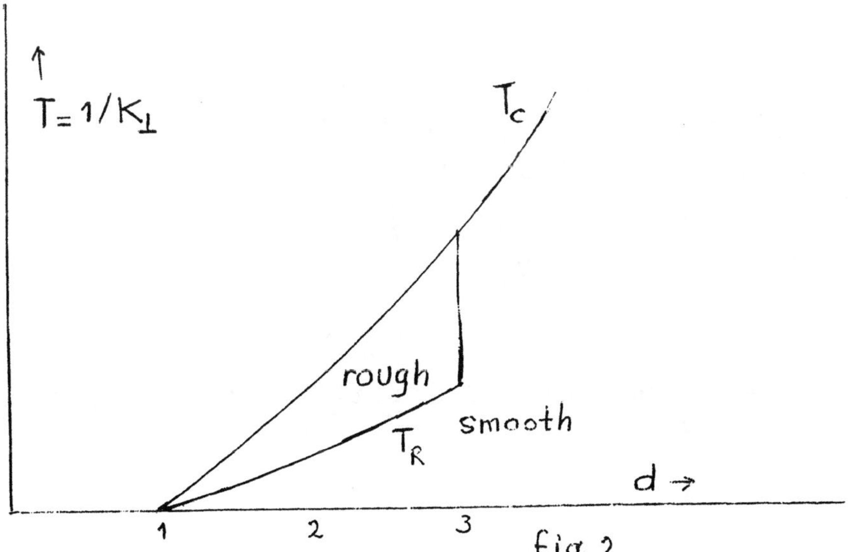

fig. 2

This treatment shows an interface phase transition but it has several shortcomings.

- The phase transition T_R for d = 3 is located at T_c for d = 2 which is of the right order of magnitude. But the transition is Ising-like whereas it should be of the Kosterlitz-Thouless type.
- $T_R \neq 0$ for d = 2 whereas we know that the rough phase exists all the way down to T = 0.

Nevertheless the exercise shows that with a delicate handling of the interface degrees of freedom the roughening transition comes within the real space renormalization approach.

Acknowledgement

The author is much indebted to Th.W. Burkhardt, F. van Dieren and H.J. Hilhorst for numerous discussions and critical comments on the subjects raised in these lecture notes.

Appendix Formulae for the Migdal transformation

In Migdal's approximation bonds are moved from the interior of a d-dimensional hypercube to the edges. We change the cube to a parallelepiped with length $b_\perp$ in $d-1$ dimensions and length $b_{/\!/}$ in the direction of the gradient of the field.

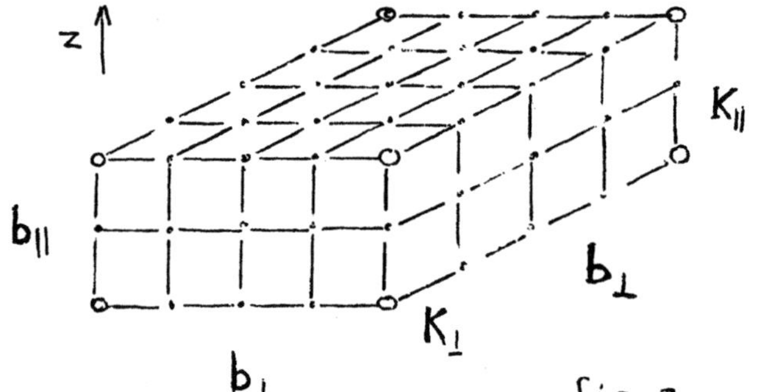

fig. 3

The bond $K_\perp$ on the edge is then multiplied by $b_{/\!/}\, b_\perp^{d-2}$ and the bond $K_{/\!/}$ obtains the strength $b_\perp^{d-1} K_\perp$. We set

$$b_{/\!/} = b_\perp^a \tag{A.1}$$

with $0 < a < 1$ and take the limit $b = 1 + dt$.
Calling the bondstrength on the edge K_1, we have

$$K_{1\perp} = K_\perp(1 + (d - 1 + a)dt)$$
$$K_{1/\!/} = K_{/\!/}(1 + (d - 1)dt) \tag{A.2}$$

The 1-dimensional chain of b elements coupled by a bond K_1 leads in zero field to a bond K' given by

$$\tanh K' = [\tanh K_1]^b \tag{A.3}$$

In the limit $b = 1 + dt$ this yields

$$K' - K_1 = K_1\, dt = B(K_1)dt \tag{A.4}$$

with

$$B(K) = (\sinh K)(\cosh K)\log(\tanh K) \tag{A.5}$$

For the parallel bond $K_{/\!/}$, dt should be replaced by adt. Combining (A.4) and (A.2) leads to the flow equations

$$\dot{K}_\perp = (d - 1 + a)K_\perp + B(K_\perp)$$
$$\dot{K}_{/\!/} = (d - 1)K_{/\!/} + a\, B(K_{/\!/}) \tag{A.6}$$

For the magnetic flow we consider the more general response $\delta H'_{i'}/\delta H_i$. We take the variant of Migdal's transformation in which no fields are moved. Then $H'_{i'}$ on the corners responds only to the field variations on the edges. Let s_i be the spin on site i of an edge. Then a variation δH_i on site i produces a field δH on the corners.

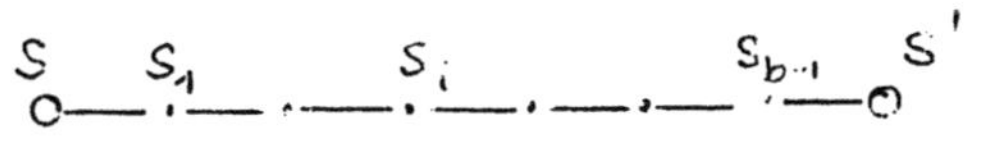

fig. 4

We have

$$e^{g+K'ss'}(\delta H's+\delta H_1's') = \sum_{\{s_1\ldots s_{b-1}\}} e^{K(ss_1\ldots s_{i-1}s_i)}\delta H_i s_i\, e^{K(s_i+s_{i+1}\ldots s_{b-1}s')} \tag{A.7}$$

where the spin-independent constant g is given by

$$e^g \cosh K' = 2^{b-1}[\cosh K]^b \tag{A.8}$$

For the value $\delta H'$ we find from (A.3), (A.7) and (A.8)

$$\delta H' = \delta H_i \cosh^2 K' \{(\tanh K)^i - (\tanh K')^2(\tanh K)^{-i}\} \tag{A.9}$$

For i = 0 this relation reduces to $\delta H' = \delta H_i$ as it should do.

For the magnetic exponent y_H we have to set $\delta H_i = \delta H$ in (A.9) and sum over i from 1 to b-1. This gives the contribution to $\partial H'/\partial H$ from one edge ending at a corner. We obtain

$$[\partial H'/\partial H]_1 = \cosh^2 K' \frac{\tanh K - (\tanh K)^b}{1 - \tanh K} + \tanh^2 K' \frac{1-(\tanh K)^{1-b}}{1-\tanh K} \tag{A.10}$$

The limit b = 1 + dt can now be taken easily. As

$$\partial H'/\partial H = 1 + y_H dt \tag{A.11}$$

we find by summing over the 2d edges ending at one corner (with 2(d - 1) edges of length $b_\perp$ and 2 of length $b_{/\!/}$)

$$y_H = -2\{(d-1)B(K_\perp) + aB(K_{/\!/})\} \tag{A.12}$$

References

1) K.G. Wilson, Phys. Rev. B4 (1971) 3174, 3184.

2) See e.g. Th.W. Burkhardt and J.M.J. van Leeuwen in "Real-Space Renormalization", Topics in Current Physics 30, Springer 1982.

3) Th. Niemeyer and J.M.J. van Leeuwen, Physica 71 (1974) 17.

4) See e.g. R.H. Swendsen in "Real-Space Renormalization", Topics in Current Physics 30, Springer 1982.

5) R.H. Swendsen, Phys. Rev. Letters 52 (1984) 1165.

6) J. Hubbard and P. Schofield, Physics Letters 40A (1972) 245.

7) M.S. Green, unpublished preprint.

8) F. van Dieren and J.M.J. van Leeuwen, to be published.

9) See e.g. J.M.J. van Leeuwen in "Fundamental problems in Statistical Mechanics III, North-Holland, 1975.

10) J.V. Sengers and J.M.J. van Leeuwen, Physica 116A (1982) 345.

11) S. Fisk and B. Widom, J. Chem. Phys. 50 (1969) 3219.

12) This suggestion is due to Th.W. Burkhardt.

13) F.P. Buff, R.A. Lovett and F.H. Stillinger, Phys. Rev. Letters 15 (1965) 621.

14) D. Jasnow and J. Rudnick, Phys. Rev. B17 (1978) 1351, Phys. Rev. Letters 41 (1978) 698.
D. Jasnow, T. Ohta and J. Rudnick, Phys. Rev. B20 (1979) 2774.

15) Th.W. Burkhardt in "Real-Space Renormalization", Topics in Current Physics 30, Springer 1982.

16) This suggestion is due to H.J. Hilhorst.

II. SPIN GLASSES AND FRUSTRATION

SPIN GLASSES AND THE RKKY INTERACTION

Michel Droz
Département de Physique Théorique
Université de Genève
1211 Genève 4, Switzerland

ABSTRACT

The main theoretical results known for spin-glasses with RKKY interactions are briefly reviewed. They are compared to results obtained for other spin-glass models. Possible scenarios for the existence or the non existence of an equilibrium phase transition are discussed.

1. INTRODUCTION

For a long time it was granted that metallic spin-glasses, whose spins interact via a Ruderman-Kittel-Kasuya-Yosida (RKKY) interaction, could be described by short-range interaction models with appropriated distribution of interactions [1]. Thus conclusions drawn from the study of short-range interaction models were applied as well to metallic spin-glasses. However, there is some increasing evidence [2,3] that metallic spin-glasses (with RKKY interactions) and insulating spin-glasses (with short-range interactions) may belong to different universality classes.

The goal of this lecture is to review briefly the main theoretical results obtained for spin-glasses with RKKY interactions. In Section 2, we recall the physical mechanism leading to the RKKY interaction. In view of the importance played by the dimensionality of the system, we give the expression for the RKKY interaction in arbitrary dimensions. The damping of this interaction is also discussed. In Section 3 we summarize the known analytical results for spin-glasses with RKKY interaction, which are basically mean field like. In Section 4, we discuss several numerical simulations performed for such systems. These simulations provide some information on the ground state properties, the structure of the phase space below the freezing temperature and the phase diagram. In Section 5 the similarities and differences between spin-glasses with RKKY interactions and other models of spin-glass (with short-range or long-range interactions) are discussed. Different scenarios leading to a true equilibrium transition or to an apparent transition due to metastability effects are discussed.

2. METALLIC SPIN-GLASSES AND THE RKKY INTERACTION

Metallic spin-glasses (as CuMn, AuFe) are formed by a non magnetic metallic matrix (Cu,Au) in which magnetic impurities are randomly distributed. For a very low impurity concentration, the impurities are not coupled; one is in the Kondo regime. For sufficiently high concentration, the magnetic impurities are coupled by the Ruderman-Kittel-Kasuya-Yosida interaction [4]. Let us briefly recall the main ingredients of the RKKY interaction. If one adds for example an Mn impurity in the copper matrix, this impurity retains part of its Hund's rule magnetisation and polarizes the spins of the conduction electrons in its neighborhood. The resultant spin polarization is not localized in the vicinity of the impurity but is oscillatory and long ranged. A second Mn atom at an arbitrary distance from the first one suffers a ferromagnetic or antiferromagnetic interaction depending upon the distance between the two atoms. The strengh of this interaction is computed as follows. The interaction atom-electron is the usual s - d interaction $H_{s-d} = J_o \vec{\sigma}\cdot\vec{S}$, ($\vec{S}$ = atomic magnetic moment, $\vec{\sigma}$ = electron's spin). One then computes, to second order in perturbation theory, the eigenfunctions and eigenvalues of the Hamiltonian $H = H_o + H_{s-d}$, where H_o describes the conduction electrons of the host (Bloch states). One then ends up with an effective interaction between the impurities of magnetic moment $\vec{S}_i$ of the form :

$$H = -\sum_{ij} J_d(R_{ij})\vec{S}_i\cdot\vec{S}_j \qquad (1)$$

where R_{ij} is the distance between the impurities i and j, and d is the dimensionality of the system. The original derivation was done for $d = 3$ [4]. However in view of the important role played by the dimensionality in the study of phase transitions, it is relevant to generalize this analysis to arbitrary dimensions. One finds for $d \geq 1$ [5]

$$J(R) = -\frac{\Pi[2^{d-3} n d \Gamma(d/2) J_o]^2}{E_F(d-1)x^{d-2}}\cdot\{J_{d/2-1}(x/2)Y_{d/2-1}(x/2) +$$

$$+ J_{d/2}(x/2)Y_{d/2}(x/2)\} \qquad (2)$$

where $x = 2Rk_F$, k_F being the Fermi wave number, E_F the Fermi energy, n the number of free electrons per atom and J and Y the usual Bessel functions.

Thus for large distances, $x >> 1$, one has

$$J(R) \sim -\frac{1}{E_F}[2^{d-2}nd\Gamma(d/2)J_o]^2 \frac{\sin(x-\pi d/2)}{x^d} \quad (3)$$

i.e. a long-range interaction with oscillations for all dimensions greater or equal to one.

Note that in the above derivation, one assumes that the conduction electrons are described by a free electron gaz. In fact, due to the presence of random impurities, the electrons are no longer free but have a finite mean free path λ. The role of this finite mean free path on the indirect exchange interaction has been considered a long time ago by de Gennes [6)] and a more complete analysis has been made more recently by de Châtel [7)]. For weak disorder, this effect can be taken into account by multiplying the expressions (2) and (3) by a damping factor $\exp - R/\lambda$. In conclusion metallic spin-glasses with RKKY interaction are characterized by a well defined interaction between spins with random positions. Although real spin-glasses correspond generally to Heisenberg spins, many works are devoted to the simpler case of Ising spins.

3. ANALYTICAL RESULTS

Due to the complexity of the interaction (2), it is obviously hopeless to get some exact results for this problem. However several attempts have been made in order to derive a self consistent mean field like approximation. In all the cases, the interaction (2) is replaced by a much more simpler form, namely

$$J(R) = \pm J_o/(k_F R)^3 \qquad (4)$$

with signs + or - at random.

The above relation is valid at large distances. Difficulties associated with the divergence of $J(R)$ for small R are avoided by introducing a cut-off r_o (depending on the impurity concentration x) and such that $J(R)$ vanishes for $R < r_o(x)$. Without cut-off, (4) may lead to infinite freezing temperature.

As we have seen in the introductory talk, it is not obvious to derive a reasonable mean field approximation for a spin-glass. Beside the Parisi solution [14] of the Sherrington-Kirkpatrick model [13] already discussed, another approach to the mean field theory has been proposed : the TAP approach [9,12]. For independent bonds with gaussian distribution, TAP obtained a self consistent equation for the magnetization. But, below the critical temperature, the solution of this equation is not unique [11], leading to new difficulties.

Moreover, the probability distribution $P(J_{ij})$ of a random distribution of magnetic impurities with RKKY interactions is significantly different from a gaussian distribution [1]. Accordingly, other mean-field approximations have to be devised for spin-glasses with RKKY interactions. Many works have been done based on the concept of internal field distribution introduced by Marshall [15] and Klein and Brout [16]. The idea is to generalize the "internal field" concept, familiar in the mean field theory of pure systems, to the case with disorder. For a fixed configuration of magnetic impurities, the local fields are well defined. The distribution of the local fields $P(h_i)$ arises from the requirement that the position of the impurities are random variables. It turns out that in order to get an implicit non linear integral equation for the distribution $P(h)$, one has to make several drastic approximations, namely :

i) The probability distribution for the positions of the spins is assumed to be uniformed.

ii) Spin and position variables are treated as mutually independent sets of random variables.

iii) The internal fields at each site are treated as independent random variables, with identical probability distributions.

This approximation is called the mean random field approximation (MRF). For Ising spins such an approach predicts a linear specific heat at low T with a slope independent of the impurity concentration [16,17]. This is mainly due to the fact that $P(h=0) \neq 0$. However, for Heisenberg spins one has $P(0) = 0$ and thus a single particle specific heat calculation is unsuccessful.

Clearly, the above approximations are not very realistic for a spin-glass. For example, the frustration is completely overlooked by this method. Due to frustration, the internal field for many spins can vanish. Several attemps have been made to improve the MRF method by including some correlations between the spins [18,19]. However when tested with the SK model [18], these approximations reproduce the SK solution without the breaking of the replica symmetry. Morandi and Corbelli [20] have analyzed critically the hypotheses used in the above generalized mean field theories.

In conclusion, one can say that there is no good mean field theory for RKKY systems, and thus it is hopeless to get reliable analytic results even in the simplest approximation.

4. NUMERICAL APPROACHES

In Section 4.1, we shall illustrate the two types of widely used numerical approaches on a simple example. In Section 4.2, we shall briefly reviewed results obtained for more realistic situations.

4.1. Simple one dimensional model

Let us consider a simple one dimensional model consisting of an open chain of N Ising spins interacting via an one dimensional RKKY interaction [21]. The Hamiltonian is

$$H = \sum_{i,j=1}^{N} J_{ij} s_i s_j - h \sum_{i=1}^{U} s_i \tag{5}$$

where
$$J_{ij} = \frac{J_o}{\rho(N)} \frac{\cos\{\alpha|x_i - x_j|\}}{|x_i - x_j| + b} = \frac{J_o}{\rho(N)} V_{ij} \tag{6}$$

b prevents the interaction to blow up when $|x_i - x_j| \to 0$. $\rho(N)$ is a normalization factor determined in such a way that the thermodynamic limit is well defined [21].

The two standard numerical approaches are the following :

A. Exact computation for small samples

The general strategy is the following :

a) Choose a value of N.

b) Choose the positions of the spins $\{x_i\}$ randomly. This define a sample of size N.

c) Compute exactly (i.e. by taking the trace over the 2^N states of the system) the physical quantities of interest for one sample.

d) Go to "b" and average the quantities computed in "c" over the samples. This gives the quenched average.

e) Go to "a" and choose another value of N. Do the job for several values of N in such a way that :

f) The results for finite N can be extrapolated to $N \to \infty$ using finite size scaling arguments.

Without going into the details (see Ref. 21) for more details), we can look at the qualitative behaviour of two quantities.

i) The ratio m/h as a function of the temperature, where m is the quenched magnetization. Despite the rather poor statistics, one sees on Figure 1 that one can distinguish two regimes. At low temperatures, one has a plateau ending at a temperature $T_c(h)$ which decreases when h increases. For $T > T_c(h)$, m/h decreases when T increases.

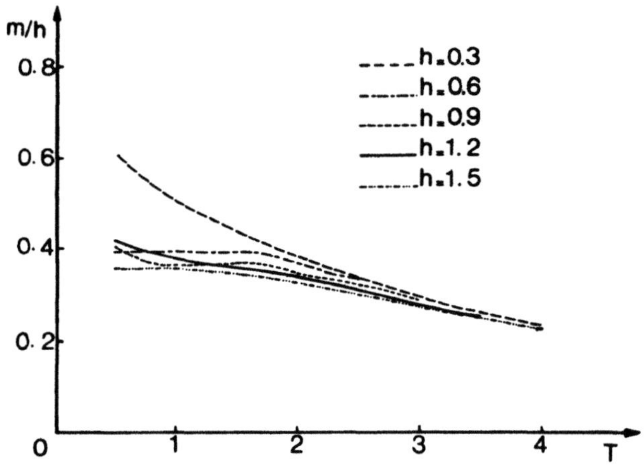

Fig. 1 : Ratio m/h as a function of the temperature for several fields. Results obtained for N = 16.

Such a behaviour is qualitatively similar to the one predicted by Parisi for the SK model (see the introductory lecture).

ii) The order-parameter. One considers the parameter Q(T) related to $\chi_q(T)$, the susceptibility associated to the Edwards-Anderson order parameter :

$$Q(T) = [N(N-1)]^{-1} < \sum_{i<j} <s_i s_j>_T^2 >_J \sim N^{-1} k_B T^2 \chi_q(T) . \qquad (7)$$

On Figure 2, Q is plotted as a function of T/T_c for several values of N. If this system exhibits a second order transition, one expects that, for $N \to \infty$, $\chi_q \sim t^{-\gamma}$ where $t = (T-T_c)/T_c$ and that for $T \lesssim T_c$, $Q(t,N) \sim |t|^{\beta}$. These two behaviours can be reproduced by assuming for Q(t,N) the following scaling form :

$$Q(t,N) = N^{-x} f(tN^y) . \qquad (8)$$

Results for N = 8, 12 and 16 are fitted with such a scaling form on Figure 3.

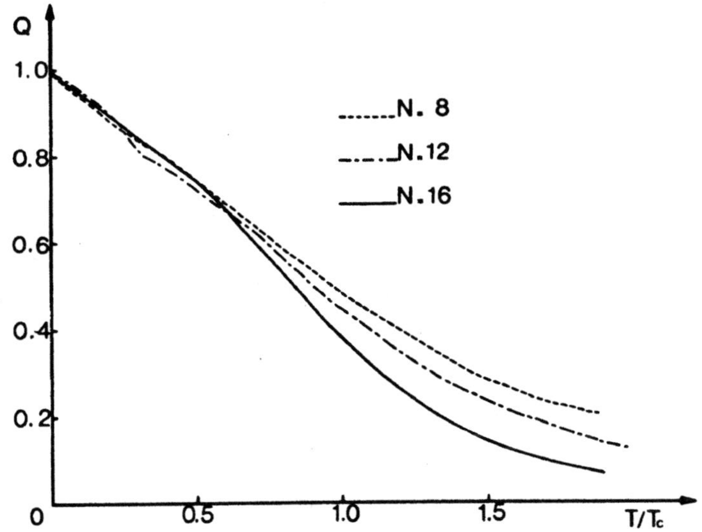

Fig. 2 : The order parameter Q as a function of T/T_c for N = 8, 12 and 16.

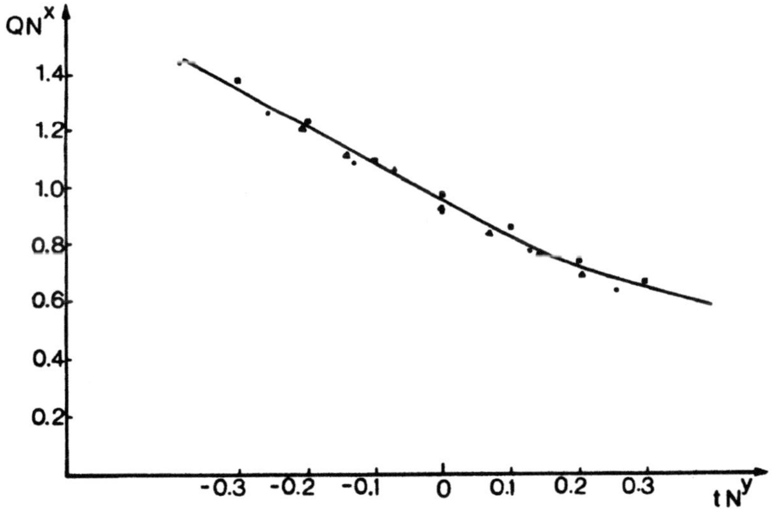

Fig. 3 : Fit of the numerical data with a form $Q(t,N) = N^{-0,31} f(tN^{0,90})$ (Δ, □ and • are for N = 8,12 and 16)

Thus one concludes that this model has a well defined second order transition in the thermodynamic limit. Note that similar exact results have been obtained for the SK model [22].

B. Monte Carlo simulation

Using standard Monte Carlo techniques [23], one can study the behaviour of this model. Two different procedures have to be distinguished.

i) Irreversible path

One cools the sample in zero external field to reach the ground state $\psi_o(h=0)$. Then one switches on the magnetic field and increases the temperature. This corresponds to the dotted line on Figure 4, where the magnetization in constant field is plotted as a function of the temperature.

ii) Field cooling or reversible path

Starting at high temperature in a field, one reduces the temperature. This corresponds to the full line on Figure 4. This behaviour looks similar to the one obtained exactly for small chains. Indeed, one can check that this path is reversible and thus correspond to the true thermodynamic equilibrium state.

The first path is irreversible and can be understood in terms of metastable states.

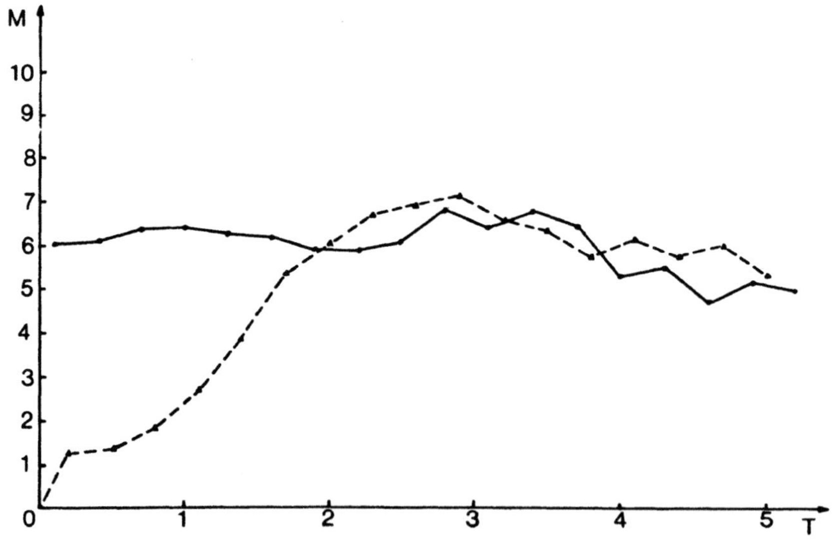

Fig. 4 : Monte Carlo magnetization at constant field as a function of the temperature. The dotted and full lines correspond respectively to the irreversible and reversible paths.

Let φ be the variable labelling the 2^N states of the system. Let $E_o(\varphi)$ the energies of these states in zero field and $E_h(\varphi)$ the energies in a field h. Let us assume that the energy profiles are the ones depicted on Figure 5. In the irreversible process, one starts in $\varphi_o(0)$ and goes to $\varphi_1(h)$ by switching on the field. But $\varphi_1(h)$ is a metastable state and the simulation will mainly probe the states in the vicinity of $\varphi_1(h)$. The real ground state in a fields is $\varphi_o(h)$ which is found in the field cooling approach.

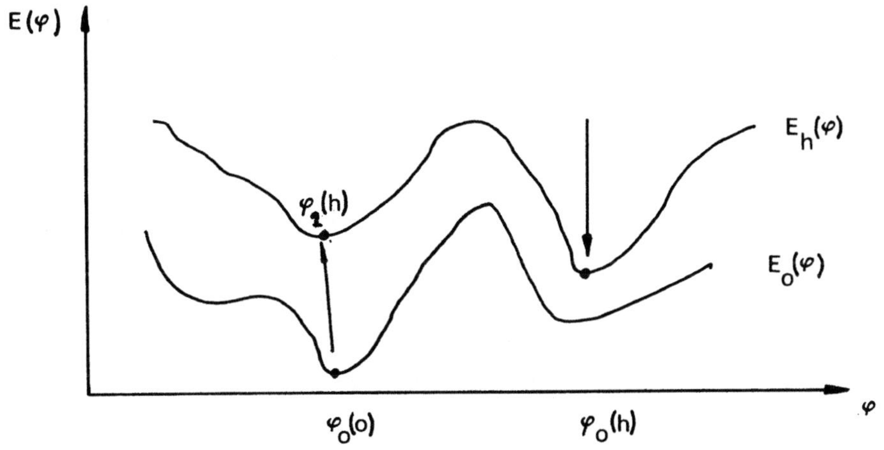

Fig. 5 : Energies of the different states of the system $\{\varphi\}$ in zero field $E_o(\varphi)$ or in a field $E_h(\varphi)$.

One sees on this simple example that the "energy structure" of the phase space is very important in order to understand the properties of the spin-glass phase.

4.2. Survey of the results of 3 dimensional RKKY systems

L.R. Walker and R.E. Walstedt have devoted a lot of efforts since 1977 to simulate the properties of a 3 dimensional RKKY system on a fcc lattice [24]. Mostly Heisenberg spins were studied, the size of the system varying from 96 to 960 spins. A detailled discussion of their work is not possible here, and we shall review only the most important aspects.

i) Low temperature properties

- Distribution of local fields $P(h) \sim h^2$ for Heisenberg spins. Thus $P(h) \to 0$ for $h \to 0$ and a single particle specific heat calculation based on P(h) would be unsuccessful.

- The elementary excitations are generalized spin-waves moderately extended in space.
- The linear specific heat, with the right concentration dependence, is reproduced.

ii) Critical properties

Here the situation is not very clear and rather puzzling. For a purely isotropic Heisenberg system, there seems to be no stable phase transition. However, a small dipolar anisotropy suffices to stabilize the transition (well marked peak in the non linear susceptibility). The role played by this anisotropy is not very clear yet.

iii) Energy barriers in the phase space

The phase space consists of hills and valleys. The picture that one could draw for Heisenberg spins with dipolar anisotropy is the following [24,25]. Below T_c, some large valleys are separated by free energy barriers whose hights diverge with the number of spins N. Inside a valley there are many small hills and secondary valleys whose hights remain finite (smaller than k_BT_c) as $N \to \infty$.

5. POSSIBLE "SCENARIOS" FOR THE (NON)-EXISTENCE OF A PHASE TRANSITION

In view of the above results and of the results obtained for short-range models [26], two different scenarios can be considered.

i) First scenario

The phase space consists of hills and valleys. Some of the hills become infinitely high when $N \to \infty$ dividing the system in several ergodic components. Inside the large valleys there are smaller hills whose heights are $\leq k_B T_c$. Such systems have a well defined equilibrium transition. Two long-range models are know to belong to this class : the SK model and the van Hemmen model [27,28]. The results of Section 4 would suggest that the three dimensional RKKY system belongs to this class too.

ii) Second scenario

The hills in the phase space remain finite when $N \to \infty$. Such a system does not have a true equilibrium phase transition. However, if the heights of the hills are large in comparison with a typical energy in the system (the short-range interaction for example), the system can be trapped into one valley (metastable state) for a time very long in comparison with the observation time. Such a system will exhibit an "apparent" or non-equilibrium transition. This is the case for two-dimensional $\pm J$ model [26].

Systems following the first or the second scenario will belong to different universality classes.

We would like to conclude this lecture by a note of caution. As we have seen, there are some evidences that 3d metallic spin-glasses follow the first scenario but we do not have yet a clear cut answer and more work need to be done.

REFERENCES

1. Binder K. and Schröder K., Phys. Rev. B14, 2142 (1976).

2. Maletta H., in "Heidelberg Colloquium on Spin Glass", Lecture Notes in Physics 192, 90, Ed. J.L. van Hemmen and I. Morgenstein. Springer Verlag, Berlin 1983.

3. Young A.P., in Ref. 2) p. 323.

4. Ruderman M.A. and Kittel C., Phys. Rev. 96, 99 (1954).
Kasuya T., Prog. Theor. Phys. 16, 45 (1956).
Yosida K., Phys. Rev. 106, 893 (1957).

5. Larsen V., Phys. Lett. 85A, 471 (1981).
Ariosa D., Master thesis University of Geneva (1981) (unpublished).

6. de Gennes P.G., J. Phys. Radium 23, 630 (1962).

7. de Châtel P.F., Jour. Magn. and Magn. Mat. 23, 28 (1981).

8. Klein M.W., Phys. Rev. 173, 552 (1968).

9. Anderson P.W., "Les Houches 1978", R. Balian et al. ed., North-Holland 1979.

10. Thouless D.J., Anderson P.W. and Palmer R.G., Philosoph. Magn. 35, 593 (1977).

11. Bray A.J. and Moore M.A., J. Phys. C13, L469 (1980).

12. Edwards S.F. and Anderson P.W., J. Phys. F5, 965 (1975).

13. Sherrington D. and Kirkpatrick S., Phys. Rev. Lett. 32, 1792 (1975).

14. de Dominicis C., in Ref. 2) p. 103.

15. Marshall W., Phys. Rev. 118, 1520 (1960).

16. Klein M.W. and Brout R., Phys. Rev. 132, 2412 (1963).

17. Klein M.W., Phys. Rev. 173, 553 (1968).

18. Klein M.W., Schowalter L.J. and Shukla P., Phys. Rev. B19, 1492 (1979).

19. Kinzel W. and Fisher K.H., J. Phys. F7, 2163 (1977).

20. Morandi G. and Corbelli C., Il Nuovo Cimento 54, 129 (1979).

21. Ariosa A., Droz M. and Malaspinas A., Helv. Phys. Acta 58, 29 (1982).

22. Young A.P. and Kirkpatrick S., Phys. Rev. B25, 440 (1982).

23. Monte Carlo Methods, Topics in Current Physics, K. Binder ed., Springer Verlag, Berlin 1979.

24. For a review of these numerous contributions, see Walstedt R.E. in Ref. 2) p. 177.

25. Walsted R.W. and Walker L.R., Phys. Rev. Lett. 47, 1624 (1981).

26. Morgenstern I., in Ref. 2) p. 305.

27. van Hemmen J.L., Phys. Rev. Lett. 49, 409 (1982).

28. Morgenstern I. and van Hemmen J.L., preprint (1984).

PHASE TRANSITIONS IN INHOMOGENEOUS ISING MODELS

J. Zittartz
Institut für Theoretische Physik
Universität zu Köln
Zülpicher Straße 77
D-5000 Köln 41

I. Introduction

In this survey we shall report recent theoretical work on inhomogeneous Ising models where Ising spins, $\mu=\pm1$, interact via nearest-neighbour couplings varying both in strength and sign. As we were interested in exactly soluble models, we first had to restrict our investigations to two dimensions and secondly we had to restrict ourselves to models with homogeneous couplings in one direction of the lattice and arbitrarily varying interactions only in second direction, namely Ising models of layered structure. These models are defined in Sect. II where we also briefly describe the method of solution and discuss the possible occurence of phase transitions. Groundstate properties and correlation functions are considered in Sect. III and IV, respectively. In the last Section V we shall describe a different route to the random-bond Ising model. In the general case the nearest-neighbour couplings are randomly distributed according to some probability distribution. So far this model has defied an exact solution. It is possible, however, to formulate a perturbation expansion about the non-random case and to calculate explicitly the first terms in the expansion in two dimensions. This then leads to a conjecture for the scaling function which describes the crossover from the non-random to the random case and determines the critical behaviour.

II. Layered Ising Models

As mentioned before we have investigated exactly soluble models /1-5/; surveys on this work have already been given in /6/. It turns out that the most general classes of models one is able to solve so far are two-dimensional Ising models of layered structure. In these models the nearest-neighbour couplings are homogeneous, i.e. translationally invariant, in one direction of the square lattice, but arbitrarily distributed in the orthogonal direction. This then results in two general classes of models, namely models with a horizontally layered structure (HL) and with a diagonally layered structure (DL). Fig.1 shows the unit cell of different couplings $K_j=\beta J_j$ for horizontal and $\bar{K}_j=\beta\bar{J}_j$ for vertical couplings with $j=1,\ldots,\nu$.

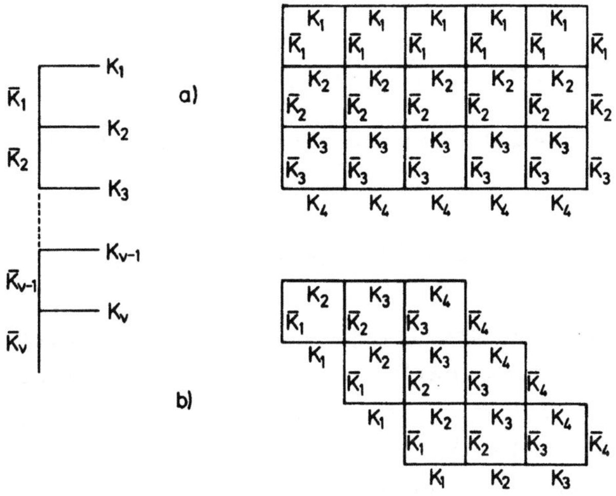

Fig. 1: Elementary cell of a layered Ising model: a) horizontally layered and b) diagonally layered structure.

These unit cells can be stacked together horizontally or diagonally resulting in layers extending in horizontal or diagonal direction. These layers have finite but arbitrary widths ν and are then repeated in the orthogonal direction periodically to make up the whole lattice of our HL or DL models. In typical models we consider widths of say $\nu=2$, 3,4, while $\nu=1$ is always the homogeneous Ising model. General results, however, apply to arbitrary ν, and, in particular, for $\nu\to\infty$ we consider models with a random coupling distribution in one direction. These latter models may be viewed as a first approximation to the situation of a spin glass where the couplings, of course, should also vary randomly in the second direction. The restriction is the prize we have to pay for being able to derive exact results.

Layered Ising models are described by Hamiltonians

$$\mathcal{H}\equiv \beta H = -\sum_j\left[K_j\sum_{h.n.n.}\mu\mu'+\bar{K}_j\sum_{v.n.n.}\mu\mu'\right] \tag{1}$$

where the sums inside the square bracket run over horizontal (h.n.n.) and vertical (v.n.n.) nearest-neighbour pairs and the external j-sum extends over horizontal or diagonal rows with the periodicity condition

$K_j = K_{j+\nu}$. Thermodynamics is derived from the partition function $Z = \text{Trace } e^{-\mathcal{H}}$.

As noticed by Toulouse /7/ the Hamiltonian (1) is invariant under the following local gauge transformations (acting on both spin and interaction variables):

$$\mu_i \rightarrow s_i\mu_i \;;\quad K_{ij} \rightarrow s_i s_j K_{ij} \quad \text{(j adjacent to i)} \tag{2}$$

for any choice of $s_i = \pm 1$. One is therefore led to introduce the frustration index

$$\phi = \prod \text{sign } K_{ij} = \pm 1 \tag{3}$$

where the K_{ij}'s are the four couplings around a plaquette. The thermodynamic properties, i.e. partition function and free energy, of the models defined by (1) are gauge invariant and do not depend on the coupling distribution $\{K_{ij}\}$ but rather on the distribution of coupling strengths $\{|K_{ij}|\}$ and of frustration $\{\phi\}$. This means that many different coupling distributions have the same equilibrium thermodynamics. For our layered Ising models it is clear from the definition (1) that the frustrated plaquettes with frustration index $\phi=-1$ are arranged in horizontal or diagonal rows.

Horizontally layered Ising models have been considered in many earlier papers /8-13/, with the restriction on only ferromagnetic couplings, i.e. couplings of the same sign. The effect of frustration is then completely missing. Only after Toulouse's introduction of the frustration concept some typical frustration models have been considered /14-18/ which also appear as special cases of our HL or DL models.

The layered structure of our models, i.e. the translational invariance either in horizontal or in diagonal direction of the square lattice, allows the application of the transfer matrix method for an exact calculation of the partition function and thus the thermodynamics. This means that the corresponding transfer operator is diagonalized by using a Jordan-Wigner transformation to fermions and Fourier transforming. The partition function Z is then solely determined by the largest eigenvalue of the transfer operator. Other eigenvalues and the eigenstates are only needed for the correlation functions.

In the horizontal case the transfer operator relates the configuration $\{\mu\}$ of spins in a horizontal row to the configuration $\{\mu'\}$ of spins in the next row, while in the diagonal case one considers configurations of spins in diagonal rows. The diagonalization of this operator is well known in the horizontal case /19/ and more easily achieved than in the diagonal case /20/. However, the results obtained in the latter case are more general and include all other models as special cases. By letting couplings go to zero or to infinity in a suitable way one obtains from the DL models /2/: 1. the HL models /1/, 2. layered models on the triangular (LT) and honeycomb lattice (LA), 3. layered models on mixed lattices, consisting of arbitrary combinations of rows of triangles, pentagons and hexagons /3/.

Layered Ising models are characterized by the global $\mu \to -\mu$ symmetry of the Hamiltonian (1). This symmetry can be spontaneously broken at low temperatures resulting in a phase transition at some finite temperature T_c. At the transition the system goes over from the completely disordered, unique high temperature phase to a 2-fold degenerate low temperature phase characterized by an ordering of the spins in complicated domains of ferromagnetic and antiferromagnetic substructures, which depend on the specific distribution of couplings and of frustration. As expected by universality this transition is always of Ising type, i.e. the critical behaviour and critical exponents are the same as in the homogeneous Ising case.

A phase transition occurs if there is a singularity in the free energy as a function of temperature. For the DL models the transition temperature is determined by the formula /2/

$$\prod_{j=1}^{\nu} \left. \frac{\sinh|K_j - s\bar{K}_j|}{\cosh(K_j + s\bar{K}_j)} \right|_{T=T_c} = 1 \, , \tag{4}$$

which must be satisfied either for s=1 or s=-1 corresponding roughly to predominant antiferromagnetic or ferromagnetic ordering, respectively. For the HL models one obtains |1|

$$\sum_{j=1}^{\nu} L_j = \left| \sum_{j=1}^{\nu} K_j \right| \tag{5}$$

where we could choose all vertical couplings positive, $\bar{K}_j > 0$, without loss of generality and L denotes their dual coupling, $\tanh \bar{K}_j = \exp(-2L_j)$. Special cases are the pure frustration models with couplings of equal strength, $|K_j| = |\bar{K}_j| = K$, such that only the signs can vary from row to row. Then all DL models have $T_c = 0$ as eq. (4) cannot be satisfied, except in the case $\nu = 1$, while for HL models we get from eq. (5) the simple formula

$$L = \frac{|\nu_+ - \nu_-|}{\nu_+ + \nu_-} K \qquad (6)$$

where $\nu_\pm$ with $\nu=\nu_+ + \nu_-$ are the numbers of horizontal rows with positive and negative horizontal couplings within the period ν. This also includes the (m,n) models, where m rows with ferromagnetic couplings are succeeded by n rows with antiferromagnetic couplings such that we have two layers of width m and n without frustration which are separated from each other by a row of frustrated plaquettes.

A transition to an ordered state at some finite temperature $T_c > 0$ occurs if one of the conditions (4)-(6) can be satisfied. For the (m,n) models this happens if the two layers have different widths, $m \neq n$; otherwise we have no transition, $T_c=0$, precisely for m=n (or $\nu_+ = \nu_-$). With decreasing temperature first the layers of finite width order internally, i.e. local ordering takes place. As the sum of vertical couplings through the separating frustrated row between successive layers adds up to zero, there would be no interaction between layers, if these would be completely ordered. This means that with decreasing temperature also the effective interaction between layers diminishes at the same time. The global ordering of adjacent layers is therefore aggravated by the internal ordering process and the interaction is maintained only by the disorder of edge spins still present at finite T. This might be called "entropy- or disorder-coupling". A global ordering, i.e. a transition, then only takes place if $m \neq n$ while $T_c=0$ if m=n.

This ordering behaviour is seen in the specific heat curves of (m,n) models shown in figs. 2 and 3 /1/. For models with m=n and no transition the curves (fig. 2) generally show a broad maximum attributed to the internal ordering within the layers. The maximum becomes more pronounced for increasing m and for $m \to \infty$ approaches asymptotically the case of the homogeneous Ising model with a logarithmic divergence. For models with $m \neq n$ (fig. 3) and a transition at some finite $T_c > 0$ we again have a broad maximum indicating the internal ordering process. The global ordering, which occurs at much smaller T_c, requires a logarithmic singularity indicated by the thin line. The amount of energy needed for this global ordering is, however, much smaller than for the internal layer ordering as the area under the logarithmic peak is hardly visible.

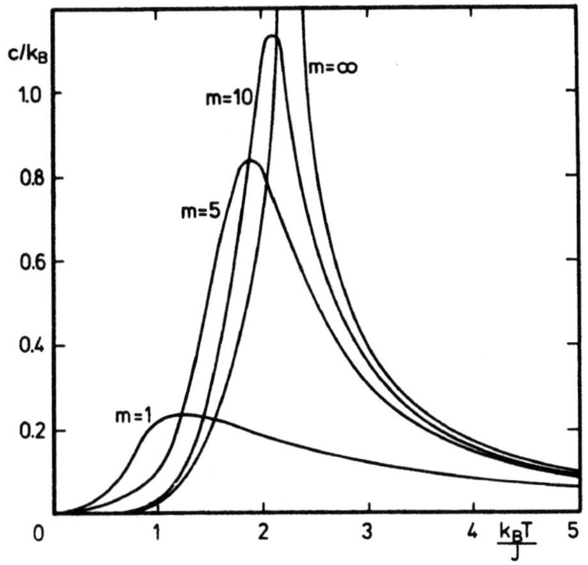

Fig. 2: Specific heat of (m,m)-models for different values of m. m → ∞ corresponds to the Ising model.

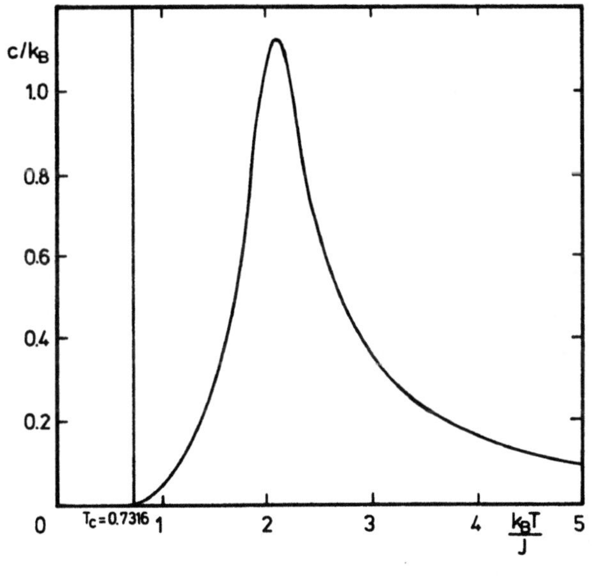

Fig. 3: Specific heat of (m,n)-models for the special case m=10, n=11. The thin line at T_c indicates the logarithmic singularity.

III. Ground-State Properties

Frustration describes and quantifies the competition between ferro- and antiferromagnetic couplings. In frustrated systems not all couplings can be satisfied at T=0, which means that there are spins or clusters of spins with a vanishing net interaction with the rest of the system, i.e. they can flip without cost in energy. This then leads to groundstate degeneracies and possibly to finite rest entropies at T=0. Pure frustration models with their high symmetry in couplings, namely couplings of equal strength, are the most liable candidates to show this behaviour. In ref. /1/ we showed that the rest entropy vanishes if and only if each row of frustrated plaquettes is separated from the next one by at least one row of nonfrustrated plaquettes. Otherwise we have finite rest entropies. For the particular case of (m,n) models we found for the rest entropy per site /1/

$$s_o = \begin{cases} G/\pi \approx 0.291, & m = n = 1 \\ (m+n)^{-1} \ln \frac{1}{2}(\sqrt{5}+1), & m \geqslant 2, n = 1, \\ 0, & m,n \geqslant 2, \end{cases} \tag{7}$$

where G is Catalan's constant.

Eq. (7) shows that the occurence of a phase transition at some finite temperature $T_c > 0$, which happens if $m \neq n$, is not simply related to the occurence of rest entropy. All combinations are possible: transition and no transition on the one hand and presence and absence of rest entropy on the other hand. However, in ref. /5/ we showed the following: if for any two groundstates one can be obtained from the other by a succession of purely local transformation, then the global symmetry $\mu \to -\mu$ cannot be broken, i.e. then $T_c = 0$. If this is not the case, then there might be a transition or not.

IV. Correlations

The bulk properties of our layered models can be inferred from the knowledge of the exact free energy and have been described in the foregoing sections. Spin correlation functions are a suitable tool to gain some insight into the microscopic properties, in particular into the details of the ordering process. In refs. /15, 21-25/ it has already been shown that correlations in frustrated systems show quite interesting behaviour. One usually studies two-spin correlations where for convenience the two spins are separated by r steps in the direction of homogeneity, i.e. parallel to the layering.

Here we shall describe the main results obtained in ref. /5/. The transfer matrix technique turns out to be also well-suited for the calculation of correlations. Formally the correlation $f(r) \equiv \langle \mu_o \mu_r \rangle$ can be written as a block Toeplitz determinant, which in all cases could be reduced to a usual Toeplitz determinant. We then could follow standard treatments described in /26/. For a general discussion it is convenient to distinguish the two cases where (a) the critical temperature is finite, $T_c > 0$, and (b) $T_c = 0$, more precisely, no transition at finite temperature.

(a) In models with $T_c > 0$ correlations show the usual Ising type critical behaviour. In the disordered high temperature region above T_c f(r) decays exponentially with a correlation length ξ diverging as $|T-T_c|^{-\nu}$ with the Ising exponent $\nu=1$. At T_c we have algebraic decay with $r^{-\eta_c}$ and $\eta=1/4$. Below T_c f(r) approaches $m(T)^2$ asymptotically, where the spontaneous magnetization or local order parameter m(T) vanishes at T_c with exponent $\beta=1/8$. This general behaviour has to be expected from universality, while the influence of inhomogeneities and frustrations is seen in details.

Below T_c we may have different local order parameters along layers of spins with different couplings. The ordering is , however, always unique, i.e. ferromagnetic or antiferromagnetic and is determined at $T>0$ by the globally dominating couplings. At zero temperature we may, however, encounter typical one-dimensional chain ordering different from the bulk ordering, e.g. an antiferromagnetic ordering in a chain with negative couplings while the rest orders ferromagnetically. Such behaviour causes the local order parameter in the chain to diminish again for $T<T_c$, to go to values smaller than one or even to zero for $T \rightarrow 0$, and to jump discontinuously at T=0. A typical example for such properties is the HL model with period $\nu=2$ /5b/.

One further effect of frustration is the fact that certain spins may be totally uncorrelated, i.e. $f(r) \equiv 0$, for any temperature $T>T_c$, i.e. in the disordered region. This occurs for instance in the fully-frustrated square lattice model (FFS) and in the chessboard model discussed in detail in ref. /5a/, for spins separated by an odd distance r in diagonal direction.

Depending on the details of the couplings the decay of correlations in the disordered region above T_c develops in many cases quite interesting oscillatory behaviour superposed on the exponential decay

$$f(r) \sim e^{-r/\xi} \cos(\Theta(T)r). \tag{8}$$

The oscillations may be commensurate with the lattice, in which case $\Theta(T)$ in eq. (8) takes on values, which are rational fractions of 2π, e.g. 0 for ferromagnetic alignment and π for usual antiferromagnetic alignment or period 2. Such behaviour is seen for instance in the HL model /5b/, where in the whole temperature range above T_c we either have ferromagnetic or antiferromagnetic correlations depending on the dominating couplings. However, for the "general square lattice" model (GS) /5c/ we may have a change from commensurate periods, which we always have for temperatures just above T_c, to incommensurate periods. This then occurs at some "disorder temperature" $T_D (> T_c)$, where correlations decay as a pure exponential and is just the same behaviour as seen in the triangular antiferromagnet /21/. Above T_D we have periods with the wavevector $\Theta(T)$ depending on temperature. This change is seen only in the $r \rightarrow \infty$ limit of the correlation f(r). As the thermodynamic quantities remain analytic, the smooth change at T_D is no phase transition.

(b) In the case $T_c=0$ frustration is so effective that the system cannot order globally at any finite temperature. While for T>0 we have exponentially decaying correlations as described before, we have to distinguish three cases at T=0.

(i) There is perfect ordering at T=0, i.e. $f(r)=\pm 1$. The ordering is usually one-dimensional chain ordering. As this is expected to be typical for models with layered structure, it is not surprising that the majority of our models with $T_c=0$ shows this behaviour, i.e. a first order transition at zero temperature.

(ii) T=0 is a critical point, i.e. $T_c=0$ in the strict sense. In that case the system has a vast groundstate degeneracy and correlations decay with $r^{-\eta} (r \rightarrow \infty)$. The critical exponent η, however, is different from the Ising value 1/4 and seems to assume the value $\eta=1/2$ universally in all investigated cases. The first known case is the triangular antiferromagnetic (FFT) /21/, the second case is the FFS model /22,5a/. However, this also occurs in the general HL model with period $\nu=2$ on a whole one-dimensional manifold in the space of the three couplings of the model /5b/. In the GS model we also find this behaviour, algebraic decay with $\eta=1/2$ accompained by oscillations with both the commensurate periods 3 (as in the FFT) and 4 (as in the FFS) /5c/.

(iii) Even at T=0 the system is disordered. This means that also at T=0 the correlations decay exponentially with a finite correlation length $\xi(T=0) < \infty$. That frustration can be so effective and the ground state degeneracy so vast in certain models as to cause this fast exponential decay is perhaps the most surprising result. So far we have found three special models with this behaviour, namely the chessboard model /5a/ which is also - and this is somewhat surprising - that model with the largest rest entropy per site, the fully frustrated honeycomb model (FFH), solved in /3/, and the frustrated "pentagon lattice" /27/.

V. Low-Concentration Expansion

So far we have considered inhomogeneous Ising models with fixed couplings varying only in one direction of the lattice. This restriction is sufficient to allow for an exact solution and the computation of correlation functions. The more general case where the couplings vary randomly also in the second direction is presumably not solvable. One therefore has to use either numerical or approximate treatments /28/.

In the general random-bond model Ising spins on a square lattice interact via nearest-neighbour couplings $K_{ij}=\beta J_{ij}$ randomly chosen with the probability distribution

$$P(K_{ij})=(1-p)\delta(K_{ij}-K)+p\delta(K_{ij}-\alpha K) \tag{9}$$

where $-1\leq\alpha\leq 1$ can be assumed without loss of generality. These systems may be thought of as Ising models with couplings K and randomly distributed impurity bonds αK with concentration p. Two special cases have received widespread attention, namely the bond-diluted ferromagnet with $\alpha=0$ /29-32/ and the $\pm J$ spin glass model with $\alpha=-1$ (see for instance /33/). In the first case a fraction p of interactions K is randomly removed from the system. For small dilution p one still expects a ferromagnetic phase transition at some finite transition temperature $T_c(p)$ as in the pure, nondiluted system. Above the percolation threshold p_c the transition temperature vanishes due to the disappearance of the largest percolating cluster. The phase diagram has been obtained by using renormalization group calculation /31/. However, there has still remained the problem of determining the nature of the transition at $T_c(p)$.

Similarly, in the $\pm J$ model ($\alpha=-1$) the ferromagnetic phase transition is still present for small enough p. However, with increasing p frustration becomes so effective that the transition is already destroyed for p larger than $p \simeq 0.12$ /34/. Though numerical analysis could rule out the existence of a spin glass state ($p=1/2$), it was not possible so far to clarify the nature of the transition at $T_c(p)$.

Unfortunately the general case $p \neq 0$ is presumably not soluble. However, it is possible to investigate the case of small p by performing a concentration expansion about the soluble pure system, $p=0$ /28/. As we have pointed out, it is essential to obtain results at least to second order in p to include impurity correlations and to clarify the difference between quenched and annealed randomness. In two dimensions where the pure Ising model is soluble, all relevant quantities can be explicitly calculated. In particular we have obtained

a) the critical temperature $T_c(p;\alpha)$ to second order in the impurity concentration,

b) the exact leading singular terms of the quenched free energy, $\bar{f}=\langle f[K_{ij}]\rangle_{av.}$, to second order in p.

The critical temperature $T_c(p;\alpha)$ determines the phase diagram of the quenched system. For $p=0$ we have the pure Ising system with a transition at $T_{c,o}=K_o^{-1}=2/\ln(\sqrt{2}+1)$. To second order in p the transition temperature is given by

$$T_c(p;\alpha)/T_{c,o} = 1-s_1p-s_2p^2 \tag{10}$$

and the coefficients $s_{1,2}$ can be expressed in closed form in terms of $T_{c,o}$ and the interaction parameter α /28/. The explicit formulas are omitted for brevity. Depending on the values of α we encounter two different cases (see fig. 4). For positive α all impurity bonds are ferromagnetic and only disorder but no frustration is present in the system. Then the critical temperature $T_c(p;\alpha)$ is finite for all values of p until for $p=1$ it reaches the known value $T_c(1;\alpha)$ for the pure Ising system with couplings αK. On the other hand the transition temperature is expected to vanish above some threshold p_c if $\alpha<0$ and a separate antiferromagnetic phase occurs near $p=1$ for $\alpha<0$. However, for clarity this phase has been left out of the figure. As can be seen frustration is more effective in reducing the critical temperature than dilution.

In fig. 4 the transition temperature T_c is plotted for several values of α from eq. (10) and numerical estimates /31,34/ are also shown for comparison as dashed curves. For small values p the agreement is very good, only near p_c we have substantial deviations where higher order corrections become relevant.

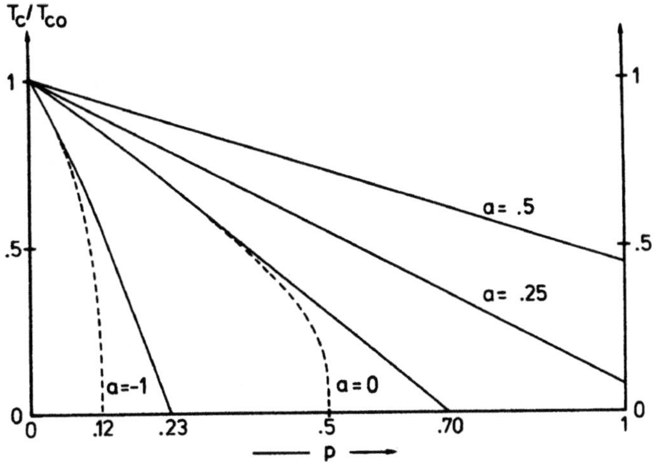

Fig. 4: The critical temperature $T_c(p;a)$ is shown for various interaction parameters a. For clarity the antiferromagnetic phase which occurs for $a<0$ in the lower right has been omitted from the picture. Numerical estimates for $a=0$ /31/ and $a=-1$ /34/ (dashed curves) are also shown for comparison.

Our second result is the expansion of the quenched averaged free energy $\beta\bar{f}$ in the vicinity of $T_c(p;a)$, i.e. near the transition. Keeping only leading terms the singular part of $\beta\bar{f}$ is given by

$$\beta\bar{f}|_{\text{sing.}} \simeq - \frac{t^2}{2b_o^2 p}[y - \frac{1}{2}y^2 + \frac{1}{3}y^3 + O(y^4)] , \tag{11}$$

to second order in p, where

$$y = pb_o^2 \frac{8}{\pi} \ln|t|^{-1} \tag{12}$$

and $t \sim T_c(p;a)-T$ goes to zero. The coefficient b_o depends on the interaction parameter a.

The expression (11) suggests that the expansion in brackets defines some universal scaling function G(y) which describes the crossover from the pure Ising model, in which case p=0 and $G(y) \Rightarrow y$, to the random model. In the latter case p is nonzero and the critical behaviour of the free energy, as $t \to 0$, is then determined according to (11) by the asymptotic behaviour of G(y) for $y \to \infty$. As all p and a dependence is absorbed in the one variable y, apart from the unimportant amplitude, this strongly suggests that we have universally the same critical behaviour along the whole T_c curve for all concentrations p and also for all values of the parameter a. This would mean that frustration, which occurs for $a<0$, in particular for $a=-1$, would have no qualitative influence on the type of phase transition, when we compare this situation with the case of a simple mixture of ferromagnetic bonds, $0<a<1$, where we have no frustration.

The exactly known terms in the expansion (11) allow to check recent suggestions in the literature on their validity. (11) rules out Fisher renormalization /35/ which holds for the annealed case /36/, but was also suggested for the quenched case /32,29/; it also rules out a power law singularity $t^{-\alpha(p)}$ with an exponent continuously varying with p, as suggested in /37/. In both cases the free energies do not fit the correct expansion (11). However, it has been suggested /38/ on the basis of an approximate mapping of the random-bond Ising model onto a zero-component Gross-Neveu model that the scaling function G(y) should be given as

$$G(y) = \ln (1+y) \tag{13}$$

whose expansion obviously fits with (11). We believe that (13) is essentially correct. In the limit $t \to 0$ ($y \to \infty$) it leads to the universal critical behaviour

$$\beta \bar{f}|_{sing.} \simeq - \frac{t^2}{2b_o^2 p} \ln|\ln|t|| \ . \tag{14}$$

Most remarkable is the fact that the type of singularity is the same for all values of the interaction parameter a, which is definitely true to second order in p. This would mean that frustration, which occurs for $a<0$, does not qualitatively change the criticalities and thus the nature of the transition as compared to the case of $a>0$, where we have no frustration, but only disorder.

Acknowledgement

I would like to thank W.F. Wolff for fruitful collaboration, stimulating discussions and help.

References

/1/ Hoever, P., Wolff, W.F., Zittartz, J.: Z. Phys. B - Condensed Matter 41, 43 (1981); Wolff, W.F., Hoever, P., Zittartz, J.; Z. Phys. B - Condensed Matter 42, 259 (1981); Hoever, P., Zittartz, J.: Z. Phys. B - Condensed Matter 44, 129 (1981).

/2/ Wolff, W.F., Zittartz, J.: Z. Phys. B - Condensed Matter 44, 109 (1981); Wolff, W.F.: Dissertation, Köln (1981).

/3/ Wolff, W.F., Zittartz, J.: Z. Phys. B - Condensed Matter 49, 139 (1982).

/4/ Hoever, P.: Z. Phys. B - Condensed Matter, 48, 137 (1982); Hoever P., Zittartz, J.: Z. Phys. B - Condensed Matter 49, 39 (1982).

/5/ Wolff, W.F., Zittartz, J.: Z. Phys. B - Condensed Matter, 47, 341 (1982); 49, 229 (1982); 50, 131 (1983).

/6/ Zittartz, J.: J. Magn. Magn. Mat. 31-34, 1105 (1983); Wolff, W.F.: J. Magn. Magn. Mat. 38, 217 (1983); Wolff, W.F., Zittartz, J. in: Lecture Notes in Physics 192, Berlin-Heidelberg-New York, Springer 1983.

/7/ Toulouse, G.: Commun. Phys. 2, 115 (1977); Vannimenus, J., Toulouse, G.: J. Phys. C10, L 537 (1977); Fradkin, E., Huberman, B.A., Shenker, S.H.: Phys. Rev. B18, 4789 (1979).

/8/ McCoy, B.M., Wu, T.T.: Phys. Rev. 176, 613 (1968); 188, 1014 (1968); B2, 2795 (1970).

/9/ Fisher, M.E.: J. Phys. Soc. Jpn. Suppl. 26, 87 (1969).

/10/ Barouch, E.: J. Math. Phys. 12, 1577 (1971).

/11/ Au-Yang, H. McCoy, B.M.: Phys. Rev. B10, 886, 3885 (1974).

/12/ Hahn, H.: J. Magn. Magn. Mat.: 7, 209 (1978); Decker, I., Hahn, H.: Physica 83A, 143 (1976); 89A, 37 (1977); 93A, 215 (1978).

/13/ Rander, K., Hahn, H.: Phys. Lett. 53A, 287 (1975); Rander, K.: Z. Naturforschung 31A, 1465 (1976).

/14/ Villain, J.: J. Phys. C10, 1717 (1977).

/15/ Villain, J., Bideaux, R., Carton, J.-P., Conte, R.: J. Phys. (Paris) 41, 1263 (1980).

/16/ André, G., Bideaux, R., Carton, J.-P., de Seze, L.: J. Phys. (Paris) 40, 479 (1979).

/17/ Longa, L., Oleś, A.M.: J. Phys. A13, 1031 (1980).

/18/ Bideaux, R., de Seze, L.: J. Phys. (Paris) 42, 371 (1981).

/19/ Schultz, T.D., Mattis, D.C., Lieb, E.H.: Rev. Mod. Phys. 36, 856 (1964).

/20/ Stephen, M.J., Mittag, L.: J. Math. Phys. 13, 1944 (1972); Temperley, H.N.V., in: Domb, C., Green, M.S.: Phase Transitions and Critical Phenomena, Vol. I, London, Academic Press 1972.

/21/ Stephenson, J.: J. Math. Phys. 5, 1009 (1964); 7, 1123 (1966); 11, 413, 420 (1970).

/22/ Forgacs, G.: Phys. Rev. B22, 4473 (1980).

/23/ Gaby, M.: J. Physique Lettre 41, 427 (1980).

/24/ Sütö, A.: Z. Phys. B - Condensed Matter 44, 121 (1981).

/25/ Peschel, I.: Z. Phys. B - Condensed Matter 45, 339 (1982).

/26/ McCoy, B.M., Wu, T.T.: The Two-Dimensional Ising Model, Harvard University Press, Cambridge, Mass. 1973.

/27/ Waldor, M.: Diplomarbeit Köln, 1984.

/28/ Wolff, W.F., Zittartz, J.: Z. Phys. B - Codensed Matter 52, 117 (1983).

/29/ Harris, A.B.: J. Phys. C7, 1671 (1974).

/30/ Domany, E.: J. Phys. C11, L 337 (1978).

/31/ Tsallis, C., Levy, S.V.: J.Phys. C13, 465 (1980).

/32/ Tamaribuchi, T., Takano, F.: Progr. Theor. Phys. 64, 1212 (1980); Tamaribuchi, T.: Progr. Theor. Phys. 66, 1575 (1981).

/33/ Binder, K., Kinzel, W. in: Lecture Notes in Physics 149, Berlin-Heidelberg, New York, Springer 1981.

/34/ Morgenstern, I., Binder, K.: Phys. Rev. Lett. 43, 1615 (1979); Phys. Rev. B22, 288 (1980).

/35/ Fisher, M.E.: Phys. Rev. 176, 631 (1968).

/36/ Kasai, Y., Syozi, I.: Progr. Theor. Phys. 50, 1182 (1973).

/37/ Suzuki, M.: J. Phys. C7, 255 (1974).

/38/ Dotsenko, V.S., Dotsenko, V.S.: JETP lett. 33, 40 (1981); J.Phys. C15, 495 (1982).

STATIC AND DYNAMIC PROPERTIES OF SHORT RANGE ISING SPIN GLASSES

Wolfgang Kinzel

Institut für Festkörperforschung der KFA Jülich,
Postfach 1913, D-5170 Jülich, W. Germany

1. Introduction

In this lecture I would like to give a short introduction and overview over the properties of short range spin glass models. Special emphasis is given to recent developments with respect to a T=0 scaling and a comparison with experiments. More detailed overviews are given in previous lecture notes [1,2,3,4]. Are spin glasses still interesting? In fact, as shown in Fig.1, the number of publications on spin glasses has reached about one per day and still does not seem to have saturated.

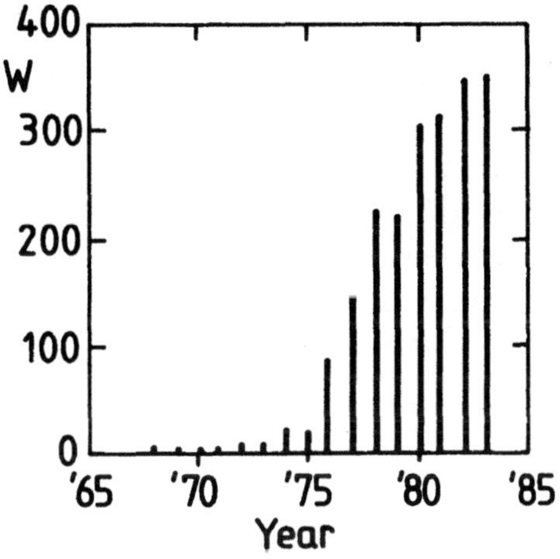

Fig. 1: Number of publications per year which have the word "spin glass" in their title or abstract

This extensive research on spin glasses may have the following reasons:

(i) spin glass properties are rather universal; for instance for the metallic crystal $AuFe_x$, for the insulator $Eu_xSr_{1-x}S$ as well as for the amorphous metal Gd_xAl_{1-x} one finds the same qualitative results in the spin glass phase;

(ii) spin glasses have new properties caused by disorder and competition of the interactions, the description of which needs new concepts.

(iii) Even after more than 2000 publications and many important results I think that spin glasses are still not well understood.

Lowering the temperature of spin glasses one usually finds that below a rather well defined freezing temperature T_f the magnetic moments freeze into randomly distributed directions. One of the main problems in spin glasses is the question: Is this transition into the spin glass phase a true static phase transition or is it a gradual freezing process far from thermal equilibrium?

To study this problem a model and a method of solving it has been proposed by Edwards and Anderson (EA) 1975 [5]. Its only relation to real materials are the two essential ingredients to spin glasses: disorder and competition. Therefore it consists of spins $S_i=\pm 1$ (we restrict ourselves to Ising variables) on a regular lattice and interactions J_{ij} which are distributed randomly by a Gaussian with width ΔJ and mean J_o or by a $\pm J$ distribution. In addition to this Hamiltonian

$$\mathcal{H} = -\sum_{ij} J_{ij} S_i S_j - H \sum_i S_i \tag{1}$$

we introduce a Glauber relaxational dynamics [3]. In the following the properties of this model are shortly reviewed

2. Mean field theory

The simplest approach to a model is usually its mean field theory. This theory is obtained from the infinite range model [6]. That is, each spin S_i interacts with each other one S_j by a coupling J_{ij} and the width ΔJ of the coupling distribution goes to zero with the number N of spins like $\Delta J \propto 1/\sqrt{N}$. It has been shown [6] that this model has a true phase transition in thermal equilibrium with a cusp both in the magnetic susceptibility χ and the specific heat C. However, while the singularity in χ may have been observed in the equilibrium (=field cooled?) susceptibility, the experimental specific heat definitely has no singularity at T_f. Nevertheless the properties of this model have stimulated many experiments, and many experimental results are in remarkable qualitative agreement with the mean field predictions.

Usually the mean field theory has a rather simple solution in terms of an order parameter. In spin glasses, however, there is still no rigorous solution of the low temperature phase. It turned out, that a whole order parameter function q(x) is necessary to describe the static properties [7] for instance one has $\chi=\frac{1}{k_B T}[1-\int_0^1 q(x)dx]$. q(x) has been obtained by the replica trick together with a selfsimilar replica symmetry breaking [8] but it also is given by a dynamic formulation where the parameter x is related to a set of time scales [9].

It is beyond the scope of this lecture to go into details of the mean field theory, but I just want to mention three recent results which firstly are very interesting and secondly may have applications to other problems with many stable and metastable states:

(i) The order parameter q(x) gives the distribution W(q) of an overlap $q_{12}=\frac{1}{N}\sum_i \langle S_i^1\rangle\langle S_i^2\rangle$ between two states $\langle S_i^1\rangle$ and $\langle S_i^2\rangle$ (= solutions of the mean field equations or states within a free energy valley), one gets [10]

$$W(q) = \frac{dx}{dq}$$

(ii) Some quantities like $q=\frac{1}{N}\sum_i \langle S_i\rangle^2$ are not self averaging, i.e. even in the thermodynamic limit $N\to\infty$ one obtains different results from sample to sample. However, energy and magnetization are self-averaging [11]

(iii) The states have a ultrametric overlap structure, i.e. if one has three states $\langle S_i^1\rangle$, $\langle S_i^2\rangle$ and $\langle S_i^3\rangle$ and corresponding overlap $q_{12}\geq q_{23}\geq q_{13}$ one finds $q_{23}=q_{13}$ [12].

3. Short range model

Since the EA model is a very crude approximation to real spin glasses (site randomness, RKKY interactions etc.) it is important to know how well this model reproduces spin glass properties. In the following we concentrate mainly on the 2-dimensional model with Gaussian distributed nearest neighbour bonds. Since there still does not exist any reliable theory, one has to investigate this model experimentally using Monte Carlo simulations [3]. In fact one finds that this model qualitatively reproduces many details of the real experiments [13,14]. I just want to mention a few results:

(i) A cusp in the time dependent susceptibility

(ii) A plateau in the field cooled susceptibility

(iii) A broad maximum in the specific heat

(iv) The existence of thermoremanent and isothermal remanent magnetization and their dependence on temperature, field and time

(v) Slow logarithmic time decay of autocorrelation functions

(vi) Apparent singular behaviour of the magnetization as a function of field at low temperatures

On the other side, there is now general agreement that at least the 2-dim model does not have a phase transition in thermal equilibrium at a nonzero temperature T_f [3].

However, it is not easy to get this result from Monte Carlo simulations. Namely in the infinite system a phase transition should show up in the decay of the correlations

$$q(t) = \frac{1}{N} \Sigma <S_i(0)S_i(t)> \qquad (2)$$

to a nonzero value $q(\infty)=\frac{1}{N}\sum_i <S_i>^2$. But it turned out that (i) q(t) decays very slowly with time between 2 and 5 dimensions such that an extrapolation to $t=\infty$ is not possible [15] and (ii) related to it, in contrast to usual phase transitions one presumably cannot distinguish between the spectrum of intrinsic relaxation times and the ergodic time scale t_e; the latter one is finite in a finite system and restores any broken symmetry, therefore q(t)=0 for times $t>t_e$.

Thus one cannot use the order parameter but one has to investigate the corresponding susceptibility

$$\chi_{EA} = \frac{1}{T}\sum_{ij} <S_iS_j>^2 \qquad (3)$$

which should diverge at any sort of a phase transition. To get thermal equilibrium in computer simulations, one has to average Eq.(3) over times which are longer than any relaxation time of the system, which may be estimated from [16]

$$\tau = \int_0^\infty q(t)dt \qquad (4)$$

It turned out, that even with today's largest computers it is not possible to go lower than $T/\Delta J \gtrsim 0.7$ in two dimensions [16]. Nevertheless the results are consistent with

$$\chi_{EA} \sim \frac{1}{T^4}\ , \ \ln\tau \sim \frac{1}{T^2} \tag{5}$$

that is, a phase transition at T=0.
These Monte Carlo results have been confirmed earlier by exact calculations of small systems. Both the quantities χ_{EA} and ζ_{EA} with

$$\langle S_o S_r \rangle^2 = A\exp(-r/\xi_{EA}) \tag{6}$$

have been calculated in thermal equilibrium using a transfermatrix method [17]. The result is consistent with

$$\xi_{EA} \sim \frac{1}{T^2} \tag{7}$$

and it is in agreement with Eq.(5) and the relation $\chi_{EA} \sim \xi_{EA}^2$. At $T/\Delta J=1$, where earlier Monte Carlo simulations found a cusp in χ, the correlation length ξ_{EA} is of the order of just two lattice spacings! Therefore one may ask, is it possible to get spin glass properties with the dynamics of such small correlated regions? In the following section the dynamics of the two dimensions Gaussian EA model is investigated with respect to this question.

4. Nature of the freezing process

In the previous section we have learned that the spin glass properties, which are observed in computer simulations, are not due to a thermal phase transition at a nonzero T_f but rather due to a dynamic freezing process far from thermal equilibrium. But still we have to understand this freezing process.

Monte Carlo simulations have shown that in a first approximation a spin glass may be considered as a kind of two fluid system [18]. Below the freezing temperature $T_f(t)$, which depends on the observation time t, a fraction q of spins is completely frozen over the time period t. The frozen spins appear in small clusters, thus ruling out previous suggestions of a percolation of frozen spins at T_f [19]. The rest of 1-q of the spins still seems to remain in thermal equilibrium such that it gives the equilibrium susceptibility $\chi=1/T$. Thus the total susceptibility should be given by

$$\chi = (1-q)/T \tag{8}$$

Comparison of Eq.(8) with the true susceptibility averaged over the same time interval indeed gave a satisfactory agreement [18].

Another problem is related to the smallness of the static correlation length ξ_{EA} at the freezing transition $T_f(t)$. In fact, an investigation of an ensemble of small rings of 4 Ising spins with Gaussian distributed interactions, which was solved exactly, showed the following [20]:

(i) For the same time and temperature scales as in the 2-d EA model the autocorrelation function and the remanent magnetization become nonzero, and in the dynamic structure factor a large zero frequency peak appears.

(ii) For somewhat lower temperatures $T \lesssim 0.3\Delta J$ the autocorrelation function decays over more than 15 order of magnitude in time. Note that also in experiments such large relaxation times are observed.

One may go a step further and estimate time dependence of T_f using the static correlation length ξ_{EA} [20]. The spin glass may be considered as an ensemble of isolated clusters of size ξ_{EA} which relax with the Arrhenius law

$$\tau = \tau_o \exp(\Delta E/T) \tag{9}$$

τ_o is the single spin relaxation time and ΔE the energy barrier for going from one ground state to another one by moving an interface through the cluster. For an usual phase transition one has $\Delta E \sim \xi_{EA}$, however, due to the large degeneracy of low lying spin glass states it was assumed $\Delta E \sim \xi_{EA}^{1/2}$ [21]. Using this and Eq.(7) in Eq.(9) one obtains

$$\tau = \tau_o \exp(\Delta J/T)^2 \tag{10}$$

The freezing occurs if the observation time t is of the order of the relaxation time τ, therefore

$$T_f \propto (\ln t/\tau_o)^{-\frac{1}{2}} \tag{11}$$

In fact Eq.(11) is consistent with the numerical result Eq.(5). Another estimate would be assume $\Delta E = J\xi^{1/2}$ and use a Gaussian distribution of J. Surprisingly such a crude approximation reproduces the observed susceptibility even quantitatively [20].

All these considerations suggest that spin glass behaviour can be understood in terms of a relaxation of small clusters of size ξ_{EA}.

5. T=0 scaling

Since the numerical estimates suggest a diverging correlation length ξ_{EA} at T=0, Eq.(7), it is tempting to scale spin glass quantities around T=0 with a single length ξ_{EA} in analogy to usual critical points [22]. One quantity of interest is the magnetization M(T,H) in thermal equilibrium

$$M(T,H) = \chi_o \cdot H - a_3 (H/T)^3 - O(H^5) \tag{12}$$

By symmetry one has $\chi_o = 1/T$, which does not involve any critical behaviour. However one has $a_3 \sim \chi_{EA}$ at $T \to 0$, therefore the nonlinear susceptibility is sensitive to a diverging ξ_{EA}. A scaling assumption would be to scale the field H by a power of temperature T^Δ, thus [21, 22]

$$\frac{M}{H}(T,H) = \frac{1}{T} g\left(\frac{H}{T^\Delta}\right) \tag{13}$$

with some scaling function g(x). Eq.(13) may also been written in the form

$$1 - T\frac{M}{H} = \tilde{g}\left(\frac{H}{T^\Delta}\right) \tag{14}$$

which includes all singular nonlinear terms of Eq.(12). Since by symmetry $\tilde{g}(x) = a \cdot x^2 + O(x^4)$, one finds by comparing Eqs.(12) and (14)

$$a_3 \sim \chi_{EA} \sim T^{2-2\Delta} \tag{15}$$

and from Eq.(5) $\Delta \simeq 3$ in two dimensions. Eq.(13) implies

$$\frac{M}{H}(T=0) \sim H^{-\frac{1}{\Delta}} \tag{16}$$

In fact, from the plateau values of the field cooled magnetization Eq.(16) was verified by computer simulations with $\Delta = 3.5 \pm 0.5$ [23]. Also the scaling relation Eq.(14) seems to hold for the nonlinear susceptibility for $T \lesssim \Delta J$ and fields H as shown in Fig.2

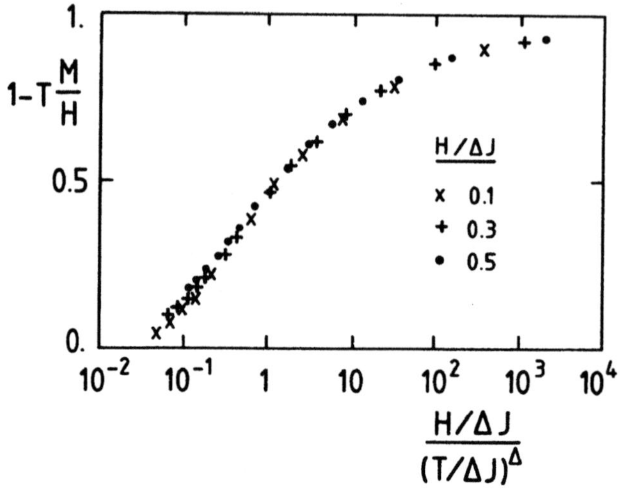

Fig. 2: Nonlinear susceptibility of the 2-dim. EA model as a function of field H and temperature T with scaling exponent $\Delta=3.5$ [from Ref.21]

Another consequence of scaling is the relation

$$\xi_{EA} = T^{-\nu} f\left(\frac{H}{T^{\Delta}}\right)_{T=0} \sim H^{-\nu/\Delta} \tag{17}$$

with $\nu \simeq 2$ from Eq.(7). The prediction of a diverging correlation length for $H \to 0$ has still to be verified.

Dynamic quantities may scale with ξ_{EA}, too. For instance for the relaxation time τ from Eq.(4) one expects

$$\ell n\tau(T,H) = T^{-\nu z}\, t\left(\frac{H}{T^{\Delta}}\right) \tag{18}$$

where according to Eq.(11) $\ell n\tau$ was assumed to be a scaling variable instead of τ itself and z is a new (dynamical) critical exponent. Eq.(18) implies

$$\frac{\ell n\tau(T,H)}{\ell n\tau(T_f,0)} = \left(\frac{T}{T_f}\right)^{-\nu z} \frac{t\left(\frac{H}{T^{\Delta}}\right)}{t(0)} \tag{19}$$

Thus curves of constant relaxation time in the H-T plane should fall on a single curve function if they are plotted on the axis H/T^{Δ} and T/T_f. In fact, this has been numerically verified in 2 and 3 dimensions [24] with $z \simeq 1$.

6. Comparison with experiments

It has already been mentioned that the two dimensional EA model describes real spin glasses qualitatively surprisingly well. However, recently more and more experiments seem to favour a true phase transition at a nonzero critical temperature T_f, whereas the model has a nonequilibrium freezing temperature $T_f(t)$ which depends on the observation time t and $T_f=0$ for $t\to\infty$ as indicated by Eq.(11).

I just want to mention three recent experiments which seem to show a phase transition at $T_f>0$ [4]:

(i) The plateau in the field cooled magnetization

(ii) The deAlmeida-Thouless H-T phase diagram

(ii) Scaling laws of the nonlinear susceptibility for $T>T_f$

However, it has been shown that also the computer simulations can reproduce these findings, therefore these experiments are not a proof of a static phase transition [4,21]:

To (i): The simulations show a plateau in the field cooled magnetization, however it diverges as in Eq.(16). The experimental data also seem to be consistent with Eq.(16), however with a very large value of $\Delta\approx 75$. Note that in the simulations one obtains a very large effective exponent Δ if one does not cool slowly enough to be in thermal equilibrium.

To (ii): Also in the model one can define a phase boundary in the field-temperature plane from the onset of irreversible effects for a given observation time t. If one scales T and H by

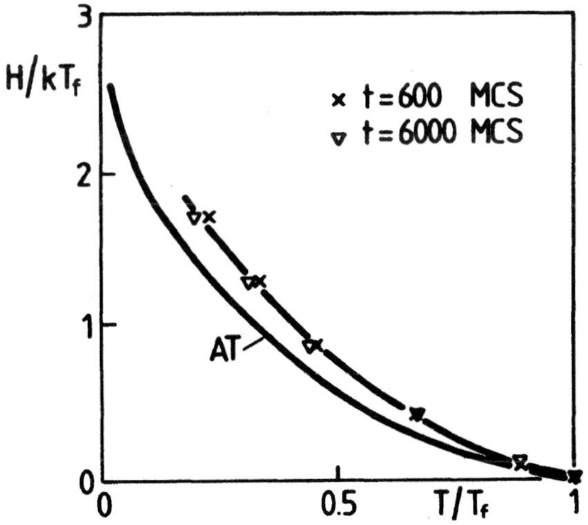

Fig. 3: Onset of irreversible behaviour for different time scales t as obtained from computer simulations of the 2-d EA model. For comparison the mean field prediction AT is shown [from Ref.23]

$T_f(t)$ one indeed obtains a boundary which is surprisingly similar to the mean field prediction [25], which gives $(T_f(H)-T_f) \propto \propto H^{2/3}$. This is shown in Fig.3. Note that in the model there are no real singularities, thus one has $(T_f(H)-T_f) \propto H^2$ for very small fields H. This crossover from $H^{2/3}$ to H^2 has recently been observed experimentally [26]

To (iii): Also in the model the data of the nonlinear susceptibility in thermal equilibrium can be squeezed onto a single "scaling function" as shown in Fig.4.

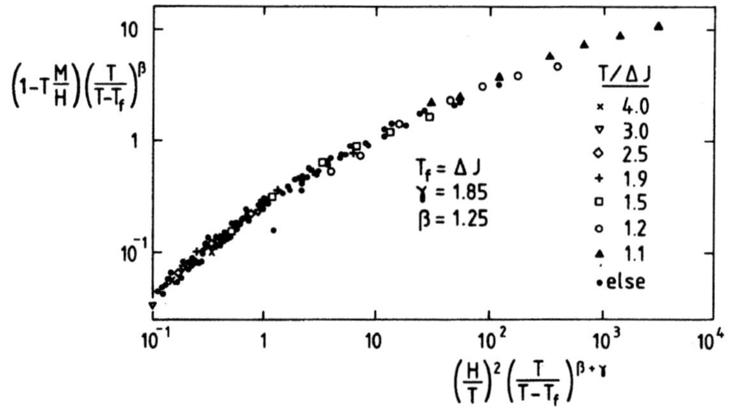

Fig. 4: Scaling plot of the nonlinear susceptibility of the 2-d EA model assuming a static transition at T_f= J and fitting "critical exponents" γ and β [Ref. 21]

$T_f=\Delta J$ has been arbritrarily chosen and the exponents β and γ have been fitted.

Since the model can reproduce some of the experimental results, one may ask whether the T=0 scaling behaviour of the model can be seen in real spin glasses. In fact, an analysis of the present experimental data gave satisfactory agreement with equations (15) and (19) [24].

7. Comparison with mean field theory

The results of the (2-dimensional) computer simulations resemble the predictions for the infinite range model in many respects. For instance one observes a cusp in the short time susceptibility, a plateau in the field-cooled magnetization and the singular field-temperature phase diagram. Therefore I recently investigated whether other mean field predictions can be observed in the short range model. Two problems have been studied [27]:

(i) The ferromagnet-spin glass phase diagram

(ii) The overlap structure of the spin glass states.

To (i): If the Gaussian distribution of the coupling J_{ij} is shifted to a nonzero average $J_o = [J_{ij}]_{av}$, the mean field theory predicts a constant spin glass transition temperature T_f up to $J_o = \Delta J$ and a double transition for $J_o > \Delta J$ [28]. In the latter case the system is in the paramagnetic phase at high temperatures T, in a ferromagnetic phase at intermediate T and in a mixed phase at low T. Note that while earlier theories predicted a zero magnetization M in the mixed phase, the Parisi theory gives $M \neq 0$ but irreversible effects (replica symmetry breaking) in this phase. Experiments seem to show that M=0 in this low temperature phase [29].

Since for the short range model the spin glass phase does not exists in thermal equilibrium, one can only define it dynamically. This has been done by computing the relaxation of

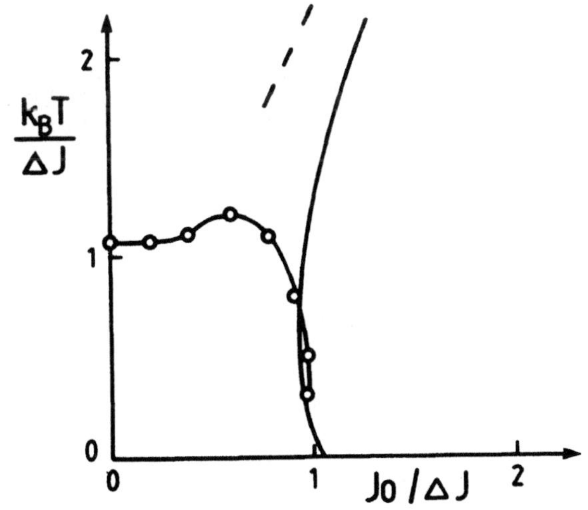

Fig. 5: Dynamically defined spin glass boundary (points) which are given by $[U_o(150)-U(150)]/U(150) = 1^o/oo$. The full line shows the exact ferromagnetic phase boundary of the corresponding annealed model. [from Ref.27]

the internal energy U(t) as a function of time (in Monte Carlo steps per spin, MCS) after starting from an initial ferromagnetic or antiferromagnetic state.

In fact Fig. 5 shows that such a dynamically defined spin glass phase is similar to the mean field predictions. While the ferromagnetic boundary exists in thermal equilibrium, it cannot be detected from Monte Carlo simulations since the relaxation times are too long, as mentioned in section 3. In Fig. 5 I have included this boundary for a different model, namely the annealed case which can be solved exactly. Surprisingly the annealed model shows what we expect for the quenched one: A destruction of the ferromagnet for $J_o \lesssim \Delta J$ and a double transition around $J_o = \Delta J$.

To (ii): To study the overlap structure I cooled the system ($J_o=0$) slowly down to T=0 and recorded the state $(S_i)_i$ which is usually not a ground state but a metastable state with an energy very close to it. Repeating this many times I recorded the distribution W(q) of the overlap q between the states $(S_i)_i$ as described in section 2. In contrast to W(q) of the mean field theory I obtained a Gaussian-like peak around q=0 and no indication of a ultrametic structure.

This shows that the low lying states, which - as in experiments - are accessible by slow cooling, do not have the structure of the mean field states.

8. Conclusions

Let us now draw some conclusions for the two-dimensional Ising model with short range random interactions and a relaxational dynamics:

- The model describes experiments qualitatively
- It has no static transition
- Nevertheless in the field H, temperature T and observation time t diagram, Fig. 6, there is a rather well defined surface below which spin glass behaviour is observed. This surface is rather singular, since one finds $T_f(H=0,t) \sim (\ell nt)^{-1/2}$ and $T_f(H,t)-T_f(0,t) \sim H^{2/3}$

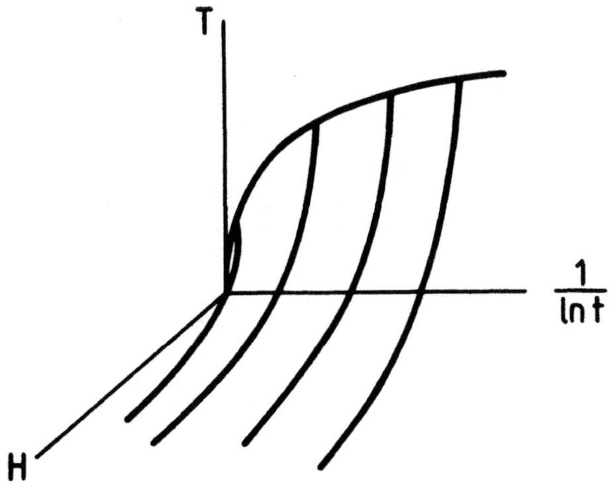

Fig. 6: Schematic view of the spin glass transition surface as a function of field H, temperature T and observation time t.

- Below $T_f(H,t)$ the spins freeze into small completely frozen clusters, the rest seems to remain in thermal equilibrium
- The freezing process can be described by a dynamics of small decoupled clusters
- The model even reproduces recent experiments which favour a static phase transition
- Only at T=0 one has a phase transition with scaling laws
- Also experimental data are not inconsistent with T=0 scaling
- The low lying metastable states do not have the structure of the mean field states

Acknowledgement
The author wishes to thank Prof. K. Binder for fruitful and continuous encouragement, support and collaboration.

REFERENCES

1. Disordered Systems and Localization, Lecture Notes in Physics, Vol. 149 (Springer 1981)
2. Heidelberg Colloquium on Spin Glasses, Lecture Notes in Physics, Vol. 192 (Springer 1983)
3. K. Binder and W. Kinzel, in Ref.1, p. 124
4. K. Binder and W. Kinzel, in Ref.2, p. 279
5. S.F. Edwards and P.W. Anderson, J. Phys. F5, 965 (1975)
6. D. Sherrington and S. Kirkpatrick, Phys. Rev. Lett. 35, 1792 (1975)
7. G. Parisi, Phys. Rev. Lett. 43, 1754 (1979)
8. see e.g. C. de Dominicis, in Ref.2, p. 103
9. H. Sompolinski and A. Zippelius, Phys. Rev. Lett. 47, 359 (1981)
10. G. Parisi, Phys. Rev. Lett. 50, 1946 (1983); A.P. Young, Phys. Rev. Lett. 51, 1206 (1983)
11. A.P. Young, A. Bray and M. Moore, to be published
12. M. Mezard, G. Parisi, N. Sourlas, G. Toulouse and M. Virasora, to be published
13. K. Binder and K. Schröder, Phys. Rev. B14, 2142 (1976)
14. W. Kinzel, Phys. Rev. B19, 4595 (1979)
15. D. Stauffer and K. Binder, Z. Physik B34, 97 (1979)
16. A.P. Young, Phys. Rev. Lett. 50, 917 (1983)
17. I. Morgenstern and K. Binder, Phys. Rev. Lett. 43, 1615 (1979); Phys. Rev. B22, 288 (1980)
18. W. Kinzel, Z. Phys. B46, 59 (1982)
19. See e.g. J.A. Mydosh in Ref. 1, p. 87
20. W. Kinzel, Phys. Rev. B26, 6303 (1982)
21. W. Kinzel and K. Binder, Phys. Rev. B29, 1300 (1984)
22. K. Binder, Z. Phys. B48, 319 (1982)
23. W. Kinzel and K. Binder, Phys. Rev. Lett. 50, 1509 (1983)
24. K. Binder and A.P. Young, to be published
25. J.R.L. de Almeida and D.J. Thouless, J. Phys. A11, 983 (1978)
26. J. Hamida, C. Paulsen, S.J. Williamson and H. Maletta, to be published
27. W. Kinzel, unpublished
28. see e.g. D. Sherrington in Ref. 2, p. 125
29. see e.g. H. Maletta in Ref. 2, p. 90

SOME MATHEMATICAL METHODS FOR FRUSTRATION MODELS

András Sütő
Central Research Institute for Physics
H-1525 Budapest 114, P.O.Box 49. Hungary

Abstract

The adoption of the Lee-Yang method (search for the zeroes of the partition function) and the Peierls (or domain wall) argument is presented for the case of Ising frustration models. The first leads to some examples of extremely strong frustration and the second to an explicit demonstration of ordering via entropy barrier.

I. Introduction

The main concern of this lecture is, as its title already indicates, some technical aspects of dealing with frustration models. Most of the powerful technics of mathematical statistical physics have been originally developed for ferromagnetic interactions and later generalized for other cases.

This happened with the Lee-Yang theorem [1] which found his extension in the Asano contraction [2, 3], making possible to study the analytical properties of the free energy also on the complex $\beta(=1/k_BT)$ plane and in other than ferromagnetic models. Further extensions followed (for a recent one and for other references see [4]) and by now the Lee-Yang theorem and its descendents together constitute a method of the statistical physics. This method consists of extending the partition function in some of its variables (typically in β or in the external field h) to a complex function and studying the distribution of the zeroes of this function. For finite systems the zeroes remain in a finite distance from the real axis but they may accumulate to it as the size of the system goes to infinity. If they do not, the infinite-volume free energy will certainly be an analytic function of the given variable along the real axis.

The Peierls argument is an invaluable tool for proving ordering at low temperatures. It was originally invented by Peierls [5] and applied to the two-dimensional ferromagnetic Ising model. After the removal of some mathematical inaccuracies [6] it also underwent some generalizations, the most powerful of them being that of Pirogov and Sinai [7] and became widely used for very different problems. The method is based on the idea of the energy-entropy balance. One assumes that at low temperatures a stable phase of the system looks like a ground state plus small perturbations added to it. Then one estimates the cost of energy and the gain in the entropy due to these perturbations and by comparing them one can sometimes establish that the ground state is stable against finite perturbations at sufficiently low temperatures.

Both the Lee-Yang and the Peierls methods are applicable to frustration models. If they are suitably adjusted to the given problem, they can yield interesting results. In the next part of this lecture I present some results obtained via the Lee-Yang method [8], showing that in some "superfrustration" models the free energy and the correlations are analytic at all temperatures, including $T=0$. The exponential decay of the correlations then follows (and also can be shown explicitly) for all $T \geq 0$. In the third part of my talk an example is given for ordering through entropy barrier [9]. Here the proof is a Peierls type argument.

It is needless to say that there exist other methods yielding rigorous results for frustration models. The direct computation of the free energy and the correlations is possible in two-dimensional periodic models. This led to a series of beautiful papers by Hoever, Wolff and Zittartz, reviewed in [10] (see also Zittartz's lecture at this School).

II. The absence of phase transitions in strongly frustrated models

Let us consider a network of triangular plaquettes built up in such a way that any vertex belongs to at most two different triangles. In the forthcoming discussion the dimensionality and periodicity of this network plays no role. Instead, we will need some restriction on its connectivity, which can be formulated in the following way. For the triangle i let $N_n(i)$ be the number of self-avoiding walks of step n through neighbouring triangles, which start at i. We impose the condition that there exist ε and α, both positive and smaller than 1, so that the inequalities

$$\sum_n \frac{N_n(i)}{3^n(1-\varepsilon)^{n-1}} \leq (1-\alpha)\varepsilon \tag{1}$$

hold true for all i. As an example, one can check that the network of triangles of the Kagomé lattice satisfies (1).

Now we define a nearest-neighbour Ising model on this network. Let $(i1)$, $(i2)$ and $(i3)$ label the three edges of the triangle i and let J_{i1}, J_{i2} and J_{i3} be the corresponding interactions. Then

$$J_{ij} = J_i \, \varepsilon_{ij} \quad , \quad j = 1,\, 2,\, 3$$

$$J_i > 0 \quad , \quad \varepsilon_{ij} = \pm 1 \quad , \quad \varepsilon_{i1}\, \varepsilon_{i2}\, \varepsilon_{i3} = -1 \; . \tag{2}$$

Hence, in our model every triangle is frustrated; the J_i's may vary arbitrarily. We will se that in this model there is no phase transition, even at $T=0$: apart from uninteresting terms, the free energy extends analytically to $T=0$ and the correlations decay exponentially at $T\geq 0$.

The starting point is the high-temperature expansion for Z_N, the partition function of N triangles. If c_i denotes $\cosh\beta J_i$ and t_{ij} stands for $\tanh\beta J_{ij}$ then

$$Z_N = \prod_1^N c_i^3 \sum_g{}' \prod_{(ij)\in g} t_{ij} \tag{3}$$

where the summation runs over closed graphs of edges. Graphs intersecting with the same set of triangles can be collected together and Z_N rewritten as

$$Z_N = \prod_1^N c_i^3 \sum_g{}'' \prod_{i\in g} (t_{ij} + t_{ik}\, t_{il}) \prod_{i\notin g} (1 + t_{i1}\, t_{i2}\, t_{i3}) \, . \tag{4}$$

Here g is now a set of loops of triangles and $j\in\{1,2,3\}$ is uniquely determined by g and i: the edges $\{(ij),\ i\in g\}$ form the unique closed graph with the property that each triangle in g is represented by a single edge in the graph.

Due to the definition (2) of the model, we have

$$\begin{aligned} t_{ij} + t_{ik}\, t_{il} &= \varepsilon_{ij}\, t_i(1-t_i) \\ 1 + t_{i1}\, t_{i2}\, t_{i3} &= 1 - t_i^3 \end{aligned} \tag{5}$$

where $t_i = \tanh\beta J_i$. Introducing the notation

$$z_i = \frac{t_i(1-t_i)}{1-t_i^3} = \frac{t_i}{1+t_i+t_i^2} \tag{6}$$

it is found that

$$Z_N = \prod_1^N (c_i^3(1-t_i^3)) \sum_g{}'' \prod_{i\in g} \varepsilon_{ij}\, z_i \, . \tag{7}$$

Let us notice that $\varepsilon_{ij}\, z_i$ is a pair correlation within the triangle i and that its magnitude does not exceed 1/3. Moreover, the summation in (7) runs over a subset of the closed graphs occurring in Eq. (3). In this subset any graph has a unique decompositon into connected closed graphs. This is the property we need for our main technical assertion:

Let us consider

$$R_N = \sum_g{}'' \prod_{i \in g} \varepsilon_{ij} \, z_i \tag{8}$$

for complex values of z_i . If

$$\sum_n N_n(i) x^n / (1-\varepsilon)^{n-1} \leq \varepsilon \tag{9}$$

holds true for all i with the same $\varepsilon (0<\varepsilon<1)$ and $x>0$ then we get the bounds

$$(1-\varepsilon)^N \leq | R_N | \leq (1+\varepsilon)^N \tag{10}$$

in the domain $\{|z_i| \leq x \text{ for all } i\}$. The proof can be done by induction (see the first paper of Ref. [8]). According to the condition (1) we imposed on the network, there exists some $\delta>0$ so that (9) can be satisfied with $x = \frac{1}{3} + \delta$ and ε, $0<\varepsilon<1$. For the sake of simplicity assume that $t_i = t$ and $z_i = z$ for all i. Then the possible zeroes of R_N are confined into the domain

$$\left| \frac{t}{1+t+t^2} \right| > \frac{1}{3} + \delta \tag{11}$$

of the complex t-plane, shown by Fig. 1.

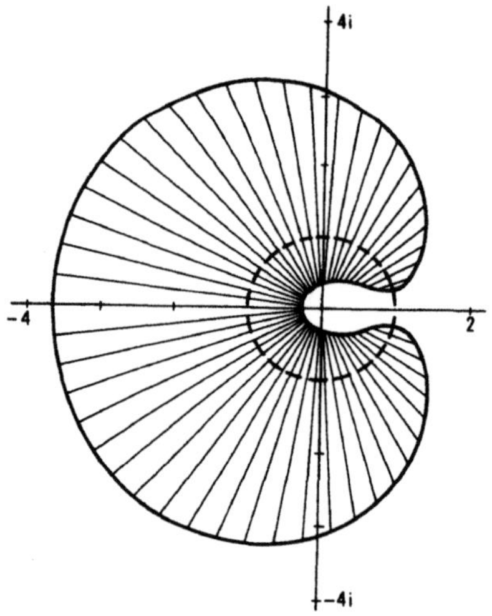

Figure 1. The cross-hatched region contains the points satisfying the inequality (11).

This domain is in a finite distance from the interval $0 \leq t \leq 1$ of the real axis which corresponds to the physically possible values of t_i . Therefore, by standard arguments, using the theorems of Hurwitz and Vitali [11], we obtain that the infinite-volume specific free energy is analytic for $0 \leq t \leq 1$, i.e. for $0 \leq T \leq \infty$, provided that it exists. The exponentially decaying upper bound for the correlations can be obtained in a very similar way (see the third paper of Ref. [8]). Models on different networks may also have these properties. For instance, the triangles can be replaced by squares; then (1) is to be modified by putting 2 instead of 3 in the denominator.

III. Ordering via entropy barrier

We study a family of two-dimensional nearest-neighbour Ising models on the square lattice, defined by the Hamiltonian

$$H = - \sum_{<ij>} \pm J \, s_i \, s_j \tag{12}$$

The signs are chosen so that the frustrated squares form pairs of columns, separated by one or more columns of non-frustrated squares:

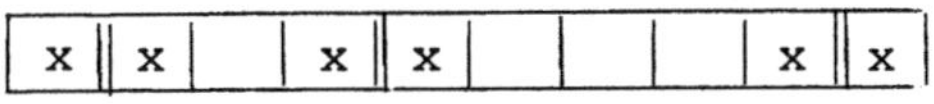

Without loss of generality we may assume that the antiferromagnetic bonds are those indicated by the double lines. The periodicity in the horizontal direction is unnecessary for our purposes; if the interaction is periodic then a direct computation of the free energy and correlations is possible ([10], [12]) and shows a phase transition at some positive temperature.

Our aim is to get an insight into the mechanism of ordering in these models. To begin with, one needs a guess about the nature of the order. It is surmised that at low temperatures in a pure phase the spins align parallely to some ground state configuration s^o, i.e. $\langle s_i \rangle \cdot s_i^o > 0$ for all i . A ground state configuration *(gsc)* is a spin configuration the energy of which cannot be decreased by changing the spins in any finite domain. In frustration models, any *gsc* has built-in defects, i.e. a set of unsatisfied bonds. In two dimensions an unsatisfied, "wrong" bond can be represented by a string of unit length drawn perpendicularly to the bond. The strings connect the centres of the frustrated squares and in a *gsc* their total length cannot be decreased by local transformations. Consider now the model defined above. In most (but not all) of the *gsc*'s every string connects neighbouring centres of frustration. The *gsc*'s are grouped into *ground states*: a ground state is a set of *gsc*'s in which any two configurations can be transformed into each other by a - maybe infinite - sequence of local transformations, each step leading through a *gsc*. In our model there exists a ground state G^+ which contains the *gsc* $s^+ \equiv +1$ and any other configuration obtained from s^+ by flipping non-neighbouring spins along the antiferromagnetically coupled chains. G^+ has a finite entropy per site: each antiferromagnetic chain contributes to it with $\ln(1+\sqrt{5})/2$ per unit length. Within G^+, s^+ is distinguished by the following fact. If one calculates the "thermodynamic" average of s_i in the ground state G^+ (by putting equal weights on every *gsc* in G^+) one finds

$$\langle s_i \rangle = \begin{cases} s_i^+/\sqrt{5} & \text{if } i \text{ is a site in an antiferromagnetic chain} \\ s_i^+ , & \text{otherwise.} \end{cases}$$

Now $G^- = \{-s \mid s \in G^+\}$ is also a ground state and $s^- \equiv -1$ is distinguished in it by the same property. These two *gsc*'s , s^+ and s^- are good candidates to play the role of s^o which characterizes the order, from two reasons. Firstly, they are the only *gsc*'s possessing the translational symmetry of the Hamiltonian (invariance with respect to vertical translation) and secondly, G^+ and G^- have the highest entropy among the ground states.

What are these other ground states ?
They can be obtained from G^+ and G^- by flipping infinite or semiinfinite vertical strips of spins. If the strip-flip is carried out on s^+ or s^-, the borders of these strips are zigzagged: every second spin along the bordering antiferromagnetic chains must also be flipped in order to obtain a *gsc*. Let G be a ground state differing from G^+ in a single strip-flip. In order to reach G from G^+ via a sequence of local transformations one has to climb through only a finite energy barrier: to turn over the spins in the first finite segment of the strip costs an energy proportional to the width of the strip and any further transformation is gratuitous. The situation is seemingly analogous to the one-dimensional case: a finite increase in the energy has to be compared to an increase of $\ln L$, where L is the vertical extension of the system, in the entropy. Let us notice, however, that once the first energy-consuming transformation is carried out, the further transformations *decrease* the entropy by an amount proportional to the length of the flipped segment. Indeed, the freedom in the alignment of the spins along the antiferromagnetic chains disappears at the borders of the flipped segment, therefore a loss of entropy is found in comparison with G^+. So far, we have considered only the ground states. At positive temperatures the argument becomes more complicated because the number of "dangerous" excitations is still exponentially increasing with their linear extension. One can nevertheless show that the probability of these excitations becomes negligible as the

temperature decreases and the reason is exactly the same that we found at zero temperature.

In order to be more definite about the argument at positive temperatures, we need the notion of a contour. Let us start with the *gsc* s^+ and allow spin flips on connected finite domains. A contour Γ is a connected line drawn on the dual lattice, separating the domains of s^- and s^+. Let us take a site i which is not situated at an antiferromagnetic chain. We would like to show that $< s_i >^+ > 0$ for sufficiently low temperatures. Now it is easy to see that

$$\frac{1}{2} < 1 - s_i >^+ \leq \sum_{\Gamma}{}' P(\Gamma) \tag{13}$$

where the summation runs over contours surrounding the site i and $P(\Gamma)$ is the probability that Γ is a maximal connected line separating domains of s^+ and s^-. The heart of the Peierls argument is the so-called Peierls condition which establishes a relation between the energy, k_Γ and length, $|\Gamma|$ of a contour Γ (the energy of a contour is the increase in the energy when one flips the spins inside Γ): provided that

$$k_\Gamma \geq c_1 \; |\Gamma| \tag{14}$$

holds for all Γ with the same $c_1 > 0$, $< s_i >^+ > 0$ can be shown for small T. The validity of this statement follows from the estimate

$$P(\Gamma) \leq e^{-2\beta k_\Gamma} \leq e^{-2\beta c_1 |\Gamma|} \tag{15}$$

and the fact that the number of contours of length ℓ , surrounding the site i, is smaller than c_2^ℓ . Clearly, in our case (14) does not hold for all Γ. We therefore have to show that

$$A_i(\varepsilon) = \sum_{k_\Gamma < \varepsilon |\Gamma|}{}' P(\Gamma) \tag{16}$$

becomes negligible if the temperature tends to zero, where, again, the contours surround i and ε is arbitratily small and fixed. The first inequality of (15) is still valid but is not sufficiently strong. It is easily seen, however, that the better estimate

$$A_i(\varepsilon) \leq e^{-4\beta J} \sum_{k_\Gamma < \varepsilon |\Gamma|}{}' 1/\nu_\Gamma \tag{17}$$

also holds. Here ν_Γ is the number of *gsc*'s which are not compatible with the contour Γ , i.e. ν_Γ is the number of transformations which can be performed without any cost of energy if (and only if) Γ is not present. Notice that the logarithm of this quantity was identified as the loss of entropy of the ground state G^+. A straightforward calculation shows that

$$\nu_\Gamma > \left(\frac{1+\sqrt{5}}{2} \right)^{(1-3\varepsilon)|\Gamma|/2} \tag{18}$$

and for the number of contours of length ℓ in the summation (16) we get

$$\begin{aligned} N_\ell(\varepsilon) &\leq c_3^{\varepsilon\ell} \quad if \quad \ell > 2/\varepsilon \\ &= 0 \quad otherwise \; . \end{aligned} \tag{19}$$

One then sees that the sum in (17) converges for sufficiently small ε and $A_i(\varepsilon)$ tends to zero with T going to zero. (Details can be found in [9].)

The entropy barrier between G^+ and G^- can be measured by the decrease of entropy due to a strip-flip. This becomes infinite in the infinite system and increases proportionally to the linear extension. We see therefore that the entropy barrier between G^+ and G^- in the present case is as effective as the energy barrier between the two ground states in the ferromagnetic model.

References

1. Yang C.N. and Lee T.D., Phys. Rev. 87, 404,410 (1952)
2. Asano T., J. Phys. Soc. Japan 29, 350 (1970)
3. Ruelle D., Phys. Rev. Letters 26, 303 (1971)
4. Lieb E.H. and Sokal A.D., Commun. Math. Phys. 80, 153 (1981)
5. Peierls R., Proc. Cambridge Phil. Soc. 32, 477 (1936)
6. Griffiths R.B., Phys. Rev. 136, A437; Dobrushin R.L., Sov. Phys. Dokl. 10, 111 (1965) and Theory Probab. Its Appl. 10, 193 (1965)
7. Pirogov S.A. and Sinai Ya. G., Teor. Mat. Fiz. 25, 358 (1975); *ibid* 26, 61 (1976); Sinai Ya. G., Theory of phase transitions: Rigorous results (Akadémiai Kiadó, Budapest 1982) pp 29-74
8. Sütő A., Helv. Phys. Acta 54, 191, 201 (1981); Z. Phys. B44, 121 (1981)
9. Sütő A., J. Phys. A14, 2733 (1981)
10. Wolff W.F. and Zittartz J., Lecture Notes in Physics 192, 252 (Proceedings of the Heidelberg Colloquium on Spin Glasses, 1983)
11. Titchmarsh E.C., The theory of functions (Oxford University Press, 1975)
12. Longa L. and Oleś A.M., J. Phys. A13, 1031 (1980)

III. RANDOM SYSTEMS

PSEUDO EQUILIBRIUM DESCRIPTION OF DISORDERED SYSTEMS

G. Sobotta
Institut für Theoretische Physik III
Ruhr-Universität Bochum
D-4630 Bochum, Federal Republic of Germany

The thermodynamical description of quenched random systems is discussed in the frame of a generalized grand canonical ensemble. This approach circumvents the usual average of the configuration dependent free energy. The differences of quenched and annealed systems are discussed utilizing exactly solvable models.

1. Introduction

If one tries to describe for instance a binary alloy of magnetic atoms M and non magnetic atoms U, $M_x U_{1-x}$, one assumes for low enough temperatures a frozen in spatial configuration of atoms c, which is given by a set of occupation numbers K_i, which are one, if site i is occupied by a magnetic atom, and zero otherwise. Instead of calculating the free energy for a given configuration c, F_c, one tries to calculate the averaged free energy[1]

$$F = [F_c] \tag{1.1}$$

where $[\cdot]$ denotes the average over an ensemble of configurations. By assuming independently distributed occupation numbers, the related distribution function reads

$$\mathbb{P}(K) = \prod_i ((1-x)\delta(K_i) + x\,\delta(K_i - 1)) \tag{1.2}$$

With a non translation invariant Hamiltonian $\mathcal{H}(K,S)$, depending on the spins S as well as on the occupation numbers K, the free energy is given by

$$F = -\beta^{-1} \sum_K \mathbb{P}(K) \ln \mathrm{Tr}_S\, e^{-\beta \mathcal{H}(K,S)} \tag{1.3}$$

with $\beta = 1/k_B T$.

The fundamental problem of the statistical mechanics of quenched disordered systems arises from this form of the free energy, which is not given by a partition function. The most popular construction of a partition function is based on the replica trick or n=o trick[2], which makes use of the identity $\ln Z_K = \frac{\partial}{\partial n} Z_K^{\ n}\Big|_{n=o}$. This transcription leads to the formal partition function

$$Z(n) := \mathrm{Tr}_{(S^\alpha)_{\alpha=1,2,\cdots,n}} e^{-\beta \mathcal{H}((S^\alpha)_{\alpha=1,2,\cdots,n})} \tag{1.4}$$

where the formal Hamiltonian is given by

$$-\beta \mathcal{H}((S^\alpha)_{\alpha=1,2,\cdots,n}) = \ln \sum_K \mathbb{P}(K) e^{-\beta \sum_{\alpha=1}^{n} \mathcal{H}(K,S^\alpha)} \tag{1.5}$$

Because for $n = 1,2,\ldots$ $Z(n)$ is the partition function of an ordered system, it seems to be reasonable to apply well-known technics for its calculation, although one has to keep in mind that the free energy is given by $F = -\beta^{-1} \frac{\partial}{\partial n} Z(n)^n\Big|_{n=o}$ and this requires an analytical continuation of $Z(n)$. Of one studies phase transitions by variational methods or renormalization group technics, it is rather questionable, whether the approximated partition function shows a reasonable behaviour.

A quite different way to find a partition function shall be discussed here. The first step to describe a quenched disordered system formally as an ordered system in equilibrium has been done by T. Morita[3] 1964. In chapter 2, we will derive for the above mentioned random site problem the thermodynamical potential for a quenched disordered system, using the information theoretical approach to statistical mechanics[4]. The equivalence of the derived thermodynamical potential and the averaged free energy will be discussed. Finally we shall compare the averages and response functions, given in the frame of both descriptions.

In chapter 3, we shall discuss the difference between quenched disorderes systems. Because the spin energy may show diverging fluctuations in the vicinity of a second order phase transition, the concept of an ideal quenched system seems to be doubtful. The possible influence of the spins on the occupation numbers shall be discussed by utelizing exactly solvable spin models.

2. Derivation of the thermodynamical potential by information theory

The first part of this chapter summarizes results, which has been derived else where[4]. We are assuming to deal with the random site model, introduced in chapter 1, being described by occupation numbers K and spins S. In the frame of information theory, we may start with the probability distribution function P (K,S) for the microscopic configuration of the magnetic atoms and of the spins, which defines the information entropy by

$$S_I = -\sum_{K,S} P(K,S) \ln P(K,S) \tag{2.1}$$

which measures our lack of knowledge, we have of a microscopic configuration of the system. To improve our knowledge, we introduce macroscopic observables, which are defined by statistical mean values. While in the canonical ensemble one has to consider the internal energy U, given by

$$\sum_{K,S} P(K,S)\,\mathcal{H}(K,S) = U \tag{2.2}$$

the grand canonical description requires also the concentration x of magnetic atoms, given by

$$\sum_{K,S} P(K,S)\, K_i = x \tag{2.3}$$

to get a complete macroscopic description. In the case of a quenched disordered system, the different non equilibrium distributions of atoms require to be distinguished by additional macroscopic variables, because, as well known, they differ in their macroscopic behaviour. Consequently, the other momenta

$$\sum_{K,S} P(K,S)\, K_{i_1} K_{i_2} \cdots K_{i_n} = f_{i_1 i_2 \cdots i_n} \tag{2.4}$$

has to be taken into account as independent macroscopic variables.

After improving with help of these complete set of macroscopic variables our lack of knowledge, the best estimate of the probability distribution function P (K,S) is obtained by maximizing our lack of knowledge or the information entropy S_I under the given constraints, introducing Lagrange multipliers. For the normalization condition, the Lagrange parameter should be α, for the energy U we introduce β , while for the momenta $f_{i_1 i_2 \ldots i_n}$ the Lagrange multipliers are $\lambda_{i_1 i_2 \ldots i_n}$. Because all conditions are linear in P (K,S), one gets

$$P(K,S) = \exp\Big(-(\alpha+1) - \beta\,\mathcal{H}(K,S) - \sum_{n=1}\sum_{i_1 i_2 \cdots i_n} \lambda_{i_1 i_2 \cdots i_n} K_{i_1} K_{i_2} \cdots K_{i_n}\Big) \tag{2.5}$$

with

$$Z = \sum_K \exp(-\mathcal{H}(K))\, Z_K = \sum_{K,S} \exp(-\beta\mathcal{H}(K,S) - \mathcal{H}(K)) \tag{2.6}$$

and

$$\mathcal{H}(K) = \sum_{n=1} \sum_{i_1 i_2 \cdots i_n} \lambda_{i_1 i_2 \cdots i_n} K_{i_1} K_{i_2} \cdots K_{i_n} \tag{2.7}$$

$\mathcal{H}(K)$ is the part of the effective Hamiltonian which contains the forces of constraints, keeping the atoms in their configurations. Z_K is the partition function for a given configuration of atoms, while the generalized grand partition function is the searched generating functional, having some more "chemical potentials[5]" than an equilibrium system.

Because the information entropy determines the thermodynamical entropy by $S = k_B\, S_I$, the discussed quenched systems has the entropy

$$S(U,\{f\}) = k_B \beta U + k_B \ln Z + k_B \sum_{n=1} \sum_{i_1 i_2 \cdots i_n} \lambda_{i_1 i_2 \cdots i_n} f_{i_1 i_2 \cdots i_n} \tag{2.8}$$

which is the thermodynamical potential as function of U and of the momenta {f}. With the temperature T, given by $1/T = \partial S/\partial U$, i.e. $\beta = 1/k_B T$, one obtaines by Legendre transformation the free energy

$$F(T,\{f\}) = U - TS = -k_B T \ln Z - k_B T \sum_{n=1} \sum_{i_1 i_2 \cdots i_n} \lambda_{i_1 i_2 \cdots i_n} f_{i_1 i_2 \cdots i_n} \tag{2.9}$$

and consequently the thermodynamical potential as function of the forces λ

$$G(T,\{\lambda\}) = k_B T \ln Z \tag{2.10}$$

For a temperature independent distribution function $\mathbb{P}(K)$, having the momenta $f_{i_1 i_2 \ldots i_n}$, the obtained thermodynamical potential $F(T,\{f\})$ is equivalent to the averaged free energy[4], because one finds

$$F(T,\{f\}) = -k_B T \sum_K \mathbb{P}(K) \ln Z_K + k_B T \sum_K \mathbb{P}(K) \ln \mathbb{P}(K) \tag{2.11}$$

where the second contribution on the right side of this equation contains the temperature independent entropy due to the atomic disorder.

To study some consequences of the proposed statistical description, let us consider a Hamiltonian with a magnetic field

$$\mathcal{H}(H;K,S) = g\mu_B H \sum_i K_i S_i^z + \mathcal{H}(0;K,S) \tag{2.12}$$

In the frame of the conventional description, let us express the thermal average by $\langle \cdot \rangle$, the configurational average by $[\cdot]$. Then, the magnetization reads

$$M = -\frac{\partial [F_c]}{\partial H} = -g\mu_B \sum_i [K_i \langle S_i^z \rangle] \tag{2.13}$$

while the susceptibility reads

$$\chi = -\frac{\partial^2 [F_c]}{\partial H^2} = \frac{(g\mu_B)^2}{k_B T} \sum_{i,j} \left([K_i K_j \langle S_i^z S_j^z \rangle] - [K_i \langle S_i^z \rangle K_j \langle S_j^z \rangle] \right) \tag{2.14}$$

In the discussed pseudo equilibrium description, let us deonte the ensemble average by $\{\cdot\}$, which is equal $[\langle \cdot \rangle]$. Then one gets

$$M = -\frac{\partial F}{\partial H} = -g\mu_B \sum_i \{K_i S_i^z\} \tag{2.15}$$

while the susceptibility is given by

$$\chi = -\frac{\partial^2 F}{\partial H^2} = \frac{(g\mu_B)^2}{k_B T} \sum_{ij} \left(\{K_i K_j S_i^z S_j^z\} - \{K_i S_i^z\}\{K_j S_j^z\} \right)$$

$$+ g\mu_B \sum_i \sum_{n=1} \sum_{i_1 i_2 \cdots i_n} \frac{\partial \lambda_{i_1 i_2 \cdots i_n}}{\partial H} \left(\{K_{i_1} K_{i_2} \cdots K_{i_n} K_i S_i^z\} - \{K_{i_1} K_{i_2} \cdots K_{i_n}\}\{K_i S_i^z\} \right) \tag{2.16}$$

Comparing the equations (2.14) and (2.16), one finds

$$\frac{(g\mu_B)^2}{k_B T} \sum_{i,j} \left([K_i K_j \langle S_i^z \rangle \langle S_j^z \rangle] - [K_i \langle S_i^z \rangle][K_j \langle S_j^z \rangle] \right)$$

$$= -g\mu_B \sum_i \sum_{n=1} \sum_{i_1 i_2 \cdots i_n} \frac{\partial \lambda_{i_1 i_2 \cdots i_n}}{\partial H} \left([K_{i_1} K_{i_2} \cdots K_{i_n} K_i \langle S_i^z \rangle] - [K_{i_1} K_{i_2} \cdots K_{i_n}][K_i \langle S_i^z \rangle] \right) \tag{2.17}$$

This identity may be used to discuss the susceptibility of a spin glass. If one follows the assumption of S.F. Edwards and P.W.Anderson[6], $q = \left[(K_i < S_i >)^2\right]$ may be regarded as the order parameter to formulate a mean field approximation. From equation (2.17) one deduces as a possible order parameter

$$p = \sum_{\delta} \left(\left[K_{i+\delta}\, K_i \langle S_i^z \rangle\right] - \left[K_{i+\delta}\right]\left[K_i \langle S_i^z \rangle\right] \right) \tag{2.18}$$

with the sum performed over nearest neighbours. This order parameter describes a short range correlation between the local density of the magnetic atoms and the magnetic moment $\mu_i^z = - g\mu_B\, K_i\, S_i^z$.

To complete this discussion, let us consider the specific heat

$$C = \frac{1}{k_B T^2} \left(\left[\langle \mathcal{H}(H;K,S)^2 \rangle\right] - \left[\langle \mathcal{H}(H;K,S)\rangle^2\right] \right) \tag{2.19}$$

which reads in the pseudo equilibrium description

$$\begin{aligned} C = {} & \frac{1}{k_B T^2} \left(\left[\langle \mathcal{H}(H;K,S)^2 \rangle\right] - \left[\langle \mathcal{H}(H;K,S)\rangle\right]^2 \right) \\ & - \sum_{n=1} \sum_{i_1 i_2 \cdots i_n} \frac{\partial \lambda_{i_1 i_2 \cdots i_n}}{\partial T} \Big(\left[K_{i_1} K_{i_2} \cdots K_{i_n} \langle \mathcal{H}(H;K,S)\rangle\right] \\ & \qquad - \left[K_{i_1} K_{i_2} \cdots K_{i_n}\right]\left[\langle \mathcal{H}(H;K,S)\rangle\right] \end{aligned} \tag{2.20}$$

3. Restricted annealed systems and ideal quenched systems

In the past there has been always discussions about the exceptional role of ideal quenched systems. Mathematically, restricted annealed systems show properties, which can be compared to annealed systems, which had been remarked by J.C. Wheeler and R.B. Griffiths[5].
M.F. Thorpe and D. Beeman[7] studied the simplest restricted model by assuming one and two particle forces of constraints in the effective Hamiltonian (equ. (2.7)).
Recently, A. Huber and R. Kühn studied a restricted annealed one dimensional Ising model[8]. While for the one dimensional Ising model without magnetic field the restricted annealed model for independently distributed occupation numbers becomes exact, if one takes into account one and two particle forces of constraints, the authors were able to show numerically, that these two forces of constraints play a dominant role in the presence of a magnetic field. - It is possible to show that there is a large class of models with a strictly finite

number of non vanishing forces of constraints.

Recently the author presented an exact method to treat d dimensional lattice models[9]. This method is a generalization of the transfer matrix method for one dimensional problems and is called the transfer tensor method, which could only be sketched very briefly here. For this purpose, let us firstly consider the one dimensional Ising model. The components of the transfer matrix ρ are given by the local Boltzmann factor

$$\rho_{S_{i-\frac{1}{2}}}^{S_{i+\frac{1}{2}}} = e^{I S_{i-\frac{1}{2}} S_{i+\frac{1}{2}}} \tag{3.1}$$

The partition function for N+1 spins and respectively N transfer matrices can be written as

$$Z_N = (11)\,\rho^N \binom{1}{1} \tag{3.2}$$

where the vectors (11) and $\binom{1}{1}$ replaces the summation over the boundary spins. Without magnetic field the transfer matrix

$$\rho = \begin{pmatrix} e^{I} & e^{-I} \\ e^{-I} & e^{I} \end{pmatrix} \tag{3.3}$$

is transformed into its diagonal representation

$$D = U \rho U^T \tag{3.4}$$

with $U = \frac{1}{\sqrt{2}} \begin{pmatrix} 1 & 1 \\ -1 & 1 \end{pmatrix}$, being independent on the value of I. From

$$D = \begin{pmatrix} D_+^+ & 0 \\ 0 & D_-^- \end{pmatrix} \tag{3.5}$$

with $D_+^+ = 2 \cosh I$ and $D_-^- = 2 \sinh I$ one obtains after the transformation $U(11) = \sqrt{2}\,(10)$

$$Z_N = 2\, D_+^{+\,N} = 2\; 2^N \cosh^N I \tag{3.6}$$

For a random system, one has to consider the position dependent transfer matrix

$$\rho(i)_{S_{i-\frac{1}{2}}}^{S_{i+\frac{1}{2}}} = e^{I(K_{i-\frac{1}{2}}, K_{i+\frac{1}{2}})\, S_{i-\frac{1}{2}} S_{i+\frac{1}{2}}} \tag{3.7}$$

but because the rotation matrix U is independent of the local interaction $I\,(K_{i-\frac{1}{2}}, K_{i+\frac{1}{2}})$, one gets for an arbitrary set of occupation numbers

$$Z_N(K) = 2 \prod_{i=1}^{N} D_+^{+}(i) \tag{3.8}$$

Using the equivalence of the averaged free energy and the pseudo equilibrium free energy

$$\begin{aligned} -k_B T \ln Z - k_B T \sum_{n=1} \sum_{i_1 i_2 \cdots i_n} \lambda_{i_1 i_2 \cdots i_n} \{K_{i_1} K_{i_2} \cdots K_{i_n}\} \\ = -k_B T \left[\ln Z_K\right] + k_B T \left[\ln \mathbb{P}(K)\right] \end{aligned} \tag{3.9}$$

one obtains a finite number of non vanishing forces of constraints, if ln Z_K and ln $\mathbb{P}(K)$ are polynomials in K of finite degree. This is fulfilled in the discussed case, because Z_K and $\mathbb{P}(K)$ factorize.

In two dimensions, let us consider the local Boltzmann factor

$$\rho^{\,S_{i+\frac{1}{2},j}\; S_{i,j+\frac{1}{2}}}_{\;S_{i-\frac{1}{2},j}\; S_{i,j-\frac{1}{2}}} = e^{-\beta \mathcal{H}_{S_{i-\frac{1}{2},j},\, S_{i,j-\frac{1}{2}};\, S_{i+\frac{1}{2},j},\, S_{i,j+\frac{1}{2}}}} \tag{3.10}$$

as components of a tensor ρ of rank 4, which is called the transfer tensor. If the full Hamiltonian is given by the sum

$$\mathcal{H} = \sum_{i,j} \mathcal{H}_{S_{i-\frac{1}{2},j},\, S_{i,j-\frac{1}{2}};\, S_{i+\frac{1}{2},j},\, S_{i,j+\frac{1}{2}}} \tag{3.11}$$

then one can write down the partition function for M X N transfer tensors as

$$Z_{MN} = \begin{array}{ccccc} & (11) & (11) & & (11) \\ (11) & \rho & \rho & \cdots & \rho \; \binom{1}{1} \\ & & & & \vdots \\ (11) & \rho & & & \\ & \vdots & & & \\ (11) & \rho & \cdots & & \rho \; \binom{1}{1} \\ & \binom{1}{1} & & & \binom{1}{1} \end{array} \tag{3.12}$$

As in one dimension, the partition function factorizes. For a random system, one gets analogously to (3.8)

$$Z_{MN}(K) = 2^{M+N} \prod_{i,j} D^{++}_{++}(K_{i-\frac{1}{2},j}, K_{i,j-\frac{1}{2}}; K_{j+\frac{1}{2},j}, K_{i,j+\frac{1}{2}}) \tag{3.16}$$

We can not discuss here the interactions, which can be treated in this way, but we want to stress that it is possible to generalize this method straightforwardly to d dimensions. For transfer tensors of finite rank, one always gets a strictly finite number of forces of constraints in the effective Hamiltonian. This result is independent on the dimension of the transfer tensor and holds therefore also for Potts spins with q states or even for continuous spins.

The discussed factorizable partition functions can not describe models with finite temperature phase transitions, because the components of D are regular functions of the local interaction, which is proportional to $\beta = 1/k_BT$. However, is seems to be quite interesting to discuss the zero temperature phase transition. This should be done for the one dimensional model, but this analysis is easily generalized to higher dimensions.

For the random interaction $I\,(K_{i-\frac{1}{2}}, K_{i+\frac{1}{2}}) = I\, K_{i-\frac{1}{2}} K_{i+\frac{1}{2}}$, one gets

$$D(i)^{+}_{+} = 2 \cosh I K_{i-\frac{1}{2}} K_{i+\frac{1}{2}} \tag{3.17}$$

Thus, from

$$\ln D(i)^{+}_{+} = \ln 2 + K_{i-\frac{1}{2}} K_{i+\frac{1}{2}} \ln \cosh I \tag{3.18}$$

one obtains the temperature depending part of $\lambda_{i-\frac{1}{2},i+\frac{1}{2}}$ by ln cosh I, where $I = \beta\, \mathbb{I} = \mathbb{I}/k_BT$ with the nearest neighbour exchange coupling $\mathbb{I}$. Obviously, one gets for $T \to 0$ the diverging force $\lambda_{i-\frac{1}{2},\, i+\frac{1}{2}} \propto 1/T$. Similar unphysical behaviour may be obtained at finite temperature phase transitions in the frame of a renormalization group analysis[10]. Other rather unphysical properties of ideal quenched systems has been observed by R.B. Griffiths[11] These Griffiths' singularities has been studied by several authors[12] and it seems to be a common feature of these unphysical properties that they are rather unobservable. This indicates that the differences between quenched and annealed systems should not be over emphasized , because for forces of constraints with

In general, this partition function can not be evaluated by elementary technics. However, if the transfer tensor has the following bi-diagonal representation D, given by

$$D_{\alpha\beta}^{\alpha'\beta'} = \delta_{\alpha}^{\alpha'}\,\delta_{\beta}^{\beta'}\,D_{\alpha\beta}^{\alpha\beta} \tag{3.13}$$

one can calculate the partition function, which becomes very simple, if D is obtained by the rotation matrix $U = \frac{1}{\sqrt{2}} \begin{pmatrix} 1 & 1 \\ -1 & 1 \end{pmatrix}$ as given by

$$D = \mathcal{U} \underset{\mathcal{U}^T}{\overset{\mathcal{U}}{\rho}} \mathcal{U}^T \tag{3.14}$$

This form should stress the diagonalization with respect to the row spins and the column spins in Z_{MN} (equ. 3.12). After the corresponding transformation of the boundary vectors, on gets

$$\begin{array}{rl} Z_{MN} = 2^{M+N} & \begin{array}{ccccc} & (10) & (10) & & (10) \\ (10) & D & D & \cdots & D\,\binom{1}{0} \\ (10) & D & & & \\ & \vdots & & & \\ & & & & \\ (10) & D & \cdots & & D\,\binom{1}{0} \\ & \binom{1}{0} & & & \binom{1}{0} \end{array} \\ & \\ = 2^{M+N} & \left(D_{++}^{++}\right)^{MN} \end{array} \tag{3.15}$$

a more realistic temperature dependence the differences between quenched and annealed systems are essentially reduces.

4. Summary.

We discussed the pseudo equilibrium description of quenched random systems. By generalizing the grand canonical ensemble it was possible to take into account frozen in spatial distributions of atoms on a lattice. The difference to the grand canonical ensemble for annealed systems is given by additional chemical potentials or forces of constraints, which are conjugated to the momenta of the occupation numbers. The obtained description is completely equivalent to the conventional description, based on the configurational average of the free energy. However, the construction of a partition function or generating functional allows to apply successful methods, which have been developed for equilibrium systems, to treat quenched disordered systems. The experience from field theoretical investigations of random ferromagnets[13], which are briefly discussed in a second article in this volume[14], seems to indicate that the pseudo equilibrium description of disordered systems may be applied also for further problems as for instance the localization phenomena.

Although this method has been discovered[3] and rediscovered[15] to perform concrete calculations, it has been considered also for more general investigations to understand the difference between annealed and quenched systems[5]. Because the forces of constraints delivers a quantitative information upon the interaction of the spins with the occupation numbers, it was possible to check to validity of an ideal quenched model. Obviously, diverging spin energy fluctuations show that the assumption of ideal quenching may be rather unphysical.

References

1. R. Brout, Phys. Rev. 115 (1959) 824
2. G. Grinstein and A. Luther, Phys. Rev. B13 (1975) 1329; V.J.Emery, Phys. Rev. B11 (1975) 239; A. Aharony, Phys. Rev. B12 (1975) 1038
3. T. Morita, J. Math. Phys. 5 (1964) 1401
4. G. Sobotta and D. Wagner, Z.Phys. B33 (1979) 271
5. J.C. Wheeler and R.B. Griffiths, Phys. Rev. 170 (1968) 249
6. S.F. Edwards and P.W. Anderson, J. Phys. F5 (1975) 965

7. M.F. Thorpe and D. Beeman, Phys. Rev. B14 (1976) 188; see also M.F. Thorpe, J.Phys. C11 (1978) 2986; H. Falk, J. Phys. C9 (1976) L 213

8. A, Huber and R. Kühn, DPG-Frühjahrstagung, Freudenstadt 1983

9. G. Sobotta, DPG-Frühjahrstagung, Freudenstadt 1983

10. G. Sobotta, J. Magn.Magn. Mat. 28 (1982) 1

11. B. Griffiths, Phys. Rev. Lett. 23 (1969) 17

12. M. Wortis, Phys. Rev. B10 (1974) 4665; Y. Imry, Phys.Rev. B15 (1977) 4448

13. G. Sobotta and D. Wagner, J. Phys. C11 (1978) 1467; see also ref. 4; for a review see G. Sobotta and D. Wagner, DPG-Frühjahrstagung, Münster 1984, to be published in J. Magn. Magn. Mat.

14. K. Westerholt and G. Sobotta, this volume

15. G. Sobotta and D. Wagner, J.Magn. Magn. Mat. 6 (1977) 92; see also ref. 5 and M. Shwartz, preprint.

ON THE VALIDITY OF THE REPLICA METHOD AND ITS APPLICATION TO ISING CHAIN IN RANDOM FIELD

András Sütő and Gergely T. Zimányi
Central Research Institute for Physics,
H-1525 Budapest 114, P.O.Box 49. Hungary

Abstract

Classical spin models in the presence of random interactions or external fields are considered. A general argument shows that the normal (non-Parisi-type) replica trick is bound to yield the correct free energy provided that this latter is an analytic function of the strength ε of the random variables. As an illustration, the free energy of the one-dimensional Ising model in random external field is calculated up to sixth order in ε by direct computation and also by the replica method, and the coincidence of the two results is demonstrated.

I. Introduction

The replica trick [1] is one of the few powerful tools to calculate the properties of disordered systems. Its use is nevertheless shadowed by the fact that the prescriptions for the computation are not well defined from a mathematical point of view. The trick exploits the identity

$$\overline{\ln Z} = \frac{d}{dx}\overline{Z^x}\Big|_{x=0} = \lim_{x\to 0}(\overline{Z^x}-1)/x \tag{1}$$

where Z is the partition function and the bar indicates averaging over the randomness. The average of the x th power of Z for general x is by no means easier to calculate than the average of the logarithm; for $x=m$ positive integer, however, $\overline{Z^m}$ is the partition function of an m-replica system, coupled by the averaging. This - at least in principle - can be calculated for general m, so that one obtains $(\overline{Z^m})^{1/N} = \lambda(m)$ and the trick consists of continuing $(\lambda(m)-1)/m$ to $m=0$ and identifying the result with $\overline{\ln Z}/N$. This is a problematic procedure because the function $\lambda(m)$ may not be unique. Indeed, there are many ways to calculate a partition function; and though the sequence of numbers $\{\lambda_m = (\overline{Z^m})^{1/N}\}_{m=0}^{\infty}$ is the same for all cases, the function $\lambda(m)$ may depend on the particular method used for the calculation. This possible ambiguity can be avoided by using nothing else than the sequence of values λ_m and trying to reproduce $\overline{Z^x}^{\,1/N}$ as a function of x out of these values. Formulating the question in this way we arrive at an interpolation problem the solution of which would consist of the following steps:

(i) Establish the analytical properties (A) of $\lim_{N\to\infty} \overline{Z^{x}}^{\,1/N}$ for complex values of x.

(ii) Show that there is a unique function $\lambda(x)$ with the properties (A) so that $\lambda(m) = \lambda_m$ for all non-negative integers m. Then

$$\lambda(x) = \lim_{N} \overline{Z^{x}}^{\,1/N} \tag{2}$$

(iii) Construct $\lambda(x)$ by using the sequence of numbers λ_m.

This program, what we call the normal replica method, is rather difficult to carry through. (It was partly done for the Sherrington--Kirkpatrick model [2], see van Hemmen and Palmer [3].) Rather than following the lines (i)-(iii) we provide a sufficient condition for the replica method to yield the correct result. This hinges about the existence of the limit in Eq. (2) and its analyticity as a function of ε, the strength of random fields or couplings. (The existence of $\lambda(x) \equiv \lambda(x,\varepsilon)$, as defined in (2), for real x and ε can be shown and will be presented elsewhere.) Our finding verifies the general expectation that the replica trick is correct at high temperatures where the analyticity in ε holds true. In case of lack of analyticity in ε, the replica method is still capable to yield the existing derivatives of the free energy with respect to ε; the divergence of these derivatives (the generalized response functions) then signals some sort of phase transitions.

In the second part of this talk the general result on the replica method is presented. In the third part we discuss the one-dimensional Ising model in a random magnetic field. We calculate the free energy up to sixth order in ε by direct computation and by applying the replica trick; then the agreement of the two results can be checked explicitely. A brief discussion of the singularity at $T=0$ finishes the lecture.

II. The analysis of the replica method by a series expansion of the free energy

We consider a classical spin system described by the Hamiltonian

$$-\beta H = K_0 + \varepsilon \cdot K_1 \; . \tag{3}$$

Here K_0 is a non-random Hamiltonian and

$$K_1 = \sum_b \xi_b \, \varphi_b(\underset{\sim}{s}) \tag{4}$$

where the summation runs over certain sets b of lattice sites, $\varphi_b(\underset{\sim}{s})$ is some function of the spin variables at the sites in b and ξ_b are independent random variables. The partition function of a system of N spins, described by the Hamiltonian (3), can be written as

$$Z = \sum_{\underset{\sim}{S}} e^{K_o} \cdot e^{\varepsilon K_1} = Z_o < e^{\varepsilon K_1} >_o \tag{5}$$

where Z_o is the partition function of the unperturbed ($\varepsilon=0$) system. Similarly,

$$\overline{(Z^x/Z_o^x)}^{\,1/N} = \overline{<exp(\varepsilon K_1)>_o^x}^{\,1/N} \tag{6}$$

On the right-hand side one has to take first the thermal average with the distribution $\sim exp\ K_o$ then the x th power then the average over the random variables ξ_b and finally the N th root. Without the thermal average, this expression would depend only on $\varepsilon \cdot x$; but even in the presence of thermal averaging the dependences on x and ε are intimately related. For N finite, the expression (6) is an analytic function of ε at $\varepsilon=0$. Our main observation is that by taking its expansion in powers of ε, the ℓ th order term in ε will be of the ℓ th order also in x . Indeed,

$$\overline{(Z^x/Z_o^x)}^{\,1/N} \equiv \Pi(x,\varepsilon)$$

$$= 1 + \sum_{\ell=1}^{\infty} \varepsilon^\ell \sum_{q=1} \binom{1/N}{q} \sum_{\substack{\ell_1 \geq 1 \\ \ell_1+\ldots+\ell_q=\ell}} \cdots \sum_{\ell_q \geq 1} \sum_{k_1=1}^{\ell_1} \cdots \sum_{k_q=1}^{\ell_q} \binom{x}{k_1} \cdots \binom{x}{k_q} B_{\ell_1\ldots\ell_q\,;\,k_1\ldots k_q}$$

$$= 1 + \sum_{\ell=1}^{\infty} a_\ell(x)\varepsilon^\ell \tag{7}$$

where a_ℓ is defined by the last equality and

$$B_{\ell_1\ldots\ell_q\,;\,k_1\ldots k_q} = \sum_{\substack{n_{11}\geq 1\cdots n_{1k_1}\geq 1 \\ n_{11}+\ldots+n_{1k_1}=\ell_1}} \cdots \sum_{\substack{n_{q1}\geq 1\cdots n_{qk_q}\geq 1 \\ n_{q1}+\ldots+n_{qk_q}=\ell_q}} \prod_{i=1}^{q} \overline{\prod_{j=1}^{k_i} \frac{< K_1^{n_{ij}} >_o}{n_{ij}\,!}} \tag{8}$$

As we see, $a_\ell(x)$ is an ℓ th order polynomial in x. Its relation to the quenched free energy can be seen by differentiating Eq. (7) with respect to x and setting $x=0$:

$$\left.\frac{\partial \pi}{\partial x}\right|_{x=0} = \frac{1}{N}\,\overline{\ell n\ Z} - \frac{1}{N}\,\ell n\ Z_o = \sum_{\ell=1}^{\infty} a_\ell'(o)\ \varepsilon^\ell\ . \tag{9}$$

Therefore, by knowing $a_\ell'(o)$ for all ℓ, the quenched free energy is also known (this latter being analytic at $\varepsilon=0$ for N finite). The calculation of a_ℓ from Eqs (7) and (8) would be rather difficult. Let us suppose, however, that one is able to calculate $\Pi(x,\varepsilon)$ for $x=m$ <u>general non-negative integer;</u> i.e. one obtains a function $\hat{\Pi}(m,\varepsilon)$ so that $\hat{\Pi}(m,\varepsilon)=\Pi(m,\varepsilon)$ identically in ε for any non-negative integer m. Though $\hat{\Pi}(x,\varepsilon)\neq\Pi(x,\varepsilon)$ may happen for non-integer x, the Taylor coefficients a_ℓ of Π can be computed from $\hat{\Pi}$. Indeed, let us observe that the $\ell+1$ numbers $a_\ell(o)=0$, $a_\ell(1),\ldots,a_\ell(\ell)$ completely determine the polynomial $a_\ell(x)$ by Lagrange interpolation. On the other hand, $\hat{\Pi}(m,\varepsilon)$ is analytic in ε at $\varepsilon=0$ and we have

$$\frac{1}{\ell!}\frac{\partial^\ell}{\partial\varepsilon^\ell}\hat{\Pi}(m,\varepsilon)\Big|_{\varepsilon=0} = \frac{1}{\ell!}\frac{\partial^\ell}{\partial\varepsilon^\ell}\Pi(m,\varepsilon) = a_\ell(m) \tag{10}$$

for all ℓ and integer m. We can therefore draw our first conclusion:

(C1) <u>For a finite system (N finite) the replica-partition functions Z^m always determine the quenched free energy, by determining its Taylor series about $\varepsilon=0$.</u>

This result does not necessarily imply that the quenched free energy can be obtained by the usual replica trick. The generally valid method is to determine $a_\ell(m)$ for $m=1,2,\ldots,\ell$ and for any ℓ through Eq. (10), then to build up the polynomial $a_\ell(x)$ by interpolation, to calculate $a'_\ell(o)$ and finally to construct the free energy from its Taylor expansion (9). The validity of the replica trick, as it is generally applied, can be checked in the following way:

(C2) <u>Consider the Taylor coefficients of $\hat{\Pi}$,</u>

$$\hat{a}_\ell(x) = \frac{1}{\ell!}\frac{\partial^\ell}{\partial\varepsilon^\ell}\hat{\Pi}(x,\varepsilon)\Big|_{\varepsilon=0} \tag{11}$$

<u>If $\hat{a}_\ell(x)$ is a polynomial for all ℓ, then the usual replica trick holds true for the whole free energy and separately for any order in ε, namely</u>

$$\frac{1}{N}\overline{\ln Z} - \frac{1}{N}\ln Z_o = \frac{\hat{\Pi}(m,\varepsilon)-1}{m}\Big|_{m=o} \tag{12}$$

and

$$a'_\ell(o) = \frac{\hat{a}_\ell(m)}{m}\Big|_{m=o} . \tag{13}$$

Indeed, if $\hat{a}_\ell(x)$ is a polynomial for all ℓ then due to (10) it coincides with the polynomial $a_\ell(x)$ for all ℓ and therefore $\hat{\Pi}(x,\varepsilon)$ coincides with $\Pi(x,\varepsilon)$. The result then follows from the first equality of Eq. (9) and from the continuity of $\hat{\Pi}$ at $x=0$.

So far, our considerations referred to finite systems: all the functions Π, a_ℓ, $\hat{\Pi}$, $\hat{a}_\ell$ depended on N, the number of spins. Our real interest is in the validity of the replica trick at $N\to\infty$ where the trick may break down. Below we formulate a sufficient condition that this does not happen. We notice that the existence of the thermodynamic limit of $\Pi(x,\varepsilon)$ can be shown for real x and ε under rather general conditions.

(C3) The conclusions (C1) and (C2) remain valid in the infinite system if the thermodynamic limit of $\Pi(x,\varepsilon)$ is an analytic function of ε at $\varepsilon=0$.

What we need is the term-by-term convergence of the Taylor series (7) as N goes to infinity. In this case $a_\ell(x) = \lim_N a_\ell^{(N)}(x)$ is also an ℓ th order polynomial of x and the subsequent considerations can be repeated without any change. Now (C3) implies that the breakdown of the replica trick indicates a singularity of $\Pi(x,\varepsilon)$ at $\varepsilon=0$. This may manifest itself, for example, in the divergence of some generalized response functions $a_\ell'(o)$ as the temperature decreases, and therefore it is related to freezing or ordering or some other kind of phase transitions. Finally we notice that in the absence of analyticity of $\Pi(x,\varepsilon)$ for the infinite system the replica method is still applicable for obtaining the existing ε-derivatives of the free energy.

III. The one-dimensional Ising model in random external field (1d-RFIM)

In this section we investigate the free energy of the 1d - RFIM, firstly by a direct method and secondly by applying the replica trick. The Hamiltonian is defined by

$$K_o = K \sum s_i\, s_{i+1}$$

$$\varepsilon K_1 = \sum H_i\, s_i = \varepsilon \sum h_i\, s_i \tag{14}$$

where h_i are independent random variables with the same distribution ρ so that $\rho(h)=\rho(-h)$ and $\int h^2 \rho\, dh=1$. It is expected that the free energy is analytic in ε at $\varepsilon=0$ for finite values of K, therefore we expect the coincidence of the two results.

In the direct calculation we start with the normal high-temperature expansion of the partition function taken with free boundary conditions. This yields

$$Z = 2^N C^{N-1} (\prod_i c_i)\, \{1+ \sum_{n=1}^{N/2} \sum_{1\le i_1<\ldots<i_n\le N} t_{i_1}\ldots t_{i_n}\, T^{i_2-i_1+i_4-i_3+\ldots+i_{2n}-i_{2n-1}}\} \tag{15}$$

where we use the notations

$$C = \cosh K,\quad S = \sinh K,\quad T = \tanh K,\quad c_i = \cosh H_i,\quad t_i = \tanh H_i\,. \tag{16}$$

In order to obtain the series of $\overline{\ln Z}$ in powers of ε, one has to take the logarithm of Eq. (15), then expand the logarithm of the bracket. This is followed by the averaging over h_i and then by a resummation of the powers of T. Using the fact that $\overline{t_i\, t_j} = \delta_{ij}\, \overline{t_i^2}$ and collecting carefully all the possible pairings of the t_i's we obtain, up to sixth order, that

$$f \equiv \lim_N \frac{1}{N}\, \overline{\ln Z} = \ln 2C + \overline{\ln c} - \frac{1}{2}\, S^2\, \overline{t^2}^{\,2} + 2S^4\, \overline{t^2}^{\,3}$$

$$= \ln 2C + \frac{1}{2}\, \overline{h^2}\, \varepsilon^2 - \frac{1}{2}\, [\frac{1}{6}\, \overline{h^4} + S^2\, \overline{h^2}^{\,2}]\varepsilon^4 + [\, \frac{1}{45}\, \overline{h^6}$$

$$+ \frac{2}{3}\, S^2\, \overline{h^2}\, \overline{h^4} + 2S^4\, \overline{h^2}^{\,3}]\varepsilon^6 + O(\varepsilon^8)\ . \tag{17}$$

Here we dropped the site index i. The free energy up to fourth order was independently obtained by Györgyi and Ruján [4].

Now we derive the same result by applying the replica method. The m th power of the partition function can be written as

$$Z^m = Tr \prod_i T_i^{\otimes m} \tag{18}$$

where

$$T_i = \begin{pmatrix} e^K & e^{-K} \\ e^{-K} & e^K \end{pmatrix} \begin{pmatrix} e^{H_i} & 0 \\ 0 & e^{-H_i} \end{pmatrix} \tag{19}$$

and $T_i^{\otimes m}$ is the m th tensor product of T_i with itself. Due to the independence of the random fields at different sites, averaging of (18) is equivalent to averaging the elements of $T_i^{\otimes m}$. This yields a $2^m \times 2^m$ matrix T which is already independent of the site index. Whence,

$$\overline{Z^m} = Tr\ T^N\ . \tag{20}$$

The elements of T are labelled by the configurations of m spins

$$T_{\underset{\sim}{s},\underset{\sim}{\sigma}} = e^{K\underset{\sim}{s}\cdot\underset{\sim}{\sigma}}\ \overline{e^{H \sum_1^m s^i}} \tag{21}$$

where $\underset{\sim}{s} = (s^1,\ldots,s^m)$. Developing the average of the exponential and keeping the terms up to sixth order we find

$$T_{\underset{\sim}{s},\underset{\sim}{\sigma}} = e^{K\underset{\sim}{s}\cdot\underset{\sim}{\sigma}}\,\{\alpha_m + \beta_m \sum_{i<j} s^i s^j + \gamma_m \sum_{i<j<k<\ell} s^i s^j s^k s^\ell$$

$$+ \delta_m \sum_{i<j<k<\ell<p<q} s^i s^j s^k s^\ell s^p s^q\,\} \tag{22a}$$

with

$$\alpha_m = 1 + \frac{m}{2}\overline{H^2} + \left(\frac{m^2}{8} - \frac{m}{12}\right)\overline{H^4} + \left(\frac{m^3}{48} - \frac{m^2}{24} + \frac{m}{45}\right)\overline{H^6}$$

$$\beta_m = \overline{H^2} + \left(\frac{m}{2} - \frac{2}{3}\right)\overline{H^4} + \left(\frac{7}{24}m^2 - \frac{5}{4}m + \frac{62}{45}\right)\overline{H^6}$$

$$\gamma_m = \overline{H^4} + \left(\frac{m}{2} - \frac{4}{3}\right)\overline{H^6}$$

$$\delta_m = \overline{H^6} \quad . \tag{22b}$$

Let us notice that in Eq. (22a) the sum subsequent to α_m is of the order of ε^2 . Therefore we can write

$$T = A + \varepsilon^2 B \tag{23}$$

and look for the largest eigenvalue, $\lambda(m)$, of T in the form

$$\lambda(m) = \lambda^{(1)} + \varepsilon^2 \lambda \tag{24}$$

where $\lambda^{(1)}$ is the largest eigenvalue of the matrix

$$A = [\alpha_m \cdot e^{K\, \mathfrak{s}\cdot\mathfrak{s}}] \tag{25}$$

It is assumed that ε is sufficiently small so that no level-crossing occurs and the deviation of $\lambda(m)$ from $\lambda^{(1)}$ is small. The detailed calculations show indeed that $\lambda \sim \varepsilon^2$. The approximate solution of the eigenvalue equation is easier to obtain if T is given in the basis formed by the eigenvectors of A. The necessary transformation can be performed by using the m th tensor product of

$$U = \frac{1}{\sqrt{2}}\begin{pmatrix} 1 & 1 \\ 1 & -1 \end{pmatrix}$$

with itself. Finally we obtain

$$\lambda(m)/(2C)^m = \hat{\Pi}(m,\varepsilon) = 1 + \frac{m}{2}\overline{h^2}\,\varepsilon^2 + \left[\left(\frac{m^2}{8} - \frac{m}{12}\right)\overline{h^4} + \frac{1}{2}m(m-1)\,S^2\,\overline{h^2}^{\,2}\right]\varepsilon^4$$

$$+ \left[\left(\frac{m^3}{48} - \frac{m^2}{24} + \frac{m}{45}\right)\overline{h^6} + m(m-1)\left(\frac{m}{2} - \frac{2}{3}\right)S^2\,\overline{h^2}\,\overline{h^4}\right.$$

$$\left. + m(m-1)(m-2)\,S^4\,\overline{h^2}^{\,3}\right]\varepsilon^6 + O(\varepsilon^8)$$

$$\equiv 1 + \hat{a}_2(m)\,\varepsilon^2 + \hat{a}_4(m)\,\varepsilon^4 + \hat{a}_6(m)\,\varepsilon^6 + O(\varepsilon^8) \tag{26}$$

where the notations $\hat{\Pi}$ and $\hat{a}_\ell$ correspond to those introduced in the previous part. Applying (C2), we see that $\hat{a}_2$, $\hat{a}_4$ and $\hat{a}_4$ are polynomials, therefore they must coincide respectively with a_2, a_4 and a_6 (c.f. Eq. (7)). Then by Eq. (13) we get

$$a_2'(o) = \frac{1}{2}\overline{h^2}, \quad a_4'(o) = -\frac{1}{12}\overline{h^4} - \frac{1}{2}S^2\overline{h^2}^2, \quad a_6'(o) = \frac{1}{45}\overline{h^6} + \frac{2}{3}S^2\overline{h^2}\,\overline{h^4} + 2S^4\overline{h^2}^3, \tag{27}$$

in a complete agreement with Eq. (17).

Let us notice, that the actual order of the polynomial $a_\ell(x)$ is only $\ell/2$. This can be understood, by inspecting Eqs (7), (8) and (14), as the consequence of the occurrence of a vanishing factor $< s_i >_o$ in all terms proportional to x^k for $k>\ell/2$.

IV. Discussion

In this lecture we have studied the possibility to reproduce the free energy of general random models from the free energies of the corresponding m-replicated systems, where m is positive integer. As a sufficient condition (which may prove to be also necessary) we found the analyticity of the free energy as a function of the strength (or variance) ε of the random variables at $\varepsilon=0$. In models with a phase transition the analyticity in ε is likely to break down below the critical temperature, therefore one may generally expect the necessity of some procedure *à la Parisi* [5] - or the choice of other methods - in this temperature region.

An interesting feature of the explicitely discussed example (1d-RFIM) is the divergence of the fourth and higher-order terms of the free energy as K goes to infinity (c.f. Eq. 17). This of course does not imply that f diverges exponentially with increasing K; it rather implies that the radius of convergence of the ε-expansion goes to zero exponentially fast as $K\to\infty$. At $K=\infty$ we expect that terms of non-even order in ε will appear, giving

$$[\, f - \ln 2C \,]_{K=+\infty} = \frac{1}{2}\overline{h^2}\,\varepsilon^2 + c_1|\varepsilon|^{2+\delta_1} - \frac{1}{12}\overline{h^4}\,\varepsilon^4 + c_2|\varepsilon|^{4+\delta_2} + \frac{1}{45}\overline{h^6}\,\varepsilon^6 + \ldots \qquad (o < \delta_i < 2) \quad . \tag{28}$$

This expectation is supported by similar findings in spin glasses at the freezing temperature where the non-linear susceptibilities diverge ([6,7]). The non-analyticity at $\varepsilon=0$ is somewhat analogous to the singularity at $p=o$ found in the model of Grinstein and Mukamel [8], though the expansion (17) is meaningless for their case.

References

1. S.F. Edwards and P.W. Anderson, J. Phys. F 5, 965 (1975)
2. D. Sherrington and S. Kirkpatrick, Phys. Rev. Lett. 35, 1792 (1975)
3. J.L. van Hemmen and R.G. Palmer, J. Phys. A 12, 563 (1979)
4. G. Györgyi and P. Ruján, J. Phys. C to appear

5. G. Parisi, Phys. Rev. Lett. 43, 1754 (1979).
J. Phys. A 13, L115; 1101; 1887 (1980)
6. G. Toulouse, in Lecture Notes in Physics, Vol. 149
(Springer 1981)
7. R. Omari, J.J. Préjean and J. Souletie, J. Physique 44, 25 (1983)
8. G. Grinstein and D. Mukamel, Phys. Rev. B 27, 4503 (1983)

HEISENBERG - ISING MAGNETS

Andrzej Pękalski
Institute of Theoretical Physics, University of Wrocław
Cybulskiego 36, 50-205 Wrocław, Poland

The best known theoretical models of a magnet are those bearing the names of Ising and Heisenberg. Both are applicable to insulators rather than metals, and have already an extensive literature. According to the universality hypothesis, they belong to two different universality classes, since the order parameter is one dimensional in the Ising model and three dimensional in the Heisenberg model. Hence, their critical behaviour should be different. This would manifest itself most clearly in the values of the critical exponents, and in the behaviour of the critical temperature, T_C, or thermodynamic functions. The most striking difference is in two dimensions (2D) where, as follows from the Mermin-Wagner theorem[1)], there is no long-range order for the Heisenberg model. It seemed therefore only natural to consider theoretically and experimentally systems where both kinds of interactions will be present.

The theories can be divided into two groups. One introduced by Dalton and Wood[2)] and called by them the anisotropic Heisenberg model, although some other authors call it also the Heisenberg-Ising model, contains a parameter $\eta \in [0,1]$ measuring degree of the anisotropy. For $\eta=1$ two neighbouring spins interact with completely isotropic forces, which corresponds to the Heisenberg model and for $\eta=0$, the interaction is completely anisotropic, which is the Ising model. The Hamiltonian may be written in the form

$$\mathcal{H} = -J \sum_{\langle ij \rangle} \left[\eta (S_i^x S_j^x + S_i^y S_j^y) + S_i^z S_j^z \right] - H \sum_i S_i^z \quad , \tag{1}$$

where J is the exchange integral and S_i^α (α=x,y,z) are components of the spin vector at the site i. H is the external magnetic field. Using the high temperature series expansion (HTSE) method Ishikawa and Oguchi[3)] determined for the square lattice the dependence of the critical temperature T_C as a function of the anisotropy parameter η, showing that it decreases continuously with increasing η. For the susceptibility critical exponent γ they found that it remains practically constant up to $\eta \sim 0.6$, and then increases rapidly to the pure Heisenberg value as $\eta \to 1$ (Fig.1).

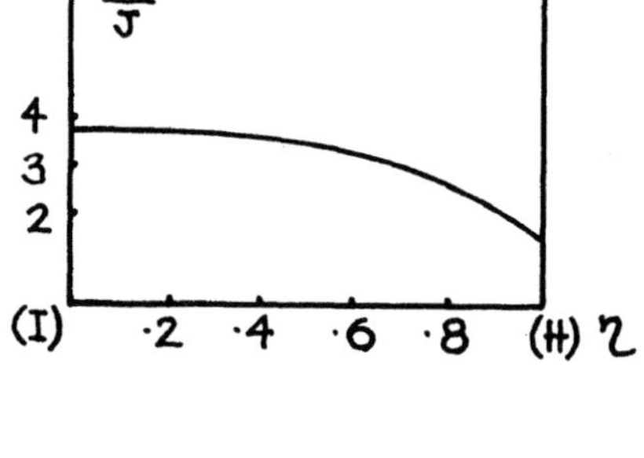

Figure 1

kT_C/J vs anisotropy parameter η for S=1 and D=2. After [3)].

The values found at the $\eta\to1$ limit are close to those predicted for a 2D Heisenberg model by Stanley and Kaplan[4], who found a singularity in the susceptibility at zero magnetization. The same continuous decrease of T_C with increasing η in 2D has been found by Stinchcombe[5] using the real space renormalization technique. Dalton and Wood[2] introduced their model to interpolate between the Heisenberg (quantum) and Ising (classical) models, and the authors believe that this model can provide a basis for comparison between the experimental results and existing theories.

In 1980 Katsura and Shimada[6] considered the same Hamiltonian (1), but now they admitted random distribution of the exchange energies J, with a distribution centered at J_o and width ΔJ. They used the effective Hamiltonian method introduced by Oguchi and Ono[7]. Firstly a single-spin Hamiltonian at a site i is constructed

$$H_i^1 = -\sum_{k=1}^{z} \lambda_{ik} S_i^z - HS_i^z \tag{2}$$

where λ_{ik} is an effective field acting on the site i, and coming from the bond between the sites i and k. Next, a pair Hamiltonian is written

$$H_{ij}^2 = -J_{ij}\left[\eta(S_i^x S_j^x + S_i^y S_j^y) + S_i^z S_j^z\right] - \sum_{k\neq i}\lambda_{ik}S_i^z - \sum_{k\neq j}\lambda_{jk}S_j^z - H\sum(S_i^z + S_j^z). \tag{3}$$

From the two Hamiltonians the reduced density matrices $\hat{\rho}_1$ and $\hat{\rho}_2$ are calculated and the self-consistency condition

$$\hat{\rho}_1 = \mathrm{Tr}_j\, \hat{\rho}_2 \tag{4}$$

is imposed. This gives a relation which, after averaging with a given distribution of the effective fields yields the equation to determine the critical temperature

$$(z-1)^{-1} = \overline{f(k)} \quad , \tag{5}$$

where bar means an average, $K=(kT)^{-1}J$, and

$$f(k) = \tanh(K - \tanh^{-1}\tanh^2\eta K). \tag{6}$$

Again the transition from anisotropic to isotropic system is continuous, and the presented dependence of $kT_C/\Delta J$ on $J_o/\Delta J$ shows that decreasing $J_o/\Delta J$ results in diminishing of $kT_C/\Delta J$. For $J_o \leq \Delta J$ the function $kT_C/\Delta J(J_o/\Delta J)$ is no longer single-valued, which is interpreted as a possibility for appearence of the spin-glass phase due to the presence of antiferromagnetic bonds coming from a long tail of the distribution.

In this model all bonds have the same character, with the same degree of anisotropy, measured by the parameter η. It is therefore not surprising that critical parameters like T_C change continuously, since varying η we have in turn: pure Heisenberg model ($\eta \in 1$), anisotropic Heisenberg ($1<\eta<0$), and pure Ising ($\eta=0$).

Another possibility of a system which is Heisenberg-like in one extreme and Ising -like in the other, can be described by a model in which part of the host bonds, say Heisenberg-like, is exchanged by Ising-like ones[8]. The system can be parametrized by p - concentration of the Heisenberg bonds. One of the plausible reasons of such a behaviour is the introduction of interstitial impurity atoms which change the local crystal field, and hence are able to change also the character of the interaction - from an isotropic (direct) to anisotropic (indirect) via the impurity atom. We have therefore a random system, which apart from p=1 and p=0 corresponds neither to the Heisenberg nor to the Ising model. We may write the Hamiltonian for our model in the form

$$H = - \sum_{<ij>} (J^H_{ij}\vec{S}_i\vec{S}_j + J^I_{ij}S^z_iS^z_j) - H\sum_{i=1}^{N} S^z_i \quad , \tag{7}$$

where

$$J^H_{ij} = \begin{cases} J^H/J \\ 0 \end{cases} \quad , \qquad J^I_{ij} = \begin{cases} 0 \\ J^I/J \end{cases} \quad , \tag{8}$$

depending whether the bond between nearest neighbouring pair of sites (i,j) is Heisenberg or Ising-like. J is a reference coupling constant. It would be very interesting to calculate the change of the critical exponents when varying p. Unfortunately, the model described by the Hamiltonian (7) cannot be solved exactly, and the effective field theories all give the classical values for the exponents. The HTSE technique permits to estimate both T_C and critical exponent γ with good accuracy, provided enough terms in the series are known.Let us then start with the HTSE. At first we have to calculate the free energy per particle, f , in a given configuration, ξ , of the bonds and then average over their distribution. This corresponds to adopting the quenched approximation, i.e. the impurities are frozen in and do not take part in creating thermodynamic equilibrium. This approach is justified since the bonds are assumed to be distributed randomly. Expanding the logarithm and taking double derivative with respect to the external field yields the power series for the susceptibility per particle[9]

$$kT\chi = \frac{X}{3} \sum_{n=0}^{\infty} a_k K^n \quad , \tag{9}$$

where

$$X = S(S+1) \tag{10}$$

and

$$a_n = \frac{3}{X}\frac{2^n}{n!} \text{ (coefficient of N in } \langle Tr(P^nQ^2)\rangle_c \text{)}. \tag{11}$$

For a given configuration ξ

$$P(\xi) = \sum_{ij} (J^H_{ij} \vec{S}_i \vec{S}_j + J^I_{ij} S^z_i S^z_j), \qquad Q = \sum_i S^z_i \ . \tag{12}$$

To calculate the coefficient a_n one has to list all expressions containing $P^n Q^2$ and calculate the traces. The procedure very quickly becomes quite complicated, and only a few (5) terms are known[8]. a_n's depend on p, the coupling constants J^I and J^H and the spin value S. Once the coefficients are known we may use any standard technique, as e.g. ratio method, to calculate the convergence radius of the series, i.e. T_C.

Just as in the anisotropic Heisenberg model investigated in 2D by Ishikawa and Oguchi, also here the Stanley-Kaplan singularity is recovered. The change of T_C with p is a continuous one, as shown in Fig.2. It is also possible to estimate γ as a function of p. It shows a minimum, the position of which depends on the values of J^I and J^H, and seems to suggest that at that particular value of p the system changes its overall character - from anisotropic (Ising-like) to isotropic (Heisenberg-like).

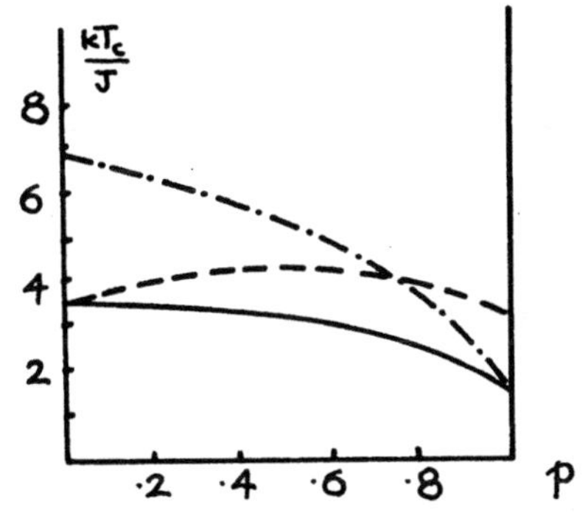

Figure 2

T_C vs concentration p of Heisenberg-like bonds for S=1 and

——— $J^I/J^H=1$

—·—·— $J^I/J^H=2$

- - - - - $J^I/J^H=0.5$

(after[7])

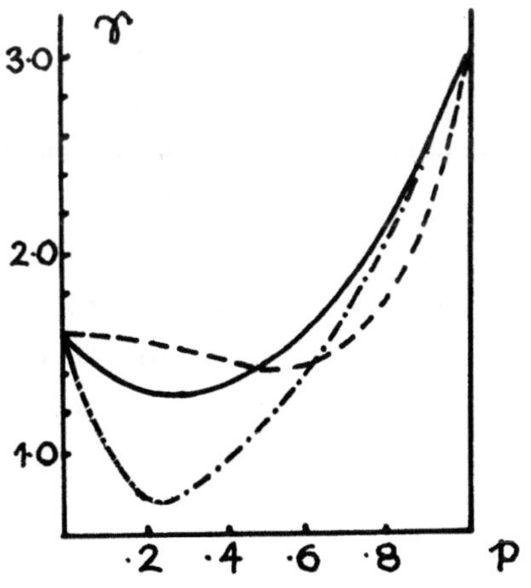

Figure 3

Critical exponent γ vs concentration p of Heisenberg-like bonds for S=1 and

——— $J^I/J^H=1$

—·—·· $J^I/J^H=2$

- - - - - $J^I/J^H=0.5$

(after[7])

One should however remember that γ measured experimentally or calculated by HTSE differs from that obtained by the renormalization group technique. In 3D we may construct similar curves with, of course, the true T_C for pure Heisenberg system. For certain values of J^I and J^H there appears a minimum on the $\gamma=\gamma(p)$ curve, as in 2D, but the effect is smaller. Here again the HTSE results indicate a continuous pass between the two pure systems.

Although the effective field theories cannot provide accurate answers concerning the critical exponents, we may use them to see whether they confirm the hypothesis of a smooth pass between the Heisenberg and Ising models. Of the not too many approximate theories suitable for dealing with the Heisenberg model, we have chosen the effective Hamiltonian method due to Oguchi and Ono[7]. Just as for the anisotropic Heisenberg model, one and two-spin Hamiltonians are constructed in the form[10]

$$H_R^1 = - z[\lambda_H p + (1-p)\lambda_I]S_R^z \tag{13}$$

$$H_{R,\delta}^2 = - 2J_H p\vec{S}_R\cdot\vec{S}_{R+\delta} - 2J_I(1-p)S_R^z S_{R+\delta}^z - (z-1)[p\lambda_H + (1-p)\lambda_I](S_R^z + S_{R+\delta}^z) \tag{14}$$

where J_H denotes Heisenberg-like and J_I Ising-like interaction. Using the self-consistency condition (4) we get an equation to determine $\Lambda_{eff}=p\lambda_H + (1-p)\lambda_I$:

$$\frac{\sinh(z-1)\beta\Lambda_{eff}}{\cosh(z-1)\beta\Lambda_{eff} - e^{-\beta J_{eff}}\cosh\beta J_H p} = \tanh(\tfrac{1}{2}z\beta\Lambda_{eff}) \tag{15}$$

where $J_{eff}=pJ_H+(1-p)J_I$.
Eq.(15) can be solved either numerically, or in the vicinity of T_C by expanding the hyperbolic functions in powers of Λ_{eff} and keeping linear terms only, since in that region the effective fields are small, and (15) reduces to

$$\frac{z-2}{z} = e^{\beta_c J_{eff}} \cosh(\beta_c J_H p). \tag{16}$$

By accounting exactly for a pair interaction between nearest neighbours, the method differentiates between the Ising and Heisenberg models. Magnetization and susceptibility have similar behaviour for both pure models, and for the intermediary model. Variation of T_C corroborates the conjecture that the system passes smoothly from isotropic to anisotropic, even when the character of individual couplings changes randomly. The above mentioned conjecture got an experimental support in a paper by Ajiro, Adachi and Maketa[11]. They considered a random mixture of two systems – $CsMnCl_3\cdot 2H_2O$, which is a quasi 1D Heisenberg antiferromanget, and $CsCoCl_3\cdot 2H_2O$ a quasi 1D Ising antiferromagnet. They obtained a lower value for the magnetization exponent β for an intermediary concentration, than for any of the pure systems. According to the authors, this is consistent with the appearance of a minimum for γ predicted by HTSE (Fig.3).Also the dependence of the critical temperature on the concentration of Co^{2+} ions is smooth. One should however bear in mind that the experiments have been carried out on quasi 1D antiferromagnets, whereas theory describes 2D ferromagnets. The resemblances are therefore encouraging but not conclusive.

If we allow now the two exchange integrals to be randomly distributed, around J_o^H and J_o^I and with the widths a and d, for the Heisenberg and Ising interactions, resp. it is possible to show[12] that the equation (15) determining T_C becomes

$$\frac{z-2}{z} = \exp\left\{\alpha\left(\frac{\alpha}{4a^2} - 1\right) + \delta\left(\frac{\delta}{2d^2} - 1\right)\right\} \cosh\left[\delta\left(\frac{\delta}{2d^2} - 1\right)\right] \tag{17}$$

where $\alpha=\beta_c J_o^I(1-p)$, $\delta=\beta_c J_o^H p$.

T_C seems to be quite sensitive to the width of, specially Ising, distributions. For two infinitely sharp distributions the system had, in 2D, $T_C=0$ only for p=1. Now $T_C=0$ also for p<1. It is possible to determine the width of both distributions for which the mixed system cannot remain magnetic, i.e. has $T_C=0$, for all values of p. This is due to introduction (through long tails of the distributions) of antiferromagnetic bonds.

The models we have so far dealt with were characterized by bond disorder. It is however possible to consider also a randomly disordered Heisenberg-Ising system, where magnetic atoms, and not bonds, are exchanged. We have therefore site disorder. The best known example is a random mixture of two 2D antiferromagnets[13] Rb_2NiF_4 (Heisenberg-like) and Rb_2CoF_4 (Ising-like). Instead of previous two there are now three kinds of interactions: Ni-Ni (Heisenberg), Co-Co (Ising) and Co-Ni (anisotropic Heisenberg or Ising). Ikeda e.a.[12] assumed $J^{zz}_{Co-Ni} \sim \sqrt{J^{zz}_{Co-Co} J^{zz}_{Ni-Ni}}$ and were able to determine the dependence of the critical temperature on the concentration, c, of Heisenberg atoms, which for $J^{zz}_{Co-Co} \sim 2J^{zz}_{Ni-Ni}$ has the form shown in Fig.4. Similar diagram has been

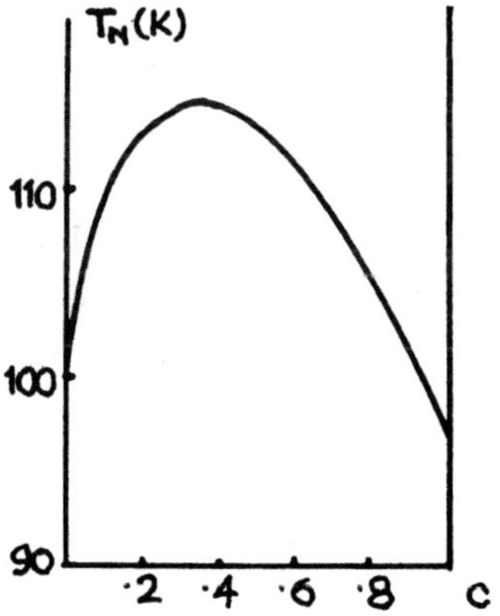

Figure 4

Critical temperature vs concentration c of Ni ions (after[12])

obtained using the effective Hamiltonian method[14]. Since the assumption that J^{zz}_{Co-Ni}

is a geometrical mean of J^{zz}_{Co-Co} and J^{zz}_{Ni-Ni}, was clearly a trial one, we have considered a more general Hamiltonian, permitting to investigate the influence of the values of each exchange integral and the anisotropy parameter :

$$H = - J_{II} \sum_{<ij>} p_i p_j S^z_i S^z_j - J_{HH} \sum_{<ij>} (1-p_i)(1-p_j)\vec{S}_i \vec{S}_j$$
$$- \tfrac{1}{2} J_{IH} \sum_{<ij>} [p_i(1-p_j) + p_j(1-p_i)][S^x_i S^x_j + S^y_i S^y_j)\eta + S^z_i S^z_j] , \qquad (18)$$

where p_i=1 if the site i is occupied by an Ising atom and is zero otherwise. η measures the anisotropy of the mixed interaction, which now, in contrast to the anisotropic Heisenberg model, is random. Its influence on T_C, for a 2D system, is illustrated in Fig.5, and as seen for η=1, i.e. fully isotropic J_{IH} interactions, the system behaves, down

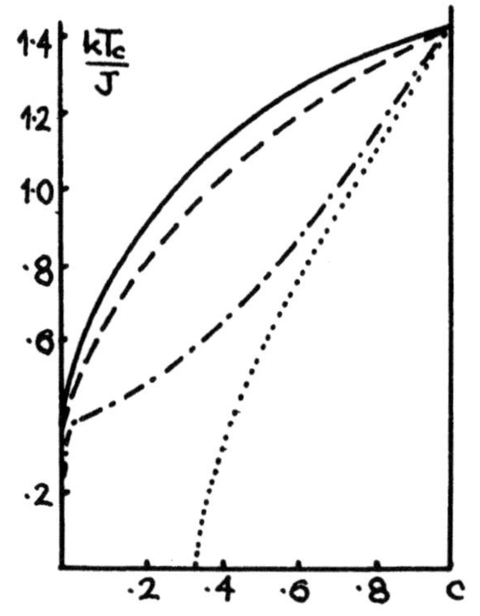

Fig. 5

Influence of the anisotropy parameter η on T_C. $J^I/J = 1$, $J^H/J = 0.5$, $J^{IH}/J = 1$.

—— $\eta = 0.1$

– – – $\eta = 0.5$

—·—· $\eta = 1.0$

······ diluted system ($J^H=J^{IH}=0$, $J^I=1$) (after[13])

to small values of c, like a diluted one. Like in the bond problem, T_C=0 only for the pure Heisenberg model. The apperance of a maximum in T_C for intermediary values of c, as in Fig.4 is associated with the value of J_{IH} - it should be larger than both J_I and J_H. Changing η shifts the location of this maximum. Although the $T_C(c)$ curves obtained experimentally[13] and theoretically[14] are quite similar, there is a difference in the systems. Rb_2NiF_4 has a small uniaxial anisotropy, hence its $T_C>0$, and the mixed system $Rb_2CoNi_{1-c}F_4$ belongs to the Ising universality class. Theoretical investigations have been carried out for a mixture of "true" i.e. isotropic Heisenberg and Ising models. As before for the bond model, the agreement is encouraging, but theoretical model differs from experimental reality.

Clearly the most interesting is the 2D case where, because of lack of long-range order in the Heisenberg model, introduction of this type of bonds resemblance dilution.

On the basis of the presented experimental and theoretical arguments, for site as well as for bond problems, it seems that the transition between the Heisenberg and Ising systems is a smooth one. In particular the critical exponents are changing continuously with concentration of exchanged bonds or atoms. Qualitative patterns following from the effective field approaches agree with those obtained experimentally and HTSE predictions concerning the $T_C(p)$ behaviour. The smooth transition with varying critical exponents is however still a conjecture, as the effective field theories cannot say anything definite about the exponents, HTSE results are based on short series and the experiments estimating the exponents are few. It seems therefore that although the above results do not invalidate the universality hypothesis, they show that theoretical and experimental work on such systems may give interesting results.

I am grateful to R.B.Stinchcombe and J.Sznajd for helpful comments and remarks.

Rererences

[1] N.D.Mermin and H.Wagner, Phys.Rev.Lett.17, 1133(1966)
[2] N.W.Dalton and D.W.Wood, Proc.Phys.Soc.90, 459(1967)
D.W.Wood and N.W.Dalton, J.Phys.C5, 1675(1972)
[3] T.Obokata, I.Ono, T.Oguchi, J.Phys.Soc.Jpn.,23, 516(1967)
T.Ishikawa and T.Oguchi, J.Phys.Soc.Jpn.,31, 1021(1971)
[4] H.E.Stanley and T.A.Kaplan, Phys.Rev.Lett.17, 913(1966)
[5] R.B.Stinchcombe, J.Phys.C14, 397(1981)
[6] S.Katsura and K.Shimada, Phys.Stat.Sol.b54, 663(1980)
[7] T.Oguchi and I.Ono, J.Phys.Soc.Jpn., 11, 2178(1966)
[8] A.Pękalski, J.Phys.C10, 4785(1977), Acta Phys.Pol.A54, 23(1978)
[9] G.S.Rushbrooke, G.A.Baker jr., P.W.Wood in "Phase Transition and Critical Phenomena" vol.III, ed.by C.Domb and M.S.Green Academic Press, London, 1974
C.Domb ibid.
[10] M.Ausloos and A.Pękalski, Solid St.Commun.26, 977(1978)
[11] Y.Ajiro, K.Adachi, M.Maketa, Solid St.Commun.37, 449(1981)
[12] A.Pękalski and M.Ausloos, Physica 101A, 507(1980)
[13] H.Ikeda, T.Abe and I.Hatta, J.Phys.Soc.Jpn., 50, 1488(1981)
[14] A.Komoda and A.Pekalski, J.Phys.C14, L1067(1981)

IV. SURFACE EFFECTS

SURFACE CRITICAL PHENOMENA IN SYSTEMS WITH A BULK PHASE TRANSITION

Theodore W. Burkhardt
Department of Physics, Temple University
Philadelphia, PA 19122/USA

Abstract: Following a discussion of the local critical behavior near isolated defects, these lectures review the critical properties of magnetic systems at free surfaces near the bulk critical temperature. Both continuous and discontinuous bulk transitions are considered. Some recent work on semi-infinite magnetic systems with smoothly inhomogeneous short-range interactions is described. A brief introduction to the wetting or interface-pinning transition is given.

1. Introduction

These lectures give a brief introduction to several aspects of surface critical behavior in magnetic systems at or near the bulk critical temperature[1-5]. A free surface may be regarded as a defect in a homogeneous bulk system. In Section 2 the local critical behavior near various types of defects in an otherwise homogeneous bulk is discussed. Section 3 reviews critical phenomena at the free surfaces of homogeneous magnetic systems. Both continuous and discontinuous (i.e., second and first-order) bulk transitions are considered. In experimental magnetic systems one would expect the spin-spin coupling strength to depend on the distance from the surface. Some recent results concerning the surface critical behavior of semi-infinite magnetic systems with inhomogeneous short-range interactions that vary smoothly with the distance from the surface are described in Section 4. Finally, Section 5 gives a brief discussion of the wetting or interface-pinning transition.

For additional information on these and other aspects of critical behavior at surfaces, the comprehensive review article[1], with unusually complete referencing, by K. Binder is recommended. For older, shorter reviews see Refs. 2-5.

2. Local Critical Behavior near Defects

In this section we consider several types of defects in an otherwise homogeneous bulk magnetic system and ask whether the critical behavior in the vicinity of the defect differs from the bulk behavior. The local magnetization will vanish as $(T_c-T)^{\beta_1}$, in general, as T approaches the bulk critical temperature T_c from below. We will be particularly interested in learning when to expect a β_1 different from the bulk exponent β.

A. Point Defect

We begin by considering a single augmented bond $K_0+\Delta K$ coupling spins 1 and 2 in an infinite Ising model in which all other couplings have the value K_0. Writing the energy of the system as $H = H_0 - k_B T\, \Delta K\, \sigma_1\sigma_2$, one sees that the thermal average of a third spin variable σ_3 is given by

$$\langle\sigma_3\rangle = \frac{\langle\sigma_3 \exp(\Delta K\, \sigma_1\sigma_2)\rangle_0}{\langle\exp(\Delta K\, \sigma_1\sigma_2)\rangle_0} = \frac{\langle\sigma_3 + (\tanh \Delta K)\sigma_1\sigma_2\sigma_3\rangle_0}{\langle 1 + (\tanh \Delta K)\sigma_1\sigma_2\rangle_0} \tag{2.1}$$

where $\langle\ldots\rangle_0$ denotes a thermal average with H_0 instead of H. Since $\langle\sigma_3\rangle_0$ and $\langle\sigma_1\sigma_2\sigma_3\rangle_0$ both tend to zero as $|T_c-T|^\beta$, whereas $\langle\sigma_1\sigma_2\rangle_0$ approaches a finite nonzero value as $T \to T_c$, we find that the local exponent β_1 is the same as β.

This result can readily be extended to a large class of spin variables and defects. As long as the number of altered bonds is finite, β_1 and β are equal. Of course, if the defect breaks the up-down spin symmetry, as in the case, for example, of a local field, the local magnetization no longer vanishes above the bulk critical temperature.

B. Defect with Dimension d-1

We now consider a d-1 dimensional plane of altered bonds with strength $K_0+\Delta K$ in a d-dimensional bulk with couplings K_0. A one-dimensional defect line in a two-dimensional bulk is represented in Fig. 1.

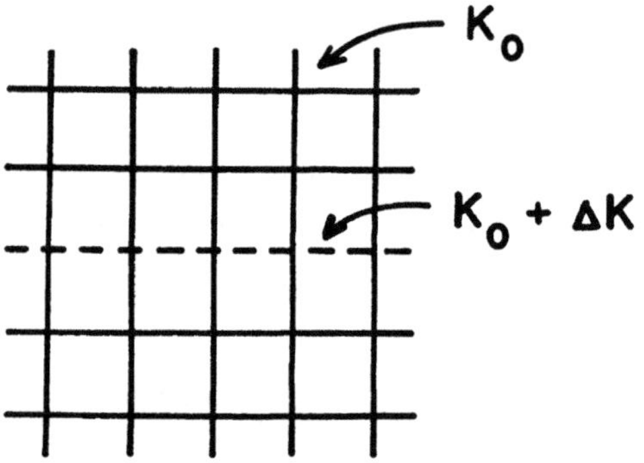

Fig. 1. Line defect in a two-dimensional system.

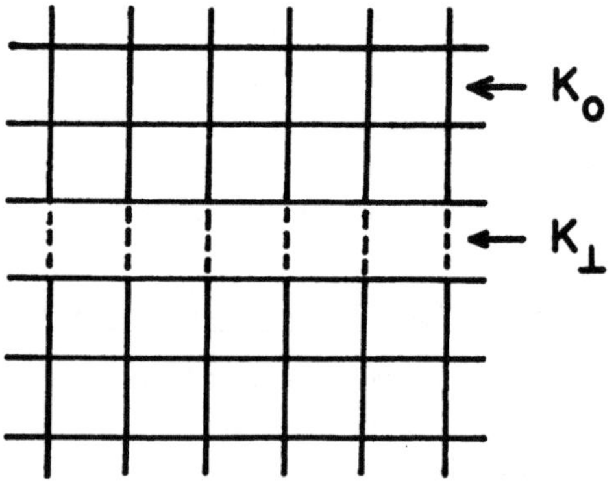

Fig. 2. Two semi-infinite systems connected by a transverse coupling $K_\perp$

To proceed we relate the crossover[6] exponent ϕ associated with the defect to the bulk critical exponent ν using scaling arguments. The sign of ϕ indicates whether or not the defect modifies the local exponent of the magnetization.

The free energy F_N of the system of N spins may be written as

$$F_N = N f_B + N^{(d-1)/d} f_D \tag{2.2}$$

where f_B and f_D represent the bulk and defect free energies. Under a scaling or renormalization-group transformation in which the lattice constant is expanded by a factor b, the number of spins N, the temperature variable $t = T-T_c$, and the quantity ΔK (assumed small) transform according to

$$N' = b^{-d} N, \qquad t' = b^{y_t} t, \qquad \Delta K' = b^{\phi y_t} \Delta K \tag{2.3}$$

The thermal index y_t and the bulk exponent ν, defined by $\xi \sim t^{-\nu}$, where ξ is the bulk correlation length, are related[6] by $y_t = \nu^{-1}$. The quantity ϕ is the crossover exponent. If ϕ is positive, ΔK grows under the rescaling. The defect is a "relevant" perturbation[6], and $\beta_1 \neq \beta$ is expected. A negative ϕ corresponds to an "irrelevant" perturbation and $\beta_1 = \beta$.

The invariance of the free energy (2.2) under the scaling transformation implies that the singular parts of f_B and f_D satisfy

$$N f_B(t) = N' f_B(t'), \quad N^{(d-1)/d} f_D(t, \Delta K) = N'^{(d-1)/d} f_D(t', \Delta K') \tag{2.4}$$

From Eqs. (2.3),(2.4) and the arbitrariness of the scale factor b, it follows that

$$f_B \sim t^{d/y_t}, \qquad f_D(t, \Delta K) \sim t^{(d-1)/y_t} \psi(\Delta K t^{-\phi}) \tag{2.5}$$

where ψ is a scaling function. From Eq. (2.5) one sees that the critical behavior of f_D as $t \to 0$ is the same for $\Delta K = 0$ and $\Delta K \neq 0$ if ϕ is negative, but not if ϕ is positive.

Following Ref. 7, we relate ϕ to the bulk exponent ν by expanding f_D in Eq. (2.5) to first order in ΔK and comparing with first order perturbation theory for the free energy, which yields $\Delta F_N = \langle H - H_0 \rangle_0 \sim N^{(d-1)/d} \Delta K \langle \sigma_1 \sigma_2 \rangle_0$. Since the singular part of the bulk energy density $\varepsilon \sim \langle \sigma_1 \sigma_2 \rangle_0$ has the leading singularity $t^{d/y_t - 1}$, one finds

$$\phi = 1 - y_t^{-1} = 1 - \nu \tag{2.6}$$

This result was first derived by Bray and Moore[8].

For typical three-dimensional magnetic systems such as the Ising and Heisenberg models[9], $\nu < 1$, so that $\phi > 0$. Thus the defect modifies the local critical behavior. For weakened couplings ($\Delta K < 0$) β_1 is presumably the same as for a free surface. More evidence for this will be given below. For enhanced couplings ($\Delta K > 0$) a nonvanishing local magnetization at the bulk critical temperature is expected.

For the percolation problem in two dimensions[10], $\nu = 4/3$, so that $\phi = -1/3$. Thus $\beta_1 = \beta$ near a defect line. For the d=2 Ising model, $\nu=1$, and $\phi=0$. The marginality is consistent with the nonuniversal exponent

$$\beta_1 = \frac{1}{2\pi^2}\left[\cos^{-1}(\tanh 2\Delta K)\right]^2 \tag{2.7}$$

derived[11,12] by Bariev for the defect line in Fig. 1.

To study whether β_1 is the same for free surfaces and internal d-1 dimensional defect planes in systems with $\nu < 1$, we couple[7] two semi-infinite systems with a weak transverse coupling $K_\perp$, as in Fig. 2, to obtain a single system with an internal defect plane. Writing $K_\perp' = b^{\phi_\perp y_t} K_\perp$, where $\phi_\perp$ is another crossover exponent, and comparing the scaling form $f_D = t^{(d-1)/y_t}\,\tilde{\psi}(K_\perp t^{-\phi_\perp})$ with perturbation theory for F_N in the small parameter $K_\perp$, one finds[7]

$$\phi_\perp = \gamma_{1,1}^{o} \tag{2.8}$$

Here $\gamma_{1,1}^{o}$ is the surface susceptibility exponent for the unperturbed system with free surfaces. This quantity is defined in Section 3, Eq. (3.8).

When $\phi_\perp$ is negative, the transverse coupling is irrelevant. For typical three-dimensional magnetic systems[9], $\nu < 1$ and $\gamma_{1,1}^{o}$ is negative (see Section 3). Since $\phi = 1-\nu$ is positive and $\phi_\perp = \gamma_{1,1}^{o}$ is negative, one expects $\beta_1 \neq \beta$ for a defect plane of weakened bonds, with β_1 the same as for a free surface. For percolation in two dimensions[10], $\nu = 4/3$; thus one anticipates $\beta_1 = \beta$ for an internal defect line and $\beta_1 \neq \beta$ for a free surface. For both the d=2 Ising model and the d=3 $n \to \infty$ vector model, $\nu=1$ and $\gamma_{1,1}^{o}=0$, as is consistent with detailed calculations of the nonuniversal behavior[11-14].

In this discussion we have only considered the critical behavior of the local magnetization. A unified renormalization-group treatment of free surfaces and defect planes that makes predictions for the critical behavior of other quantities as well is given in Ref. 7.

C. Defect with Dimension d^*

Using the methods of the previous subsection, one can readily show that the crossover exponent corresponding to a d^* dimensional hyperplane of modified bonds is given by[7,11]

$$\phi = 1 - (d-d^*)\nu \tag{2.9}$$

From the values of ν for three-dimensional magnetic systems[9] such as the Ising and Heisenberg models, one sees that ϕ is negative in the case of a one-dimensional defect line, so that $\beta_1=\beta$. Marginal behavior is expected in the Gaussian model with $d-d^*= 2$, since $\nu = 1/2$.

3. Critical Phenomena at Free Surfaces

A. Continuous Bulk Transition

Let us now consider the free-surface critical behavior[1-5] of systems with homogeneous short-range couplings and a continuous (second-order) bulk transition. In such systems the surface spins order at the bulk critical temperature in a transition known as the "ordinary" transition[1-5].

Exact results for the ordinary transition of the d=2 Ising model with free surfaces are discussed in the book of McCoy and Wu[15]. The local magnetization m_1, the correlation function $g_{\parallel}(r)$ of surface spins separated by r (at $T=T_c$), and the surface energy density ε_1 have the critical behavior shown in Eqs. (3.1). The corresponding behavior of the bulk quantities m, g, and ε is shown for comparison. Due to the missing neighbors, the surface quantities are less singular.

$$m_1 \sim t^{1/2} \qquad m \sim t^{1/8} \tag{3.1a}$$

$$g_{\parallel}(r) \sim r^{-1} \qquad g(r) \sim r^{-1/4} \tag{3.1b}$$

$$\epsilon_1 \sim t^2 \ln t \qquad \epsilon \sim t \ln t \tag{3.1c}$$

Exact results have also been obtained for the ordinary transition of the spherical[16] and $n \to \infty$ vector[8] models. (Although the bulk critical behavior of the two models is the same, the surface critical behavior is not.)

For $d > 4$ mean-field theory[1,3,5,8] gives the critical exponents of the n-vector model correctly. In mean-field theory (see Sections 3B and 5)

$$m_1 \sim t \qquad\qquad m \sim t^{1/2} \tag{3.2a}$$

$$g_{\parallel}(r) \sim r^{-d} \qquad\qquad g(r) \sim r^{-(d-2)} \tag{3.2b}$$

Scaling theories[1-6] relate a host of surface critical exponents to a small number of scaling indices. Under a scaling or renormalization-group transformation, the bulk and surface temperature and magnetic field variables t, h and t_1, h_1, respectively, transform as

$$t' = b^{y_t}\, t \qquad\qquad h' = b^{y_h}\, h \tag{3.3a}$$

$$t_1' = b^{y_{t_1}}\, t_1 \qquad\qquad h_1' = b^{y_{h_1}}\, h_1 \tag{3.3b}$$

In analogy with the discussion in Section 2B, Eqs. (3.3) and the transformation $f_B(t,h) = b^{-d} f_B(t',h')$ for the singular part of the bulk free energy per spin imply the scaling form[6]

$$f_B(t,h) = t^{d/y_t}\, \psi_B(h t^{-y_h/y_t}) \tag{3.4}$$

In the same manner one obtains

$$f_s(t,h,t_1,h_1) = t^{(d-1)/y_t}\, \psi_s(h t^{-y_h/y_t},\, t_1 t^{-y_{t_1}/y_t},\, h_1 t^{-y_{h_1}/y_t}) \tag{3.5}$$

for the singular part of the surface free energy.

The surface free energy involves two scaling indices y_{t_1} and y_{h_1} that do not appear in the bulk free energy. Dietrich and Diehl[5,17] have proved that $y_{t_1} = -1$ for the n-vector model in d=4-ε dimensions to all orders in ε. Bray and Moore[8] conjectured that y_{h_1} is determined by bulk scaling indices according to the formula $y_{h_1} = (d-y_t)/2$. Although this formula agrees with exact results for the d=2 Ising and d-dimensional $n \to \infty$ vector models and the ε expansion for the n-vector model to first order[5,8,18] in ε, it fails in second order[19,20] and in the 1/n expansion[21] and is thus not generally valid.

From Eq. (3.5) one sees that the critical exponents characterizing the surface magnetization, surface susceptibility, and surface energy density are related to the scaling indices as follows:

$$m_1 \sim \frac{\partial f_s}{\partial h_1} \sim t^{\beta_1^0}, \qquad \beta_1^0 = (d-1-y_{h_1})/y_t \tag{3.6a}$$

$$\chi_{1,1} \sim \frac{\partial^2 f_s}{\partial h_1^2} \sim t^{-\gamma_{1,1}^0}, \qquad \gamma_{1,1}^0 = (2y_{h_1} - d + 1)/y_t \tag{3.6b}$$

$$\epsilon_1 \sim \frac{\partial f_s}{\partial t_1} \sim t^{(d-1-y_{t_1})/y_t} = t^{2-\alpha} \tag{3.6c}$$

In Eq. (3.6c) we have used the result $y_{t_1} = -1$ of Dietrich and Diehl[17] and the conventional definition[6] of the exponent α, according to which the bulk energy density has singularity $t^{1-\alpha}$. Note that an irrelevant[6] scaling index y_{t_1} contributes to the leading singularity in ϵ_1.

The critical exponent $\eta_\parallel^0$ of the pair correlation function of surface spins, defined by

$$g_\parallel(r) = \frac{f(r/\xi)}{r^{d-2+\eta_\parallel^0}} \tag{3.7}$$

where ξ is the bulk correlation length, may be related to the scaling indices in Eqs. (3.5) and (3.6) by using the relation

$$\chi_{1,1} \sim \int d^{d-1}r \; g_\parallel(r) \sim t^{-\gamma_{1,1}^0} \tag{3.8}$$

A standard calculation[1-5] gives

$$\gamma_{1,1}^{o} = (1-\eta_{\parallel}^{o})\nu, \qquad \eta_{\parallel}^{o} = d - 2y_{h_1} \tag{3.9}$$

The scaling index y_{h_1} for the ordinary transition of the n-vector model was calculated to first order in $\varepsilon=4-d$ by Lubensky and Rubin[18] and to second order in ε by Diehl and Dietrich[19] and Reeve and Guttmann[20]. The second-order prediction for β_1^o is

$$\beta_1^{o} = 1 - \frac{3}{2(n+8)}\epsilon - \frac{3(n+2)(12-n)}{8(n+8)^3}\epsilon^2 \tag{3.10}$$

With n=1 and $\varepsilon=1$, one obtains a result for the d=3 Ising model that agrees nicely with predictions of other methods.

$\beta_1^o = 0.80 \pm 0.02$	ε-expansion[19,20]	
0.78 ± 0.02	series[22]	
0.78 ± 0.02	Monte Carlo[23]	(3.11)
0.795	real-space renormalization[24]	

[Although the scaling laws such as Eq. (3.9a) are exactly satisfied to second order in ε, they are, in general, only approximately satisfied by the numerical values of the critical exponents obtained by setting $\varepsilon=1$ in the second-order expressions. The uncertainty in the ε-expansion result in Eq. (3.11) was estimated[1] by calculating β_1^o in various ways using the scaling laws.]

Experimental work on surface critical exponents is rather scarce. The most definitive work is probably the recent experiment on Ni by Alvarado et al[25-26]. They report $\beta_1 = 0.825^{+0.025}_{-0.040}$. For n=3 (Heisenberg model) and $\varepsilon=1$, the ε expansion gives β_1^o = 0.85 ± 0.03. Since corrections to the leading singularity are presumably nonnegligible in the temperature range $|T-T_c|/T_c \gtrsim 10^{-2}$ considered and surface anisotropy should also affect the observed exponent[27], additional experimental work would still be very useful.

Dietrich and Wagner[28] have recently suggested X-ray and neutron scattering at grazing angles, where there is total internal reflection, as an experimental technique for measuring surface critical exponents.

We now consider semi-infinite spin systems with couplings J_B in the interior and

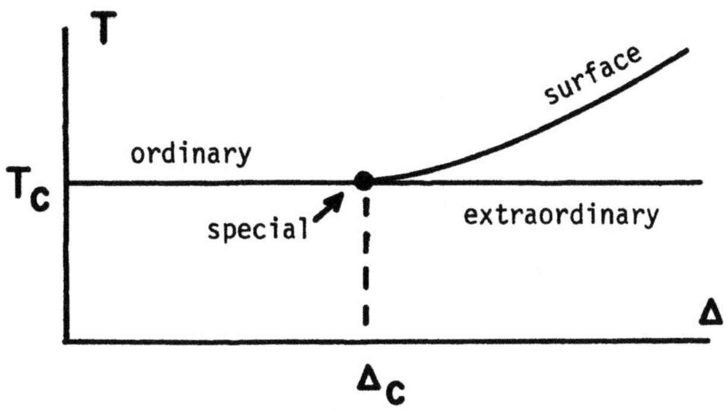

Fig. 3. Phase diagram in the variables Δ,T.

$J_{\parallel} = (1+\Delta)J_B$ on the boundary. If the bulk dimension d is sufficiently great (d>2 for the Ising model, d>3 for the n vector model with n>2), the phase diagram shown in Fig. 3 is expected[1-5]. For small Δ the surface spins order at the bulk transition temperature T_C, i.e., the surface critical behavior is "ordinary". For Δ greater than a critical value Δ_c, the surface spins undergo a transition with d-1 dimensional character above the bulk critical temperature. This is known as the "surface" transition. The line of "extraordinary" transitions $T=T_C$, $\Delta > \Delta_c$ marks the onset of bulk ordering with the surface spins already ordered. The singularities of the surface free energy at the extraordinary transition are extremely weak. The lines of ordinary, surface and extraordinary transitions intersect at the multicritical point $T=T_C$, $\Delta=\Delta_c$ corresponding to the "special" or "surface-bulk" transition.

The approach of the line of surface transitions to the special multicritical point is characterized by a crossover exponent ϕ^{sp} defined by writing $\Delta-\Delta_c \sim (T-T_c)^{\phi^{sp}}$. For the d=3 Ising model the following results have been obtained for the exponents β_1^{sp} and ϕ^{sp} of the special transition:

$\beta_1^{sp} = 0.22 \pm 0.03$	$\phi^{sp} = 0.68$	ε-expansion[29,30]	
0.175 ± 0.025	0.56 ± 0.04	Monte Carlo[23]	(3.12)
0.160	0.545	real-space renormalization[24]	

There is little hope of measuring the exponents of the special transition in magnetic systems, since the enhancement Δ is not a controllable parameter. Leibler and Peliti[31] have discussed the possibility of observing the special transition in fluids with impurities. In principle ϕ^{sp} for the d=3 n-vector model with n=0 can be determined in experiments in which polymers with one end attached to an attractive wall are absorbed as the temperature is lowered. Above the absorption temperature T_a, the fraction of the total number N of monomers at a finite distance from the wall vanishes as $N \to \infty$. Below T_a the fraction is nonzero. According to theoretical predictions of de Gennes[32] and of Eisenriegler et al.[33], the number N_1 of monomers at the surface varies as

$$N_1 \sim N(T_a - T)^{1/\phi^{sp} - 1}, \qquad T \lesssim T_a \tag{3.13a}$$

$$N_1 \sim N^{\phi^{sp}}, \qquad T = T_a \tag{3.13b}$$

Numerical simulations of polymer adsorption described in Ref. 33 yield $\phi^{sp} = 0.59 \pm 0.02$. The ε expansion gives 0.67.

B. Discontinuous Bulk Transition

According to recent work of Lipowsky[34], surface magnetic ordering may vary continuously at a discontinuous bulk transition. A kind of continuous melting of the ordered phase occurs on approaching T_c from below that begins at the surface. Analogous behavior of the interface separating coexisting phases has been discussed by Lajzerowicz[35].

Lipowsky's findings are based on Landau mean-field theory[1,3,5,8,17-20], to which we now turn. For a semi-infinite system, the free-energy functional has the form

$$F[m] = \int_0^\infty dz \left[\frac{1}{2}\left(\frac{dm}{dz}\right)^2 + f(m)\right] + \frac{1}{2}c\,m(0)^2 - h_1 m(0) \tag{3.14}$$

where $f(m)$ is a local function of the magnetization density $m(z)$. For $c > 0$ the term $\frac{1}{2}cm(0)^2$ depresses the magnetization at the surface and simulates the effect of missing spins. Making the variation $m(z) \to m(z) + \delta m(z)$ and setting $\delta F = 0$, we obtain the differential equation and boundary condition

$$\frac{d^2 m}{dz^2} = \frac{df(m)}{dm}, \qquad \frac{dm(0)}{dz} = c\,m(0) - h_1 \tag{3.15}$$

satisfied by the mean-field magnetization profile. Since the differential equation is formally equivalent to Newton's equation of motion, a single integration leads to the "energy conservation" equation

$$\frac{1}{2}\left(\frac{dm}{dz}\right)^2 - f(m) = -f(m_B) \tag{3.16}$$

Here we impose the boundary conditions $m \to m_B$, $dm/dz \to 0$, as $z \to \infty$.

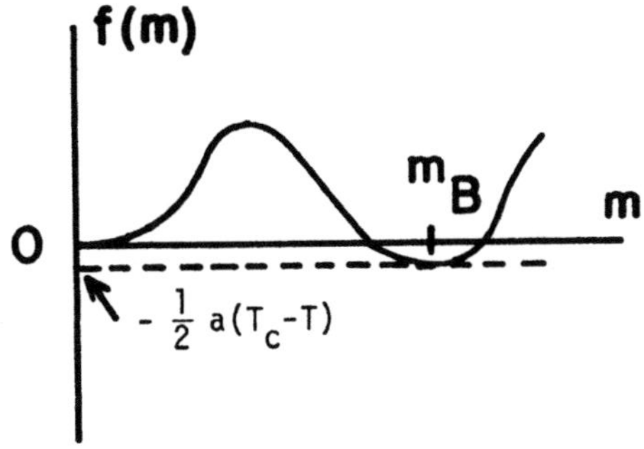

Fig. 4. The function f(m) for a first-order transition

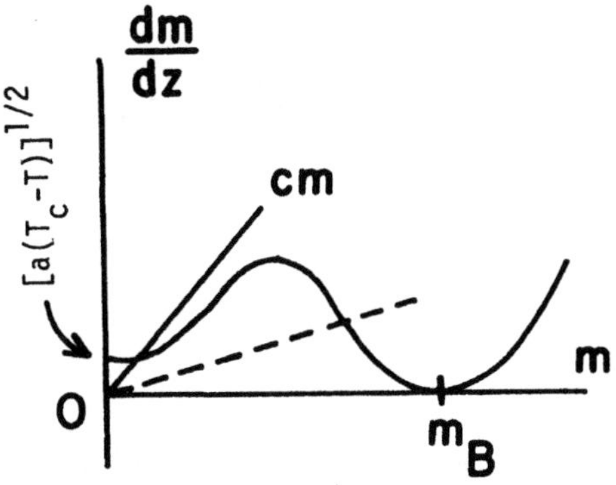

Fig. 5. Phase-space plot corresponding to Fig. 4.

For a first-order transition f(m) typically has the form shown in Fig. 4 for $T \lesssim T_c$. In Fig. 5 the equation

$$\frac{dm}{dz} = \left\{2\left[f(m) - f(m_B)\right]\right\}^{1/2} \tag{3.17}$$

which follows from Eq. (3.16), and the boundary condition of Eq. (3.15b) are depicted. The intersection of the curved and straight lines determines the boundary magnetization $m_1 = m(0)$. Since the right side of Eq. (3.17) has the form $[a(T_c-T)+bm^2]^{1/2}$ for $0 < m \ll m_B$, one sees that

$$m(0) \sim (T_c - T)^{1/2}, \qquad c^2 > b \tag{3.18a}$$

$$m(0) \sim \text{constant}, \qquad c^2 < b \tag{3.18b}$$

Thus for $c^2 > b$, the surface magnetization vanishes continuously, whereas the bulk magnetization vanishes discontinuously.

The magnetization profile for $c^2 > b$ is shown schematically in Fig. 6. It is straightforward to show that the characteristic distance ℓ varies as $\ell \sim |\ln(T_c-T)|$ as T approaches T_c from below.

Evidence beyond mean-field theory for a first-order bulk transitions with continuous surface behavior is provided by the LEED experiment on Cu_3Au of Sundaram et al.[36] and a Monte Carlo simulation[37] of the system. An effective interface model[38]

Fig. 6. Magnetization profile for $c^2 > b$, $T \lesssim T_c$.

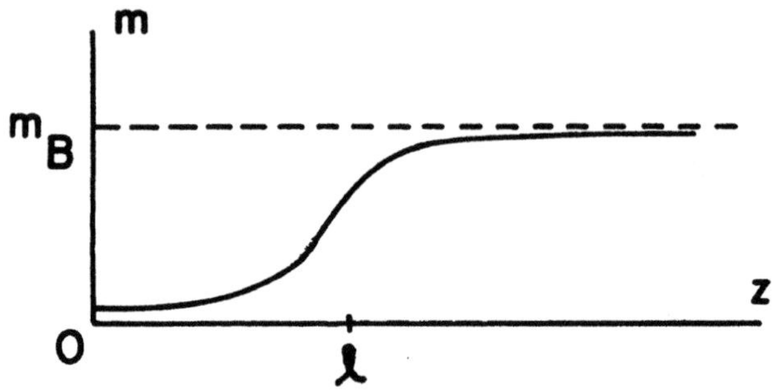

suggests that fluctuations modify the mean-field surface critical exponents.

For further details on these recent developments the reader is refered to Refs. 34 and 39.

4. Surface Critical Behavior in Smoothly Inhomogeneous Systems

Most theoretical studies of surface critical behavior in magnetic systems have considered semi-infinite models parametrized by two coupling constants, a surface coupling $K_{\parallel}$ and a bulk coupling K_B, as in Fig. 3. In real physical systems, one would expect the coupling strengths to depend on the distance z from the surface. If the difference $K(z)-K_B$ vanishes for $z \gg \ell$, where ℓ is a finite characteristic penetration length, essentially the same surface critical behavior as in the two-coupling case $K_{\parallel}, K_B$ is expected, since ℓ is negligible in comparison with the bulk correlation length sufficiently close to the bulk critical temperature T_c.

In this section we consider short-range couplings K(z) that vary asymptotically as

$$K(z) = K_B + A z^{-y}, \qquad z \gg 1 \tag{4.1}$$

where the lattice constant is 1. It is assumed that the couplings near the surface are not so strong as to produce a surface magnetization above the bulk critical temperature. Then the asymptotic form (4.1) determines the universal surface critical behavior, as will be argued below.

Elastic deformations of the lattice induced by the surface could conceively produce a variation in coupling strengths corresponding to Eq. (4.1). Since $K=J/k_BT$, the variation could also result from an inhomogeneous local temperature.

Hilhorst and van Leeuwen[40,41], Blöte and Hilhorst[42], and Burkhardt and Guim[41,43] have obtained exact results for the semi-infinite two-dimensional Ising model with coupling constants that vary as in (4.1). At the bulk critical temperature T_c one finds the following behavior:

1. For $y > 1$ the inhomogeneity does not penetrate sufficiently into the system to alter the ordinary critical behavior. The surface magnetization m_1 vanishes at $T=T_c$ for all finite A. The pair correlation function $g_{\parallel}(r)$ of surface spins falls off as r^{-1} for large r, just as in the homogeneous semi-infinite case A=0.

2. For $y < 1$ there is a nonvanishing surface magnetization m_1 at the bulk critical temperature for $A > 0$. As $A \to 0$, m_1 vanishes as $A^{1/[2(1-y)]}$. For either sign of A, $g_{\parallel}(r)$ decays according to the anomalous exponential form

$$g_{\parallel}(r) \sim \exp\left[-(r/\xi_{\parallel})^{1-y}\right], \qquad \xi_{\parallel} \sim |A|^{-1/(1-y)} \tag{4.2}$$

3. In the marginal case $y=1$, m_1 is nonzero at bulk criticality for $A > A_c$, where A_c is a positive critical value. As $A \to A_c$ from above, m_1 vanishes as $(A-A_c)^{1/2}$. The correlation function $g_{\parallel}(r)$ falls off as $r^{-\eta_{\parallel}}$, with the nonuniversal exponent

$$\eta_{\parallel}(A) = |1-A/A_c| \tag{4.3}$$

At $A = A_c$, where $\eta_{\parallel}$ vanishes, $g_{\parallel}(r)$ decays as $(\ln r)^{-1}$.

Enhanced surface couplings in conjunction with infinite-range bulk correlations produce the nonvanishing surface magnetization at the bulk critical temperature T_c for $y < 1$, $A > 0$ and $y = 1$, $A > A_c$. The bulk correlation length is finite for $T > T_c$, and $m_1 - 0$. Thus as T passes through T_c from below, the bulk magnetization vanishes continuously, whereas the surface magnetization vanishes discontinuously.

In obtaining these results, Refs. 40,41 and 43 make use of an exact mapping based on the star-triangle transformation. A Pfaffian approach is used in Ref. 42. Results for $T \neq T_c$ as well as $T = T_c$ in the marginal case $y = 1$ are obtained in Ref. 42. In Ref. 44, exact results are obtained for an inhomogeneous Gaussian model with coupling constants that vary as in (4.1).

In the remainder of this section, we show that many of the exact Ising and Gaussian results can be explained and generalized to any inhomogeneous semi-infinite system with a divergent bulk correlation length using simple renormalization-group or scaling arguments. This was first pointed out by Burkhardt[45] and Cordery[46].

In the scaling approach, the short-range couplings K(z) are assumed to transform locally under rescaling of lengths, as in real-space renormalization[47]. The renormalization transformation $K_B' = R(K_B)$ for an infinite bulk system[6,47] is replaced by

$$K'(z/b) = R(K(z)), \qquad z \gg 1 \tag{4.4}$$

for K(z) which vary sufficiently slowly with z. Here b is the length-rescaling factor.

Linearizing Eq. (4.4) about the bulk fixed point K_B^* gives[6,47]

$$K'(z/b) - K_B^* = b^{y_t}\left(K(z) - K_B^*\right) \tag{4.5}$$

The thermal scaling index y_t and the bulk exponent ν are related by $y_t = \nu^{-1}$. Substituting $K(z)-K_B^* = Az^{-y}$, $K'(z)-K_B^* = A'z^{-y}$ into Eq. (4.5), one finds that the parameter A transforms as

$$A' = b^{y_t - y} A = b^{\nu^{-1} - y} A \tag{4.6}$$

From (4.6) one sees that the inhomogeneity grows under the rescaling, i.e., is a "relevant" parameter[6,47] that modifies the ordinary surface critical behavior, for $y < \nu^{-1}$, but is "irrelevant" for $y > \nu^{-1}$ and "marginal" for $y = \nu^{-1}$. This is clearly consistent with the exact Ising results, where $\nu = 1$.

In accordance with Eq. (4.6) and standard scaling analyses[6,47], it follows that the pair correlation function of the surface spins transforms as

$$g_{\parallel}(r,A) = b^{-2(d-1-y_{h_1})}\, g_{\parallel}(r/b,\, b^{y_t - y} A) \tag{4.7}$$

at the bulk critical temperature. The scaling index y_{h_1} is the index for the ordinary semi-infinite transition introduced in Eq. (3.3b). Since the scale factor b is arbitrary, Eq. (4.7) implies

$$g_{\parallel}(r,A) = r^{-2(d-1-y_{h_1})}\, F_y(r^{y_t - y} A) \tag{4.8}$$

From Eq. (4.8) one again sees that if $y > y_t = \nu^{-1}$, the critical behavior is the same for $A = 0$ and $A \neq 0$. There is an algebraic decay of the form $g_{\parallel}(r) \sim r^{-(d-2+\eta_{\parallel}^{0})}$, with $\eta_{\parallel}^{0} = d - 2y_{h_1}$, as in Eq. (3.9b). For $y < y_t$ the large-r behavior of $g_{\parallel}(r)$ depends on the asymptotic form of the scaling function $F_y(x)$ for large $x = (r/\xi_{\parallel})^{y_t - y}$, where

$$\xi_{\parallel} \sim |A|^{-1/(y_t - y)} = |A|^{-\nu/(1-\nu y)} \tag{4.9}$$

The quantity $\xi_\parallel$ is a finite surface correlation length at the bulk critical temperature. When $y = 0$, $\xi_\parallel(A)$ is proportional to the usual bulk correlation length.

An ordinary exponential decay $\exp(-r/\xi_\parallel)$ in the homogeneous semi-infinite case $y = 0$ corresponds to

$$F_y(x) \sim \exp(-x^{1/y_t}) \tag{4.10}$$

for large x and $y = 0$. Conjecturing that Eq. (4.10) holds for all $y < y_t$, not just $y = 0$, we obtain the anomalous exponential form

$$g_\parallel(r) \sim \exp\left[-(r/\xi_\parallel)^{1-y/y_t}\right] \tag{4.11}$$

The exact Ising (y_t=1) results (4.2) and analogous results for the Gaussian[44] model (y_t=2) are both consistent with Eqs. (4.9) and (4.11).

At bulk criticality the surface magnetization $m_1(A)$ transforms as

$$m_1(A) = b^{-(d-1-y_{h_1})} m_1(b^{y_t-y} A) \tag{4.12}$$

in analogy with Eq. (4.7). Since b is arbitrary, it follows that m_1 vanishes as

$$m_1 \sim A^{(d-1-y_{h_1})/(y_t-y)} = A^{\beta_1^o/(1-\nu y)} \tag{4.13}$$

for $y < y_t$ as A tends to zero from above. In Eq. (4.13), β_1^o is the exponent for the ordinary transition defined by Eq. (3.6a). For the d=2 Ising model β_1^o = 1/2 and ν = 1. Thus Eq. (4.13) is consistent with the exact Ising result given just above Eq. (4.2).

It is easy to devise approximate real-space renormalization transformations that satisfy the locality condition (4.4) and thus correctly predict the relevance, marginality and irrelevance of the inhomogeneity for $y < y_t$, $y = y_t$, $y > y_t$, respectively. One such transformation, based on the Migdal-Kadanoff method, is applied to the two-dimensional Ising model in Ref. 48. The approximation yields a nonuniversal exponent $\eta_\parallel$ in the marginal case $y = y_t$ and a nonvanishing surface magnetization for A greater than a threshold value A_c, in qualitative agreement with the exact results.

5. Wetting or Interface-Pinning Transition

In this section a very brief introduction to recent work on the wetting or interface-pinning transition is given. Water Selke discusses one aspect of wetting in his contribution to this volume. Some of the early work on this topic is due to Cahn[49], Ebner and Saam[50], de Oliveira and Griffiths[51], and Sullivan[52]. For a review of the field with extensive references two articles[53,54] of Pandit, Schick and Wortis are recommended. Recent experimental work is discussed in Refs. 55-59.

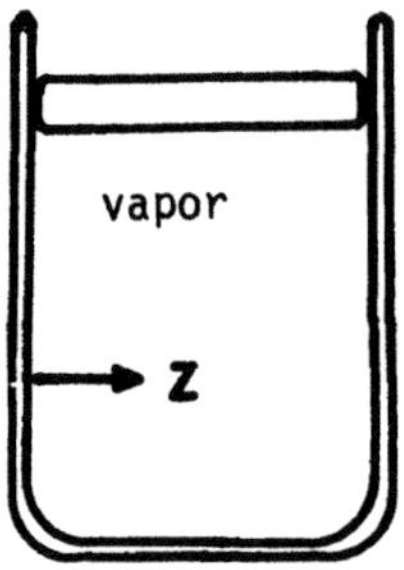

Fig. 7. Schematic representation of wetting experiment.

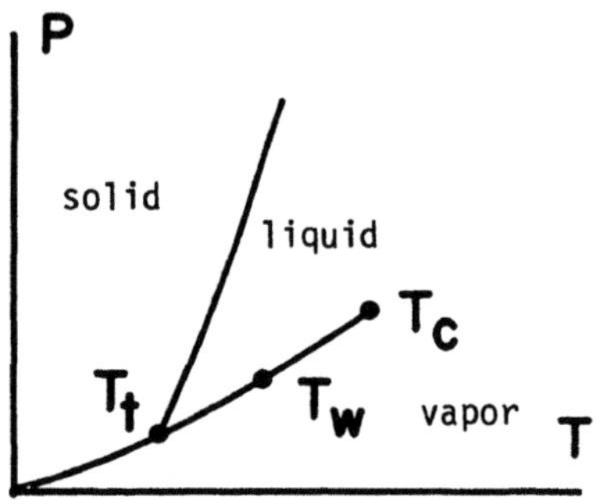

Fig. 8. Phase diagram for a one-component fluid with a wetting transition.

Consider a vapor in a closed container with a movable piston as shown in Fig. 7. The van der Waals interaction of the vapor molecules with a wall falls off as z^{-3}, where z is the distance from the wall. If this interaction is sufficiently attractive, the density of the vapor is greater than the bulk density in the vicinity of the wall. We define the excess surface density by

$$n_s = \int_0^\infty dz\, [n(z) - n_B] \tag{5.1}$$

Let us lower the piston at a constant temperature T, with $T_t<T<T_c$, where T_c and T_t are the critical and triple-point temperatures, indicated in Fig. 8. The pressure increases, and the vapor-liquid coexistence line in Fig. 8 is approached vertically from below. If n_s diverges on approaching the coexistence line from below, there is said to be complete wetting. If n_s remains finite, one speaks of incomplete or partial wetting. There is a wetting transition with transition temperature T_w, as in Fig. 8, if the point labelled T_w separates regions of complete and incomplete wetting. This section reviews some of the work which has been carried out in order to better understand this transition.

In experimental work and in lattice gas models, n_s increases through layer formation. Thus the wetting, layering and interface-roughening transitions are interrelated.

De Oliveira and Griffiths[51] and Pandit et al.[53] have discussed the variety of phase diagrams found for Ising lattice-gas models of the liquid-vapor system in molecular-field theory. The attractive substrate interaction corresponds to a surface magnetic field in the Ising model that falls off as z^{-3}.

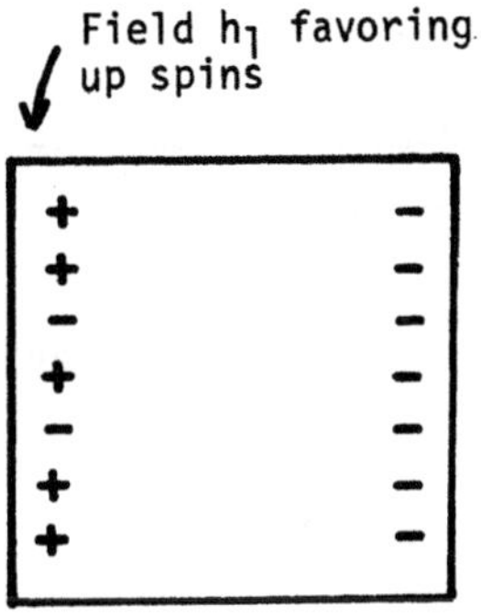

Fig. 9. Boundary conditions considered by Abraham.

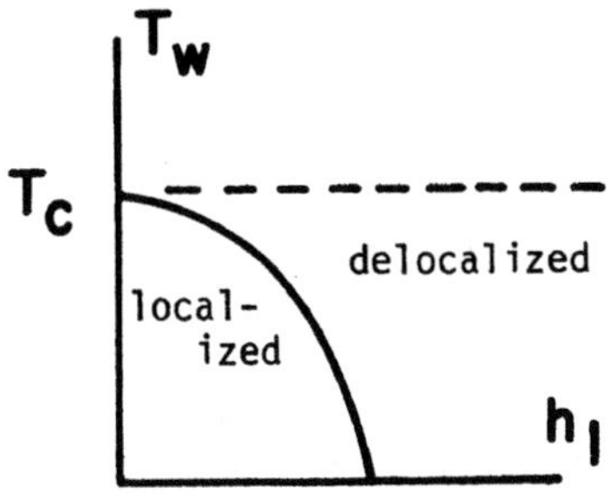

Fig. 10. Phase diagram of the localization-delocalization transition of Ref. 60.

Abraham[60] has obtained exact results for the d=2 Ising model with the boundary conditions indicated in Fig. 9. Fixing the boundary spins downward on the right edge creates a region of negative magnetization, corresponding to the vapor phase, for $T < T_C$. Along the left edge the spins are subject to a surface field h_1 that favors up spins. This model exhibits a transition that is referred to as the interface-pinning, interface localization-delocalization, or wetting transition. Below the curve $T_W(h_1)$ in Fig. 10, the interface separating the phases of predominantly positive and negative magnetization is bound, i.e., the average distance ℓ of the interface from the left edge is finite. Above the curve, ℓ is infinite, i.e., there is an infinitely thick liquid layer wetting the left edge. As T approaches $T_W(h_1)$ from below at constant h_1, ℓ diverges as $(T_W(h_1)-T)^{-1}$. The interface specific heat is discontinuous on crossing the curve. The roughening transition also takes place at $T_W(h_1)$ in this model. The bound and unbound interfaces are smooth and rough, respectively, i.e., the root-mean-square width of the interface is finite for $T < T_W(h_1)$ and infinite for $T > T_W(h_1)$.

Well below T_C, one would expect SOS(solid on solid) models, which neglect overhangs of the interface and "islands", to yield a reasonable description of the localization-delocalization transition. Several authors[61-66] have shown that in two bulk dimensions these models have transitions with the same critical properties as found by

Fig. 11. SOS model for the localization-delocalization transition. It is energetically favorable for the domain wall to pass to the left of the dashed line, where it is subject to the pinning potential.

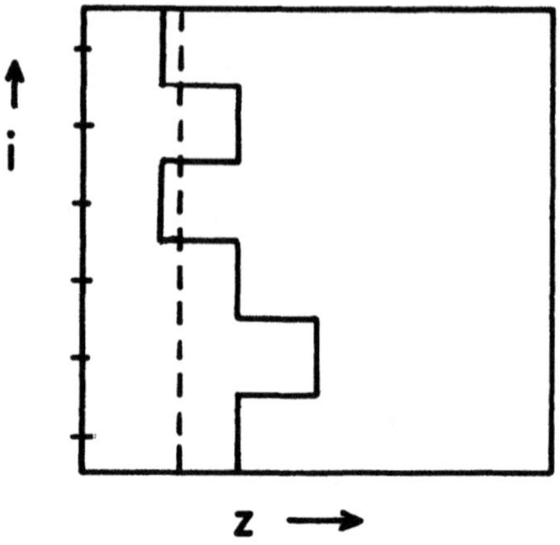

Abraham. One such model, which corresponds to Fig. 11, has the partition function[61]

$$Q = \int_0^\infty dz_1 \ldots \int_0^\infty dz_N \prod_i \exp\left[-K|z_{i+1} - z_i| - V(z_i)\right] \tag{5.2}$$

Due to the pinning potential V(z) (a square well, for example), it is energetically favorable for the interface to stay close to the left edge.

One can readily calculate Q using transfer-matrix methods. In the large-N limit, $(\ln Q)/N \to \lambda_0$, where λ_0 is the largest eigenvalue of the transfer matrix

$$T(z,z') = \exp\left[-K|z-z'| - V(z')\right] \tag{5.3}$$

The eigenvalues and eigenfunctions of T(z,z') satisfy

$$\int_0^\infty dz'\, T(z,z')\psi(z') = \lambda\psi(z) \tag{5.4}$$

Using the relation $(-d^2/dz^2 + K^2)\exp(-K|z-z'|) = 2K\delta(z-z')$, one can convert the integral equation (5.4) to the differential equation

$$\left[-\frac{d^2}{dz^2} - \frac{2K}{\lambda}\exp(-V(z)) + K^2\right]\psi(z) = 0 \tag{5.5}$$

with the boundary condition $d(\ln \psi(0))/dz = K$. Equation (5.5) has the form of a Schrödinger equation with a temperature-dependent potential. It is easily shown that a localized interface corresponds to a bound ground state of the Schrödinger equation and a delocalized interface to an unbound ground state. For an attractive potential which falls off as $z^{-\sigma}, \sigma \leq 2$, the interface is always bound[70]. For a more detailed discussion see Ref. 61.

We note that the transition remains qualitatively unchanged[62-65] if the z_i are restricted to discrete values 0,1,2,... instead of varying continuously from 0 to ∞. The correspondence between the interface problem and the Schrödinger equation also follows[66] from the path-integral formulation of quantum mechanics.

Brézin et al.[67] and Kroll and Lipowsky[68] have discussed the wetting or interface-pinning transition in continuum field theories with Hamiltonians of the form

$$\mathcal{H}/k_BT = \int d^{d-1}r \left[\frac{1}{2}(\nabla z(\vec{r}))^2 + V(z(\vec{r})) \right] \tag{5.6}$$

The pinning potential V(z) is represented by the difference of two Gaussian functions. The wetting transition is found to be continuous in d = 3, and the unbound interface is rough. We refer to these papers for further details.

Luck et al.[69] have carried out finite-size scaling calculations for SOS models with discrete rather than continuous z. In bulk dimension d=3, they find that for weak pinning, the delocalization and roughening temperatures T_D and T_R satisfy $T_D < T_R$, and the delocalization transition is discontinuous. In the strong-pinning regime $T_D = T_R$, and the transition is continuous. These field-theoretical and finite-size scaling calculations hold for finite-range pinning potentials. Very recently Nightingale et al.[70] have argued that any pinning potential V(z) that falls off as $V(z) \sim z^{-\sigma}$, $\sigma > 0$ completely suppresses the critical (continuous) wetting transition in d=3.

There are some contradictions in the literature on this rapidly developing field which, at the time of writing, remain to be resolved.

Before leaving the subject of wetting, let us turn to the description of the transition in continuum mean-field theory[1,49-54,71]. This approach neglects interface fluctuations (thermally-excited capillary waves) which are responsible for the rough interface in $d \leq 3$. Brézin et al.[71] have argued that the mean-field description is essentially correct for d > 3.

In the mean-field theory of wetting presented in Ref. 71, one solves Eqs. (3.15) with the usual free energy functional

$$f(m) = \frac{1}{2}a(T-T_c)m^2 + \frac{1}{4}bm^4 \tag{5.7}$$

corresponding to a second-order bulk transition at $T = T_C$. The bulk magnetization m_B, which follows from $f'(m_B) = 0$, is given by

$$m_B = \left[\frac{a}{b}(T_c - T)\right]^{1/2} \tag{5.8}$$

The boundary condition $m(z) \to -m_B$ as $z \to \infty$ (corresponding to the vapor phase far from the wall) is also imposed on the solution to the differential equation.

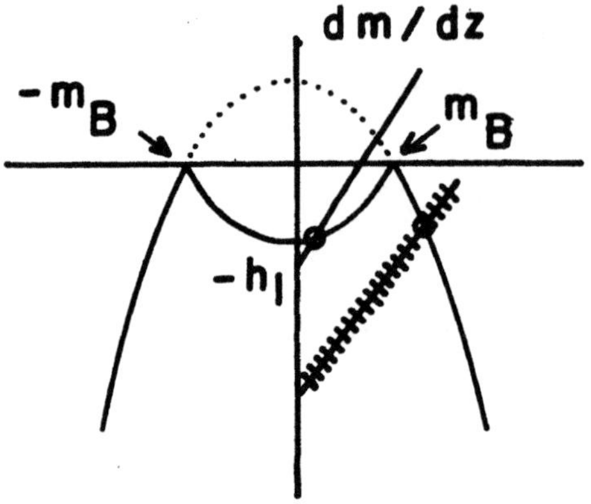

Fig. 12. Phase-space plot corresponding to Eqs. (3.15b) and (5.9).

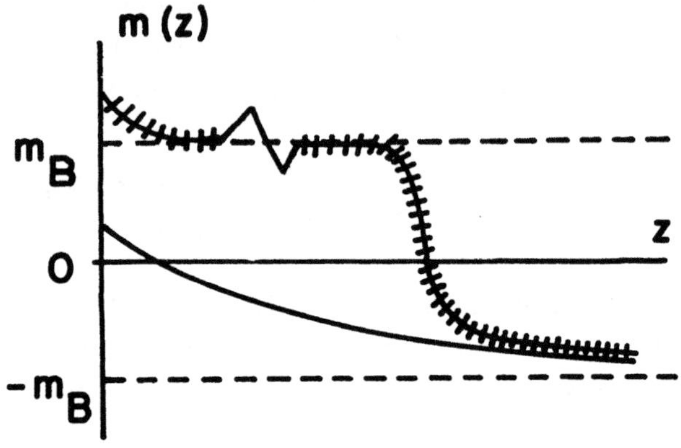

Fig. 13. Magnetization profiles corresponding to Fig. 12. The profiles with and without cross hatching are localized and delocalized, respectively.

Combining Eqs. (3.16),(5.7) and (5.8), we are led to the differential equation

$$\frac{dm}{dz} = -\sqrt{\frac{b}{2}}\,\left|m^2 - m_B^2\right| \tag{5.9}$$

This equation and the boundary condition (3.15b) correspond to the phase-space plot of Fig. 12. The intersection of the straight and curved lines determines the surface magnetization m(0). The straight lines with and without cross hatching in Figs. 12 and 13 correspond to $h_1 < cm_B$ and $h_1 > cm_B$, respectively.

The criticality condition for the wetting transition is given by $h_1 = cm_B$, or

$$T_W = T_c - \frac{b}{ac^2}h_1^2 \tag{5.10}$$

The corresponding phase diagram looks qualitatively the same as in Fig. 10. It is easily shown that the characteristic distance ℓ of the interface from the edge of the system diverges as

$$\ell \sim |\ln (T_w - T)| \tag{5.11}$$

as T approaches T_w from below at constant h_1.

For further details concerning the mean-field approach to the wetting transition, we refer to Refs. 49-54,71. Results which go beyond mean-field theory have already been discussed in the paragraph containing Eq. (5.6).

Acknowledgement: I thank Professor Andrzej Pękalski and the other members of the organizing committee of the 20th Winter School for a most enjoyable and stimulating two weeks in Karpacz. Ihnsouk Guim kindly helped with the preparation of this manuscript.

References

1. K. Binder in Phase Transitions and Critical Phenomena, edited by C. Domb and J. L. Lebowitz (Academic, London, 1983), Vol. 8.
2. M. E. Fisher, J. Vac. Sci. Technol. 10, 665(1973).
3. K. Binder and P. C. Hohenberg, IEEE Trans. Mag. 12, 66(1976).
4. M. Wortis and N. M. Švrakić, IEEE Trans. Mag. 18, 721(1982).
5. H. W. Diehl, J. Appl. Phys. 53, 7914(1982).
6. M. E. Fisher, Rev. Mod. Phys. 46, 597(1974).
7. T. W. Burkhardt and E. Eisenriegler, Phys. Rev. B24, 1236(1981).
8. A. J. Bray and M. A. Moore, J. Phys. A10, 1927(1977).
9. J. C. Le Guillou and J. Zinn-Justin, Phys. Rev. Lett. 39, 95(1977); Phys. Rev. B21, 3976(1980).
10. J. L. Black and V. J. Emery, Phys. Rev. B23, 429(1981).
11. R. Z. Bariev, Zh. Eksp. Teor. Fiz. 77, 1217(1979) [Sov. Phys. JETP 50, 613(1979)].
12. B. M. McCoy and J. H. H. Perk, Phys. Rev. Lett. 44, 840(1980).
13. E. Eisenriegler and T. W. Burkhardt, Phys. Rev. B25, 3283(1982).
14. R. Abe, Prog. Theo. Phys. 65, 1237(1981); K. Yamamoto and R. Abe, Prog. Theo. Phys. 66, 1947(1981).
15. B. M. McCoy and T. T. Wu, The Two-Dimensional Ising Model (Harvard, Cambridge, 1973).
16. M. N. Barber, D. Jasnow, S. Singh and R. A. Weiner, J. Phys. C7, 3491(1974).
17. S. Dietrich and H. W. Diehl, Z. Physik B43, 315(1981).
18. T. C. Lubensky and M. H. Rubin, Phys. Rev. B11, 4533(1975).
19. H. W. Diehl and S. Dietrich, Phys. Lett. 80A, 408(1980); Z. Phys. B42, 65(1981).
20. J. S. Reeve and A. J. Guttmann, Phys. Rev. Lett. 45, 1581(1980); J. Phys. A14, 3357(1981).
21. K. Ohno and Y. Okabe, Phys. Lett. 95A, 41(1983); Prog. Theo. Phys. 70, 1226(1983).
22. S. G. Whittington, G. M. Torrie and A. J. Guttmann, J. Phys. A12, 2449(1979).
23. K. Binder and D. P. Landau, Phys. Rev. Lett. 52, 318(1984).
24. T. W. Burkhardt and E. Eisenriegler, Phys. Rev. B17, 318(1978).
25. S. F. Alvarado, M. Campagna and H. Hopster, Phys. Rev. Lett. 48, 51(1982).
26. S. F. Alvarado, M. Campagna, F. Ciccacci and H. Hopster, J. Appl. Phys. 53, 7920 (1982).
27. H. W. Diehl and E. Eisenriegler, Phys. Rev. Lett. 48, 1767(1982).
28. S. Dietrich and H. Wagner, Phys. Rev. Lett. 51, 1469(1983).
29. J. S. Reeve, Phys. Lett. 81A, 237(1981).
30. H. W. Diehl and S. Dietrich, Phys. Rev. B24, 2878(1981).
31. S. Leibler and L. Peliti, Phys. Rev. B29, 1253(1984).
32. P. G. de Gennes, J. Physique 37, 1445(1976).
33. E. Eisenriegler, K. Kremer and K. Binder, J. Chem. Phys. 77, 6296(1982).
34. R. Lipowsky, Phys. Rev. Lett. 49, 1575(1982); Z. Physik B51, 165(1983).

35. J. Lajzerowicz, Ferroelectrics 35, 219(1981).
36. V. S. Sundaram, B. Farrell, R. S. Alben and W. D. Robertson, Phys. Rev. Lett. 31, 1136(1973).
37. V. S. Sundaram, R. S. Alben and W. D. Robertson, Surf. Sci. 46, 653(1974).
38. R. Lipowsky, D. M. Kroll and R. K. P. Zia, Phys. Rev. B27, 4499(1983).
39. R. Lipowsky and W. Speth, Phys. Rev. B28, 3983(1983).
40. H. J. Hilhorst and J. M. J. van Leeuwen, Phys. Rev. Lett. 47, 1188(1981).
41. T. W. Burkhardt, I. Guim, H. J. Hilhorst and J. M. J. van Leeuwen, to be published.
42. H. W. J. Blöte and H. J. Hilhorst, Phys. Rev. Lett. 51, 2015(1983).
43. T. W. Burkhardt and I. Guim, Phys. Rev. B29, 508(1984).
44. T. W. Burkhardt and I. Guim, J. Phys. A15, L305(1982).
45. T. W. Burkhardt, Phys. Rev. Lett. 48, 216(1982); Phys. Rev. B25, 7048(1982).
46. R. Cordery, Phys. Rev. Lett. 48, 215(1982).
47. Th. Niemeijer and J. M. J. van Leeuwen in Phase Transitions and Critical Phenomena, edited by C. Domb and M. S. Green (Academic,London,1976), Vol. 6.
48. T. W. Burkhardt and I. Guim, J. Phys. A16, L721(1983).
49. J. W. Cahn, J. Chem. Phys. 66, 3667(1977).
50. C. Ebner and W. F. Saam, Phys. Rev. Lett. 38, 1486(1977).
51. M. J. de Oliveira and R. B. Griffiths, Surf. Sci. 71, 687(1978).
52. D. E. Sullivan, Phys. Rev. B20, 3991(1979).
53. R. Pandit, M. Schick and M. Wortis, Phys. Rev. B26, 5112(1982).
54. M. Wortis, R. Pandit and M. Schick in Melting, Localization, and Chaos, edited by R. Kalia and P. D. Vashishta (North Holland,New York,1982).
55. M. R. Moldover and J. W. Cahn, Science 207, 1073(1980).
56. O'D. Kwon, et al., Phys. Rev. Lett. 48, 185(1982).
57. D. W. Pohl and W. I. Goldberg, Phys. Rev. Lett. 48, 1111(1982).
58. J. W. Schmidt and M. R. Moldover, J. Chem. Phys. 79, 379(1983).
59. J. L. Seguin, J. Suzanne, M. Bienfait and J. A. Venables, Phys. Rev. Lett. 51, 122 (1983).
60. D. B. Abraham, Phys. Rev. Lett. 44, 1165(1980).
61. T. W. Burkhardt, J. Phys. A14, L63(1981).
62. J. T. Chalker, J. Phys. A14, 2431(1981).
63. S. T. Chui and J. D. Weeks, Phys. Rev. B23, 2438(1981).
64. D. M. Kroll, Z. Physik B41, 345(1981).
65. J. M. J. van Leeuwen and H. J. Hilhorst, Physica 107A, 319(1981).
66. M. Vallade and J. Lajzerowicz, J. Physique 42, 1505(1981).
67. E. Brézin, B. I. Halperin and S. Leibler, Phys. Rev. Lett. 50, 1387(1983).
68. D. M. Kroll and R. Lipowsky, Phys. Rev. B26, 5289(1982).
69. J. M. Luck, S. Leibler and B. Derrida, J. Physique, in press.
70. M. P. Nightingale, W. F. Saam and M. Schick, Phys. Rev. Lett. 51, 1275(1983).
71. E. Brézin, B. I. Halperin and S. Leibler, J. Physique 44, 775(1983).

INTERFACIAL ADSORPTION IN MULTI-STATE MODELS

Walter Selke
Institut für Festkörperforschung der KFA,
5170 Jülich, Postfach 1913, Federal Republic of Germany

Abstract: A review on interfacial adsorption in two-dimensional multi-state models is presented including q-state Potts models, the Blume-Capel model and the three-state chiral clock model: New critical properties are found using Monte Carlo techniques as well as scaling and analytic arguments; the occurence of non-boundary states at the interface provides the clue to suggest a SOS (or Müller-Hartmann and Zittartz)-type approximation of the interface free energy to map phase diagrams of two-dimensional three-state models. - In realistic models for chemisorbed monolayers interfacial adsorption can be observed at domain boundaries in (3×1) phases.

1. Introduction

Most of the studies on interfaces have been confined to two phase systems such as the liquid-vapor system experimentally and the spin-$\frac{1}{2}$ nearest neighbour Ising model theoretically [1]. A more complex situation may be encountered at interfaces between two phases (α,β) in the presence of additional phases ($\gamma,\ldots$). For example, in a two-component fluid system in equilibrium with its vapor phase a thin film of the heavier fluid may intervene at the interface between the lighter fluid and the vapor [2]. This wetting effect is caused by an interplay of interface tensions, repulsion between interfaces, and gravitional forces which has been described within the classical, phenomenological framework [3].

We discovered [4] a similar effect ("interfacial adsorption" [5]) in multi-state models, i.e. lattice models where the variables, S_i, can have at least three values, $S_i=\alpha,\beta,\gamma,\ldots$. In particular, two-dimensional (2d) models have been studied including q-state Potts models, $q=3,4,\ldots$, [4,5,6], the Blume-Capel model [7,8], the three-state chiral clock model [4,6,9] as well as lattice gas models for adsorbates which may be mapped onto multi-state models (such as the ANNNI model in a field [10] and a model for hydrogen chemisorbed on Fe (110) [11, 12]). In all cases an interface is introduced by fixing variables on opposite boundaries in two different states, say, α and β, see Fig.1. At the interface an excess of non-boundary states, $\gamma\ldots$, occurs, possibly in form of droplets [4,5] or in a layer-like fashion [7,9]. To study the interfacial adsorption quantitatively the "net adsorption of non-boundary states at the interface" is defined by comparing their numbers in systems with and without interfaces. The adsorption is found to exhibit interesting new critical behaviour as one approaches the bulk transition temperature, T_c, or possibly a distinct wetting temperature below T_c.

The article is organized as follows: First the models are introduced and the phenomenon of interfacial adsorption is described. In the following section critical properties are discussed which have been obtained using Monte Carlo methods as well as scaling and analytic arguments. The interfacial adsorption provides the clue for the modification of a SOS-type approximation for the interface free energy

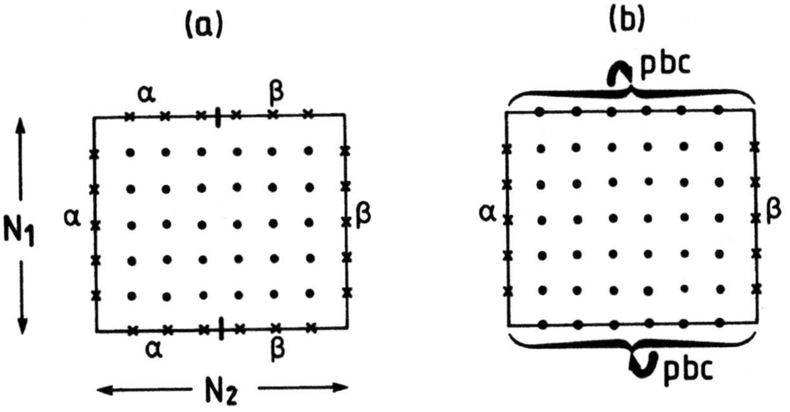

Fig. 1: Systems of size $N_1 \times N_2$ with interfaces.

(originally [13] devised to map phase diagrams of 2d two-state models) to treat 2d three-state models, as reviewed in the fourth section. Finally preliminary results on wetting phenomena in models for adsorbate systems are presented which may encourage future experimental work. A brief summary concludes the article.

2. Models and Interfaces

We consider two-dimensional multi-state models where discrete variables, S_i, situated on lattice sites, i, can have at least three values, $S_i = \alpha, \beta, \gamma \ldots$. An interface may be introduced by fixing the variables at opposite boundaries in two different states, say, α and β (we fix the variables either only at the left and right boundaries and use periodic boundary conditions at the top and the bottom (b) or we also fix the variables at the lower and upper boundaries to pin the interface in the middle of the system (a); see Fig.1). By examining typical Monte Carlo equilibrium configurations it is readily seen that at the interface more of the non-boundary states, γ, ..., are generated than is the case in the absence of an interface, i.e. by fixing the variables at opposite boundaries in the same state, say, α. To describe this phenomenon quantitatively we define the net adsorption per unit length of the interface as

$$W = \frac{1}{N_1} \sum_{n,i} (<\delta_{n,S_i}>_I - <\delta_{n,S_i}>_O) \qquad (1)$$

The sum is taken over all lattice sites, i, and non-boundary states, n=γ,... ; the subscripts I and O refer to systems of size $N_1 \times N_2$ with (α:β) and without (α:α) interfaces. W may be considered as a measure of the thickness of an idealized adsorption layer at the interface which contains only non-boundary states.

We studied the net adsorption, W, for various multi-state models (for the case of adsorbate models, see Section 5) on square lattices: (A) The q-state Potts model [14] with the Hamiltonian

$$H = -J \sum_{<ij>} \delta_{S_i,S_j} \qquad S_i = 1,2,\dots q \qquad (2)$$

The sum runs over nearest neighbours <ij>; J>0. The model is known to exhibit second-order transitions for q=3,4 and first-order transitions for larger integers of q. - In this model we take $\alpha \hat{=} 1$ and $\beta \hat{=} 2$. A typical Monte Carlo configuration for the three-state model close to T_c is shown in Fig.2.

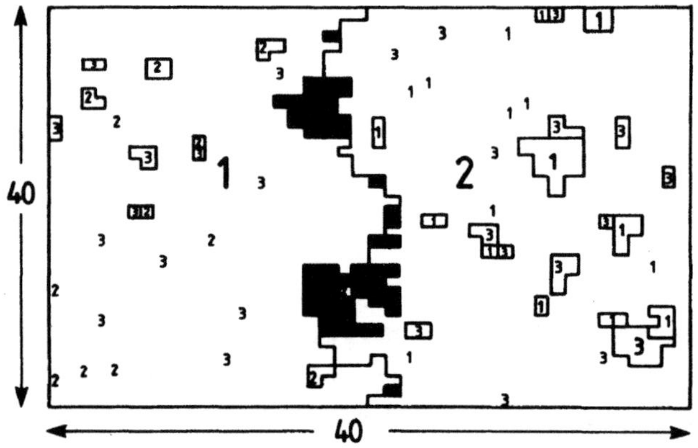

Fig. 2: Typical equilibrium configuration for the 3-state Potts model at temperature $T=0.97T_c$ showing an interface. The non-boundary states (3) are shown blackened, if adsorbed at the interface.

(B) The Blume-Emery-Griffiths model [15]

$$H = -J \sum_{<ij>} S_i S_j + D \sum_i S_i^2 - K \sum_{<ij>} S_i^2 S_j^2 \qquad S_i = -1,0,1 \qquad (3)$$

where J,D,K>0. This spin-1 Ising model reduces to the Blume-Capel model [16] if K=0. Its phase diagram is depicted in Fig.3. For $D/J < (D/J)_t \approx 1.94$ it displays a line of second order transitions which goes over into a line of first order transitions if $2>(D/J)>(D/J)_t$. $(D/J)_t$ denotes the tricritical point. - The variables at the boundaries are fixed in the two non-zero states, ±1. At zero temperature, for $D/J < 1$, the interface is a straight line separating "+1" and "-1" domains. It is interesting to note that for D/J>1, however, the ground-state energy is lowered by an insertion of a $S_i=0$ monolayer at the interface. A typical situation close to the tricritical point is shown in Fig.4.

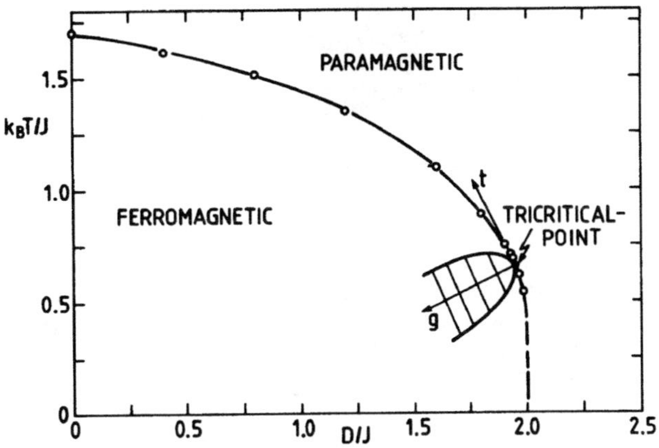

Fig. 3: Phase diagram of the 2d Blume-Capel model. The tricritical point is at $D/J \approx 1.94$. The scaling axes, t and g, are expected to be approximately perpendicular to each other. - See Ref. [7,8].

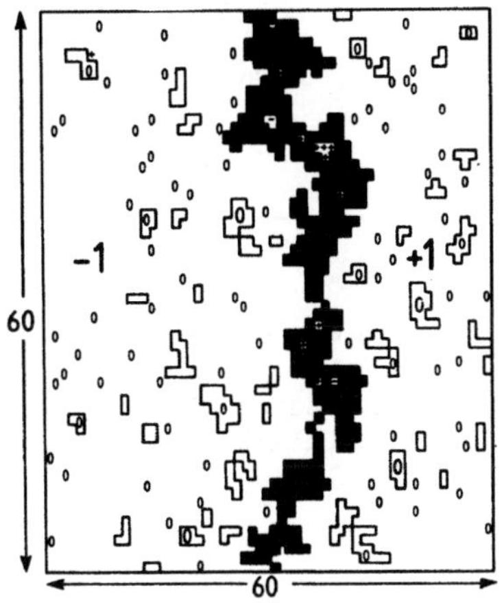

Fig. 4: Typical Monte Carlo equilibrium configuration close to the tricritical point ($D/J=1.92$, $k_BT/J=0.64$). The non-boundary states (0) adsorbed at the interface are shown blackened.

(C) The three-state chiral clock model [17]

$$H = -J \left(\sum_{<ij>}^{x} \cos\left(\frac{2\pi}{3}(S_i-S_j)\right) + \sum_{<ij>}^{y} \cos\left(\frac{2\pi}{3}(S_i-S_j+\Delta)\right) \right) \quad S_i=0,1,2 \qquad (4)$$

where the sums refer to nearest neighbours along the x- and y-axis of the system; J>0 and $0 \leq \Delta \leq 1$. At $\Delta=0$ the model reduces to the usual three-state Potts model. The phase diagram $(k_BT/J, \Delta)$ is symmetric about the line $\Delta=\frac{1}{2}$. The model is known [9,17,18] to display a floating incommensurate or modulated phase in between the low temperature ferromagnetic (or chiral phase, if $\Delta>\frac{1}{2}$) and the high temperature paramagnetic phase, see Fig.5. Possibly the floating phase is present only up to a Lifshitz point, $\frac{1}{2} \leq \Delta \leq \Delta_L \approx 0.4$ - but it has been also argued

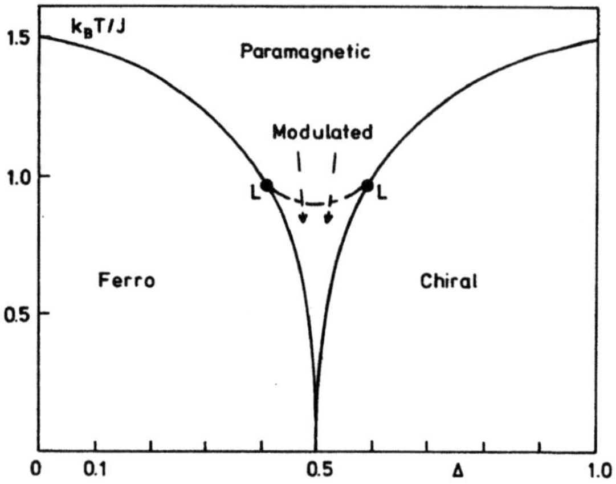

Fig. 5: Phase diagram of the 3-state chiral clock model. It is an open question whether a very narrow modulated phase extends up to Δ=0. See Ref.[18].

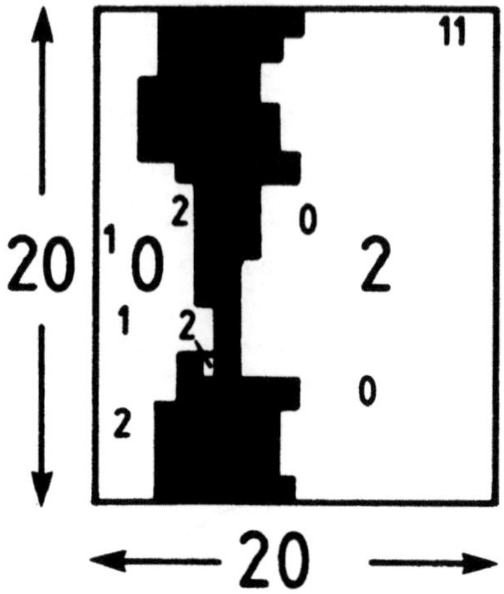

Fig. 6: Typical equilibrium configuration at Δ=0.225 slightly below the melting temperature. The non-boundary states (1) are shown blackened, if adsorbed at the interface.

that a very narrow strip extends up to Δ=0, see, e.g.,[19] for a discussion of this subtle problem. - In this case we take $\hat{\alpha}=0$ and $\hat{\beta}=2$. In the ground state, for $\Delta < \frac{1}{4}$, the interface is a straight line separating "0" and "2" regions. For $\Delta>\frac{1}{4}$, however, the insertion of a "1" layer of arbitrary thickness becomes favourable, i.e. complete wetting occurs. But even for $\Delta < \frac{1}{4}$ a "1" layer is included at the interface close to the melting temperature of the ferromagnetic phase, see Fig.6.

In the first two models the net adsorption, W, is found, using Monte Carlo techniques, to exhibit the following temperature dependence, for finite lattices (usually of size N×N): W increases, as one raises the temperature, up to its maximal value close to the bulk transition temperature T_c (presumably approaching T_c, as $N\to\infty$) and then decreases sharply, see Fig.7. The increase demonstratesthat the cost of energy to generate the non-boundary states at the interface is compensated by a gain in entropy. The latter decrease reflects the vanishing of the interface above T_c, which we know must occur in the thermodynamic limit. The, possibly, critical properties of W are discussed in the next section for these two models as well as for the chiral clock model

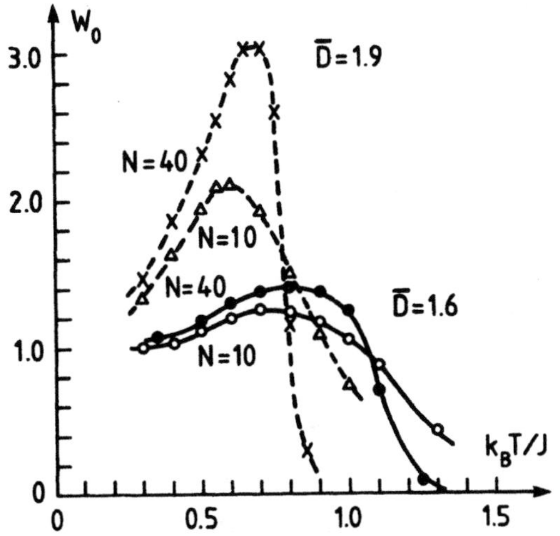

Fig. 7: The net adsorption, W_0, at two values of $D/J=\bar{D}$ for systems of size (N×N) in the two-dimensional Blume-Capel model

where the interfacial adsorption becomes critical, even for $0 < \Delta < \frac{1}{4}$, already below the bulk transition temperature [9,6].

3. Critical Properties

For the q-state Potts model and the Blume-Capel (BC) model the net adsorption, W, appears to become maximal at the bulk transition temperature, T_c, as $N\to\infty$, see e.g., Fig.8. To study possibly critical behaviour we recorded (using standard Monte Carlo techniques [20]) the size dependence of the maximal adsorption, $W^{max}(N)$, or of the adsorption at T_c, $W(T_c,N)$, $6\leq N \leq 100$, as well as the possibly critical temperature dependence, $W(t_c=(T_c-T)/T_c)$ for fixed size, typically N=40 and 60.

(A) The Monte Carlo results for the 2d BC model can be summarized as follows [7,8]:

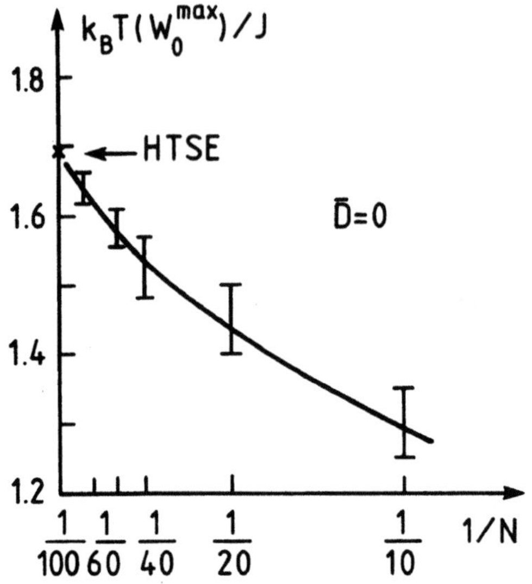

Fig. 8: Size dependence of the position of the maximum of the adsorption, W_0^{max}, at D=0 for systems of size N×N. As N→∞, the bulk transition temperature (known from high temperature series expansion) is approached

(i) At the continuous transition $(D/J < (D/J)_t \approx 1.94)$:

$$W^{max}(N) = W^{max}(\infty) - c(D/J)/N$$

where $W^{max}(\infty)$ is finite, see Fig.9. Accordingly, the temperature dependence is non-critical.

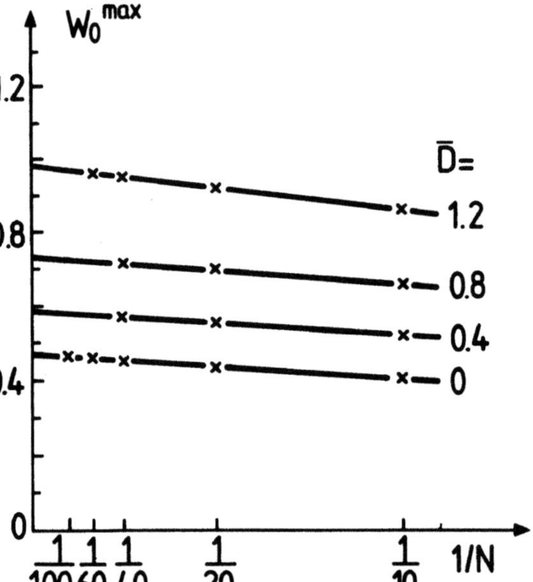

Fig. 9: Size dependence of the maximal value of the net adsorption, W_0^{max}, for chosen values of the ratio $\bar{D}$=D/J at the continuous transition

(ii) At the tricritical point ($D/J \approx 1.94$):
In order to avoid crossover effects [7] the tricritical point should be approached perpendicular to the tangent of the transition line, i.e. in terms of the scaling fields t and g, see Fig.3, at t=0. Then one obtains [8]

$W(t_c) \sim t_c^{-\omega}$ with $\omega = 0.44 \pm 0.02$. , see Fig. 10

To establish the critical dependence one presumably has to use very large systems and one has to know the position of the tricritical point very accurately [8], see below.

(iii) At the first order transition:

$W(t_c) \sim t_c^{-\omega}$ with $\omega = 0.33 \pm 0.03$, see Fig.11.

$W(T_c,N) \sim N^a$ with $a = 0.7 \pm 0.05$.

The Monte Carlo findings can be explained using scaling and analytic arguments [8]. The scaling argument are based on the identity

$$W = k_B T \; \partial f_{int}/\partial D \tag{5}$$

where f_{int} is the interfacial tension (or free energy). For the continuous Ising-like transition we expect $f_{int} \sim t_c$ [21] from which one obtains, because D is a nonordering field with a projection on t_c, $W \sim \partial f_{int}/\partial t_c \sim$ constant. - Near the tricritical point the interface free energy is suggested to scale as

$$f_{int} = g^{\mu_t} \; F_{int}(t/g^{\phi_t}) \tag{6}$$

with $\mu_t = 2-\alpha_t-\nu_t = 5/9$ and $\phi_t = 4/9$ [22]. The scaling ansatz, eq.(6), used the scaling form of the bulk free energy [22] and assumes, following Widom [1], that the relevant length of the intrinsic structure of the interface is the correlation length, ξ. Since D has a projection on g, see Fig.3, we get from equations (5) and (6) at t=0 the asymptotic form $W \sim g^{-4/9}$, in excellent agreement with the Monte Carlo result (note that $g \sim t_c$). Finite size scaling for the maximal adsorption yields $W^{max} \sim N^{4/5}$ (because $N \sim \xi \sim g^{-5/9}$ at the maximum value of W). Presumably very large systems and accurate knowledge of the position of the tricritical point are necessary to confirm this prediction in a Monte Carlo study [8]. - The power-law divergence of W at the first order transition may be understood in connection with interface delocalization transitions [23]: Within Landau theory logarithmic divergencies have been obtained at first order transitions for interfaces of two-state systems [24] and "surface induced disorder" [25]. For the latter effect in two dimensions a power-law behaviour with the exponent 1/3 has been established using collective interface coordinates [26]. Indeed,

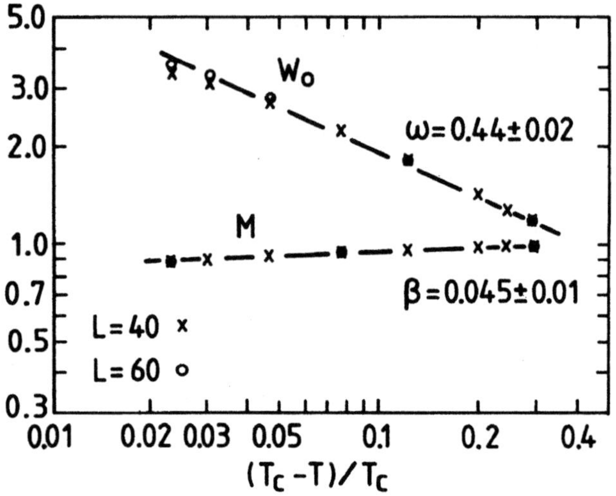

Fig. 10: To show the critical temperature dependence of the net adsorption, W_o, and the magnetization as one approaches the tricritical point of the Blume-Capel model along the path t=0, see Fig. 3. Systems of size L×L are used.

the analogous method gives ω=1/3 in our case, too [8], in excellent agreement with the Monte Carlo estimate. To understand this value one might also compare the BC model to an exactly soluble three-state model of SOS-type [27,8,28].

(B) For the two-dimensional q-state Potts model our Monte Carlo findings for the critical behaviour of the interfacial adsorption can be summarized (and explained) as follows:

(i) At the continuous transition (q=3,4) [4,5]:

$$W(T_c,N) \sim N^{a_q} \qquad q = 3,4$$

$$W(t_c) \sim t_c^{-\omega_q} \qquad q = 3,4$$

with exponents consistent with a scaling argument [5] (based on the observation that the non-boundary states are components of the order parameter and the assumption that the intrinsic structure of the interface is governed by the bulk correlation length) which yields $\omega_q=\nu_q-\beta_q$ and $a_q=1-(\beta_q/\nu_q)$. From the known bulk critical exponents [14], ν_q and β_q, one obtains ω_3=13/18, ω_4=7/12,

$a_3=13/15$, and $a_4=7/8$. - It should be mentioned that the effect of the interfacial adsorption in Potts models on depinning [29] and roughening [30] transitions has been studied quite recently [31] which leads to interesting quantitative, but no qualitative changes compared with the behaviour of the spin-$\frac{1}{2}$ Ising model.

(ii) At the first order transition ($q>4$):
Preliminary Monte Carlo data for $q=20$ (where the correlation length is presumably small and the typical features at first order transitions might be observable) suggest, see Fig.11,

$$W(T_c,N) \sim N^a \text{ with } a=0.9\pm0.1;\ 10\leq N\leq 80$$

$$W(t_c) \sim t_c^{-\omega} \text{ with } \omega=1.1\pm0.1;\ 3\cdot10^{-2}\lesssim t_c \lesssim 2\cdot10^{-1};\ N=60$$

Note, however, that extremely long runs, at least a few 10^5 Monte Carlo steps per site, are necessary to obtain stable and apparently reliable data. But because both the dependence on t_c and N show some curvature as one approchaes T_c or increases N, the above exponents should be viewed with great care concerning the truely asymptotic behaviour. Unfortunately a really convincing physical argument on that behaviour is still lacking. Possibly a 1/q-expansion might clarify the problem.

(C) In the chiral clock model the adsorption at the 0:2 interface is known to diverge at zero temperature for $\Delta\geq\frac{1}{4}$ (complete wetting)

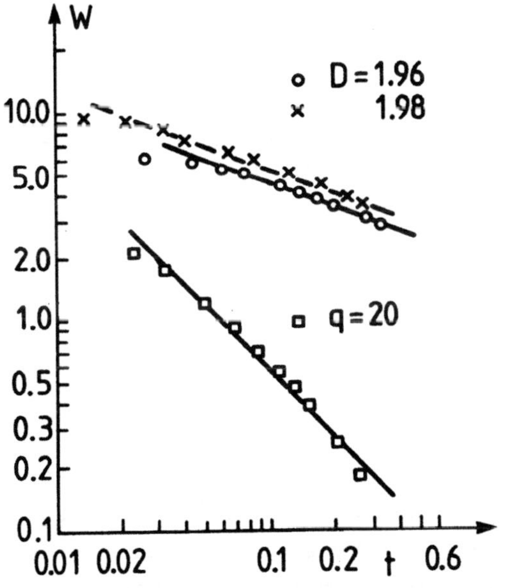

Fig. 11: To show the critical temperature dependence of the net adsorption at first order transitions for both the Blume-Capel model ($\overline{D}=1.96$ and 1.98) and the 20-state Potts model ($t=(T_c-T)/T_c$).

and is believed to diverge at a non-zero wetting temperature, T_W, below the bulk transition temperature, T_c, for $\Delta < \frac{1}{4}$ [9]. T_W should merge in T_c, as $\Delta \to 0$, i.e. for the pure Potts case. The net adsorption, W, has been argued to diverge linearly in $t_W=(T_W-T)/T_W$, as $t_W \to 0$ [9]. - Our Monte Carlo results confirm this picture. In particular, we observe that the wetting line

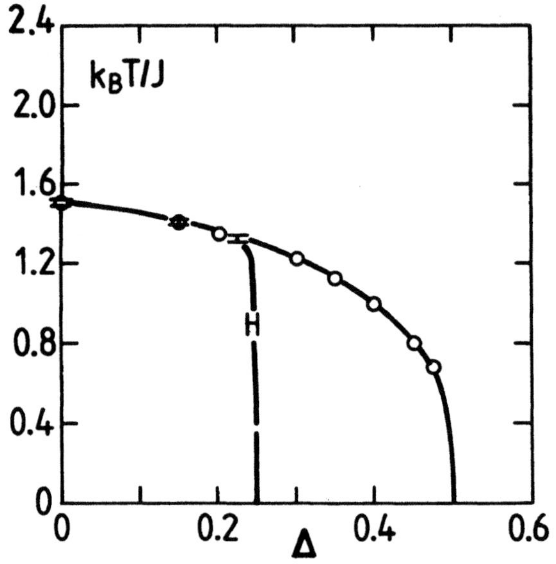

Fig. 12: Wetting line (⊢⊣) in the 3-state chiral clock model. The modulated or flaoting phase is not shown.

rises very sharply from the point ($\Delta=\frac{1}{4}$, T=0) and follows the melting line of the ferromagnetic phase very closely as Δ is decreased, see Fig.12. Indeed, it is very difficult to separate bulk and wetting transitions in a simulation.

4. SOS-type Approximation

In this section we review results on a SOS-type approximation to calculate interface free energies of two-dimensional three-state models. Our approach [4,7] modifies the method of Müller-Hartmann and Zittartz [13], orginally devised to map phase diagrams of two-state models [13,32]. - The interface is introduced into the three-state system by fixing the spins on opposite boundaries to take values α and β, respectively. To obtain good estimates for the phase boundaries those configurations where a layer of the non-boundary state, γ, is allowed to intervene at the interface must be taken into account in the calculation of the interface free energy [4]. The calculation is rendered tractable neglecting both overhangs at the interface [13] and islands of boundary states inside the intervening layer.

The notation we use to describe the interface is shown in Fig.13. Two integers are needed to specify the position of the interface in each column. $-\infty \leq m_i \leq \infty$ (i=1, ...N) denotes the position of the boundary of the "α" domain relative to a reference line in the middle of the system and $0 \leq n_i \leq \infty$ the width of the intervening "γ" layer.

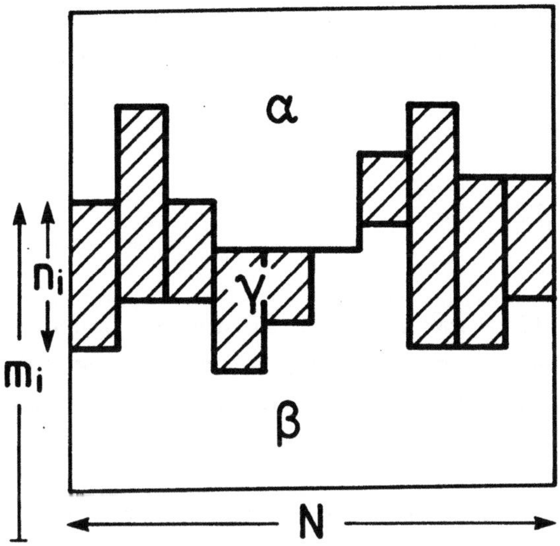

Fig. 13: Schematic diagram of interface region defining the notation used in our SOS-type approximation.

Then the free energy of the interface is defined in the usual way by

$$\sigma = -k_BT \lim_{N\to\infty} \frac{1}{N} \ln \sum_{\{m_i, n_i\}} \exp(-E/k_BT) \tag{7}$$

where E is the total energy of a given interface configuration which can be calculated easily. The summation over all configurations in equation (7) can be performed using the transfer matrix method. In the thermodynamic limit σ is obtained from the largest eigenvalue, λ_o,

$$\sigma = - k_B T \ln \lambda_o \qquad (8)$$

Usually the numerical calculation of λ_o is not difficult and has been done for the 3-state chiral clock model (hence also for the 3-state Potts model) [4] as well as the BC model and some other cross-sections of the Blume-Emery-Griffiths model [7]. In all cases the melting lines of the ferromagnetic phases could be determined quite accurately from the condition that σ vanishes at the transition temperature. In the three-state Potts model we found in addition that σ, as it follows from (7), (8), agrees very well with Monte Carlo results for the interface free energy at all temperatures (apart from the critical exponent which describes the vanishing of σ: the SOS-type approximation always leads to a linear dependence in $t_c=(T_c-T)/T_c$)) [4]. - Therefore we conclude that interface adsorption is indeed the essential effect to be included in a description of interfaces in multi-state models.

5. Chemisorbed Monolayers

Interfacial adsorption might occur at domain boundaries of gases adsorbed at surfaces. For example, He physisorbed on graphite is expected to be in the universality class of the two-dimensional three-state Potts model, and the three states are believed to correspond to the three degenerate $(\sqrt{3}\times\sqrt{3})$ overlayer structures [33]. Accordingly one might predict critical interfacial adsorption at the boundaries of such structures as one approaches the bulk transition temperature. (On wetting in physisorbed monolayers see also [34]).

In the following we shall analyse this intriguing problem in the context of chemisorbed gases which can be described by lattice gas models. The Hamiltonian can be written as

$$H = -(\varepsilon+\mu)\sum_i c_i - \sum_{i\neq j} \varphi_{ij}\, c_i c_j - \sum_{i\neq j\neq k} \varphi_t\, c_i c_j c_k \qquad (9)$$

where i,j, and k refer to the preferred adsorption sites forming a Bravais lattice; c_i is the occupation variable with $c_i=0$, if site i is empty, and $c_i=1$, if the site is occupied. The pair interactions between the adsorbed atoms or molecules, "adatoms", are denoted by φ_{ij};

three-body forces between "nearest triangles" of adatoms are given by φ_t. The binding energy gained, if an atom is adsorbed, is ε; μ is the chemical potential. Obviously one can transform Hamiltonian (9) into a spin-$\frac{1}{2}$ Ising model by introducing spin variables $S_i=1-2c_i$

$$H = - \sum_{i\neq j} J_{ij} S_i S_j - \sum_{i\neq j\neq k} J_t S_i S_j S_k - H \sum_i S_i \tag{10}$$

where the exchange constants J_{ij} and J_t can be easily expressed in terms of φ_{ij} and φ_t; the concrete transcription depends on the lattice structure. The field, H, is determined by ε, μ, φ_{ij} and φ_t.

Good candidates for displaying adsorption at domain boundaries are models which describe overlayer structures of, at least, three-fold degeneracy. A simple example is the (3×1)phase where the adsorption sites in two adjacent rows, perpendicular to the direction of modulation, say, the x-axis, are predominantly occupied (empty) while the sites in the following row are predominantly empty (occupied). Such phases have been observed experimentally, e.g., in H/Fe(110) [35] and O/Pd(110) [36]. Adequate lattice gas models have been proposed and analysed in detail for both systems [11,10]. These models show, indeed, the phenomenon of interfacial adsorption. Preliminary results for the Ising versions of the models will be presented here. A full account will be published elsewhere [12].

In the model for H/Fe(110) pair interactions between adatoms up to third nearest neighbours on a centered rectangular lattice (J_1, J_2, and J_3 in the notation for the Ising model) are considered; in addition three-body interactions are included. The phase diagram in the ($k_BT/|J_2|$, H) plane is shown in Fig. 14. The three degenerate (3×1) structures may be denoted by "A" (..++-..), "B" (..+-+..), and "C" (..-++..). To study adsorption at domain boundaries, such as A: C or B:A, one may map the spin-$\frac{1}{2}$ Ising model onto a multi-state model: In a given spin configuration spins on three successive lattice sites along the x-direction are combined and identified with one of eight possible states, A,B,C,..H. Hence the original system of size $N_1 \times N_2$, N_2 refers to the x-direction, is transformed into a new system of size $(N_1 \times (N_2/3))$. Then an interface can be introduced into the new 8-state system exactly in the same way as has been done in Section 2, i.e. by fixing the (new) variables on opposite boundaries in different states. Obviously there are six possibilities to study adsorption at domain boundaries in the (3×1) phase, A:B, A:C, B:C, C:B, C:A, and B:A.

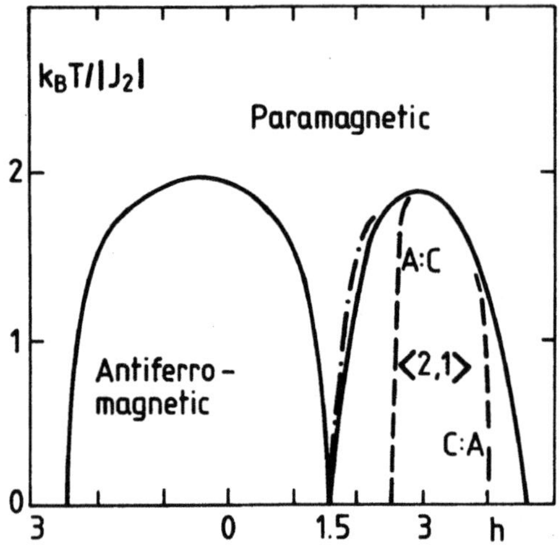

Fig. 14: Phase diagram of the Ising version of the model for H/Fe(110). The boundary of the floating phase (-.-) and the wetting lines are schematic. <2,1> corresponds to the (3×1) phase . - See Ref. [11].

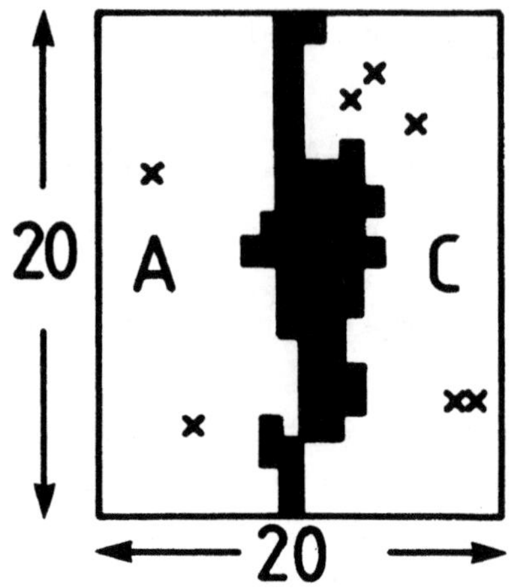

Fig. 15: Typical equilibrium configuration at $h=H/|J_2|=3$ and $k_BT/|J_2|=1.775$. Non-boundary states are shown blackened or as crosses (x). The 8-state model is used.

The interfaces A:C and C:A are of particular interest, because there the ground state energy can be lowered at fields beyond critical fields, $h_{A:C}=2.5$ and $h_{C:A}=4$, by insertion of a B layer of arbitrary thickness. Thus complete wetting occurs. In analogy to the situation in the three-state chiral clock model we expect wetting lines rising from $h_{A:C}$ and $h_{C:A}$ at zero temperature, see Fig. 14. At these lines the net adsorption, as defined in equation (1), in the 8-state model should diverge. Preliminary results of a Monte Carlo study [12] suggest that such lines can, indeed, be located as sketched in Fig. 14. For example, the net adsorption, W(T, h fixed) at the A:C interface shows a very pronounced maximum close to the melting temperature at h=3 ($N_1 \times N_2 \hat{=} 10\times30$ and 20×60), while there is no hint of critical adsorption at h=3.7. Vice versa, there is clearly no critical adsorption for small fields, e.g. h=3, at the C:A interface, in contrast to findings at larger fields, $h \lesssim h_{C:A}$. This suggests that the wetting lines rise up to the melting line of the (3×1) phase, but coincide with the melting line only in a small range (or merely at one point?). More definite results will be presented elsewhere [12]. So far our main conclusion is that interfacial adsorption does occur at domain boundaries in the (3×1) phase of this lattice gas model for H/Fe(110). In Fig. 15 a typical Monte Carlo equilibrium configuration with an A:C interface demonstrates this fact.

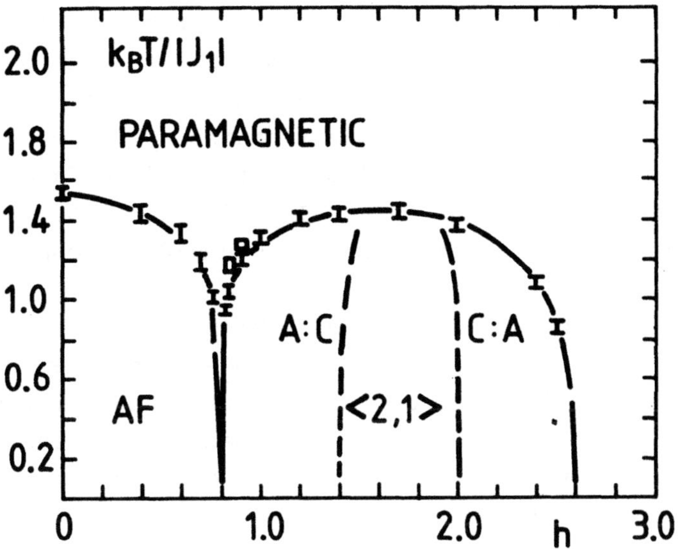

Fig. 16: Phase diagram of the two-dimensional ANNNI model in a field. <2,1> denotes the (3×1) phase, AF the (2×1) phase. The wetting lines, A:C and C:A, are shown schematically; the behaviour close to the melting line of the (3×1) phase has still to be clarified. - See Ref. [10].

The phase diagram of O/Pd(110) for coverages up to one half is similar to the one of H/Fe(110). The suggested lattice gas model is even simpler [36,10], because three-body interactions are neglected. Its Ising version is the two-dimensional ANNNI model in a field, a model with competing nearest and next nearest neighbour interactions along one axis [10,37]. Again, two wetting lines in the (3×1) phase are expected to describe critical adsorption at A:C and C:A interfaces, as sketched in Fig. 16. The lines may be located using, e.g., Monte Carlo techniques for the corresponding 8-state model.

6. Summary

In this article I have reviewed recent results on interfacial adsorption in various two-dimensional multi-state models. New critical properties have been found which can be understood by scaling and analytic arguments. The new critical exponents which describe the interfacial adsorption depend on the class of the bulk transition. They are related either to the bulk correlation length or specific interface features. - The occurence of non-boundary states at interfaces provides the clue to an approximate calculation of the interface free energy to map phase diagrams of 3-state models quite accurately as demonstrated for the chiral clock model and the Blume-Capel model.

In addition, preliminary results of a Monte Carlo study on interfacial adsorption in realistic lattice gas models for chemisorbed monolayers, H/Fe(110) and O/Pd(110), are presented. In particular, critical interfacial adsorption is expected at domain boundaries in (3×1) phases. It is hoped that these findings on a new interface effect may encourage experimental work in the future.

I wish to thank W. Pesch, J. Yeomans, D.A. Huse, D. Kroll, I. Sega, and K. Binder for useful cooperation on many of the results presented in this contribution.

References

1. Widom, B.: In: Phase transition and critical phenomena. Domb, C., Green, M.S. (eds.), Vol.II, p.79. New York: Academic Press 1972
2. Moldover, M., Cahn, J.: Science 207, 1073 (1980)
3. Cahn, J.W.: J. Chem. Phys. 66, 3667 (1977); Widom, B.: J. Chem. Phys. 68, 3878 (1978); de Gennes, P.G.: J. de Physique Lett. 42, 377 (1981)
4. Selke, W., Pesch, W.: Z. Physik B 47, 335 (1982)
5. Selke, W., Huse, D.A.: Z. Physik B 50, 113 (1983)
6. Selke, W.: Surf. Sci. (1984)
7. Selke, W., Yeomans, J.: J. Phys. A 16, 2789 (1983)
8. Kroll, D.M., Selke, W., Huse, D.A.: J. Phys. A (1984)
9. Huse, D.A., Fisher, M.E.: Phys. Rev. Lett. 49, 793 (1982); Huse, D.A., Szpilka, A.M., Fisher, M.E.: Physica A 121, 363 (1983); Huse, D.A., Fisher,M.E.: Phys. Rev. B 29, 239 (1984)

10. Rujan, P., Selke, W., Uimin, G.V.: Z. Physik B 53, 221 (1983)
11. Selke, W., Binder, K., Kinzel, W.: Surf. Sci. 125, 74 (1983); Kinzel, W., Selke, W., Binder, K.: Surf. Sci. 121, 13 (1982)
12. Sega, I., Selke, W., Binder, K.: in preparation
13. Müller-Hartmann, E., Zittartz, J.: Z. Physik B 27, 261 (1977)
14. Wu, F.Y.: Rev. Mod. Phys. 54, 235 (1982)
15. Blume, M., Emery, V.J., Griffiths, R.B.: Phys. Rev. A 4, 1071 (1971)
16. Capel, H.W.: Physica 32, 966 (1966); Blume, M.: Phys. Rev. 141, 517 (1966)
17. Huse, D.A.: Phys. Rev. B 24, 5180 (1981); Ostlund, S.: Phys. Rev. B 24, 398 (1981)
18. Selke, W., Yeomans, J.M.: Z. Physik B 46, 311 (1982); Duxbury, P.M., Yeomans, J., Beale, P.D.: J. Phys. A (1984)
19. von Gehlen, G., Rittenberg, V.: Nucl. Phys. B (1984); Howes, S.F.: Phys. Rev. B 27, 1762 (1983); Haldane, F.D.M., Bak, P., Bohr, T.: Phys. Rev. B 28, 2743 (1983); Schulz, H.J.: Phys. Rev. B 28, 2746 (1983)
20. Binder, K., (ed.): Monte Carlo Methods in Statistical Physics. Topics in Current Physics. Vol.7. Berlin, Heidelberg, New York: Springer 1979
21. Onsager, L.: Phys. Rev. 65, 117 (1944)
22. Fisher, M.E.: Rev. Mod. Phys. 46, 597 (1974); Lawrie, I.D., Saarbach, S.: In: Phase Transitions and Critical Phenomena. Domb, C., Lebowitz, J.L. (eds.), Vol. X. New York: Academic Press
23. Kroll, D.M., Lipowsky, R.: Phys. Rev. B 28, 6435 (1983)
24. Lajzerowicz, J.: Ferroelectr. 35, 219 (1981)
25. Lipowsky, R.: Phys. Rev. Lett. 49, 1575 (1982)
26. Lipowsky, R., Kroll, D.M., Zia, R.K.P.: Phys. Rev. B 27, 4499 (1983)
27. Abraham, D.B., Smith, E.R.: Phys. Rev. B 26, 1480 (1982); Huse, D.A.: private communication
28. Fisher, M.E.: J. Stat. Phys. (1984)
29. Abraham, D.B.: Phys. Rev. Lett. 44, 1165 (1980)
30. Weeks, J.D.: In: Ordering in Strongly Fluctuating Condensed Matter Systems. Riste, T. (ed.), p. 293, New York: Plenum 1980
31. Schmidt, I., Pesch, W.: to be published; on interfaces in Potts models see also: Schmittmann, B.: J. Phys. A 15, 3571 (1982)
32. For applications see: Burkhardt, T.W.: Z. Physik B 29, 181 (1978); Hornreich, R.M., Liebmann, R., Schuster, H.G., Selke, W.: Z. Physik B 35, 91 (1979); Doczi-Reger, J., Hemmer, P.C.: Physica A 109, 541 (1981); Kroemer, J., Pesch, W.: J. Phys. A 15, L 25 (1982); Slotte, P.: J. Phys. A 15, L 507 (1982); Temesvari, T.: J. Phys. A 15, L 625 (1982)
33. For a review see: Schick, M.: Prog. Surf. Sci. 11, 245 (1981)
34. Huse, D.A.: preprint
35. Imbihl, R., Behm, R.J., Christmann, K., Ertl, G., Matsushima, T.: Surf. Sci. 117, 257 (1982)
36. Ertl, G., Küppers, J.: Surf. Sci. 21, 61 (1970)
37. Fisher, M.E., Selke, W.: Phys. Rev. Lett. 44, 1502 (1980); for reviews see: Bak, P.: Rep. Prog. Phys. 45, 587 (1982); Fisher, M.E., Huse, D.A.: In: Melting, Localization, and Chaos. Kalia, R.K., Vashista, P. (eds.), p.259. Elsevier 1982; Selke, W.: In: Modulated Structure Materials. Tsakalakos, T. (ed.), Advanced Study Institute, Crete, 1983.

ELECTROMAGNETIC SCATTERING IN GRANULAR MEDIA

M. Ausloos

Institut de Physique B5,
Université de Liège
4000 Sart Tilman/Liège 1, Bëlgium.

I. INTRODUCTION

Heterogeneous materials are attracting much theoretical interest because they occur quite naturally and thus require per se scientific investigations, but also because of the growing importance taken by composite materials in technology. Several new theoretical studies have been performed on "random materials" in which e.g. one of the constituents takes random places in another. Most often one imagines particles embedded in a matrix. In the typical case of metallic particles in an insulator, such investigations lead to theories pertaining to percolation phenomena.[1] These allow better understanding of random (heterogeneous) media in describing structure and statistics of particle clusters.

Notice that important theoretical progress has been made in the percolation case. However except for describing the "static electrical conductivity" not much can be gained from such investigations. In particular, propagating electromagnetic waves and frequency dependent percolation are still studied from very approximate theories which we will recall. The same can be said of acoustic properties although work on acoustic waves in porous media have recently received much attention.[2]

The lack of elaborate theories, say beside mean field like approaches, is probably due to the lack of information on material structure. Explicit description and introduction of clustering effects in order to predict physical properties appears to be difficult theoretical work. We have attempted to provide some extension of such a limited approximation,[3] and will describe its positive aspects, and its actual shortcomings below. We will emphasize the need for a better description of clusters, hence for more exactly calculated dielectric function. In order to do so we will present a general theory based on the solution of Maxwell's equations allowing to calculate the optical absorption spectrum of aggregates.[4-7]

In section II, we will review the main theories for calculating optical properties, present their limits and show that they imperfectly reproduce experimental results. In section III, Maxwell's equations will be solved in the case of isolated clusters, -as an attempt to provide a step toward the obtention of electromagnetic properties of heterogeneous materials.

In section IV, we will show how to take into account polarization field statistics in order to calculate a dielectric function, and we will indicate further areas of

investigations.
We need to say that a complete bibliography is impossible here. Furthermore some simplification will be necessary in order to maintain the content of this contribution to an introductory level, -and also because more elaborate treatments have not yet been either presented or reduced to simple explanatory statements. The literature is abundant however. We hope that this introduction will attract those at all levels who are ready to work on a fascinating, opened, if not "too new" subject.

II. EFFECTIVE DIELECTRIC FUNCTION

An effective dielectric function (EDF) is usually introduced through two different theories. One is the Maxwell-Garnett theory (MGT)[8] (prefered for describing the optical properties of granular systems); the other is usually called the effective medium theory (EMT) (and is more often used in the study of percolation). The latter was proposed by Bruggeman[9] and generalized by Wood and Aschcroft[10]. We recall here that Gubernatis[11], has indicated that the MGT and the EMT respectively correspond to the average T matrix approximation (ATA) and to the coherent potential approximation (CPA) in the theory of disordered alloys. It is, in our opinion very important to stress that the usual MGT and EMT in fact have the same main defect that the usual ATA and CPA : they are all based on a single site approximation (SSA) and thus (totally or partially) neglect correslation between electrical or polarization fields as well as between scattering events. To go beyond the usual EMT meets difficulties. Several review articles or topical papers can be found in ref.12-14 where numerous cases are discussed. An interesting "unified EDF theory" by Sheng[15], althoug reproducing well some experimental results, also lacks the inclusion of correlation effects. Let us finally notice that formal theories of rigorous bounds for various properties of composites can be found in recent work of Bergman,[16] Milton,[17] McCoy[18] and Beran[19] where other references can be found.
From a practical point of view, it is only fair to mention that the comparison between theory and experimental data is usually as follows:transmission (or reflexion) spectra are compared to theoretical predictions without as said before much information on the material internal structure. This leaves many free parameters to play with. As we will show it, more systematic studies taking into account light wavelength, particle size distribution, particle concentration, light polarization may be in order nowadays.
Let us recall that it is possible to predict optical properties of an homogeneous medium from the knowledge of its complex index of refraction $\hat{n}=n-ik$, where n is the refraction index and k the extinction coefficient. For an isotropic material $\hat{n}$ is related to the complex dielectric function by $\hat{\varepsilon}=\hat{n}^2$. The question whether one can define an EDF for an heterogeneous material such that all optical properties be deduc-

ted from classical geometrical optics has a positive answer in the limit where the diffused light intensity is negligible with respect to that of the undiffused beam. Indeed one knows that the light diffused by a single particle decreases when the particle (linear) size to light wavelength ratio is small. Therefore theories leading to an EDF will mainly be valid when the latter ratio is small. Although we have in our work studied particles (in clusters) having such a ratio of the order of unity, we do not claim that we have found an EDF valid in such a case.

Let us first recall the basic differences between MGT and EMT; such theories are in fact restricted to a long wavelength approximation since they assume that the fields are uniform in each particle. One has thus to find a relation $\vec{D}_{eff}=\varepsilon_{eff}\vec{E}_{eff}$ such that from ε_{eff} one can define typical quantities.

The absorption coefficient $A(\omega)$ is related to the dielectric function by

$$A(\omega) = (\omega/c)\, n^{-1}(\omega)\ \mathrm{Im}\ \varepsilon_{eff}(\omega) \tag{1}$$

where ω is the light frequency and c is the speed of light. Similarly, we recall the definition of the reflectivity

$$R(\omega) = \frac{(n(\omega)-1)^2 + k^2(\omega)}{(n(\omega)+1)^2 + k^2(\omega)} \tag{2}$$

In fact, one has $A(\omega)=(2\omega/c)k(\omega)$.

The effective field is the volume average of uniform fields(inside $\vec{E}_i$,and outside $\vec{E}_o$ the particle)$\vec{E}_{eff}=f\vec{E}_i+(1-f)\vec{E}_o$, where f is the volume fraction of particles.

In MGT, the heterogeneous medium is supposed to be composed of independent (metallic) spherical inclusions surrounded by a (dielectric) matrix. The MGT also assumes that the average polarization field is locally equal to the concentration average of the constituents (and matrix) polarization field. The usual relation

$$\vec{P}(\vec{r}) = \left[\varepsilon_i(\vec{r},\omega)-1\right] \vec{E}(\vec{r}) \tag{3}$$

is also supposed to hold whatever constituent, matrix, or effective medium is at point $\vec{r}$. The assumption about spherical inclusions then allows one to write the field boundary conditions on the surface of an inclusion in terms of the applied field supposed to be uniform in the matrix (M) of dielectric constant ε_M, i.e. $E_o=E_M$ and $\varepsilon_o=\varepsilon_M$ lead to

$$\vec{E}(R_i) = (3\varepsilon_M/(\varepsilon_i+2\varepsilon_M))\vec{E}_o \tag{4}$$

where R_i is the radius of the inclusion; this supposes in fact a uniform polarization of the inclusion, or implies a long wavelength approximation $(kR_i \ll 1)$.
In so doing one obtains the Maxwell-Garnett average dielectric constant ε_{av} from

$$\frac{\varepsilon_{av}-\varepsilon_M}{\varepsilon_{av}+2\varepsilon_M} = \sum_i f_i \frac{\varepsilon_i-\varepsilon_M}{\varepsilon_i+2\varepsilon_M} \tag{5}$$

which we have written in the more general case of N types of inclusions i present with a concentration f_i in the sample; we have also $\Sigma f_i+f_M=1$ defining f_M as the percentage of matrix in the sample. Landauer has remarked however that the latter equation presupposes that the <u>same</u> effective field acts on all polarizable entities,[12] or in other words that there is no polarization variation in the sample. Notice that Eq.(4) has the form of the Clausius-Mosotti relation, i.e. ,

$$(\varepsilon-1)/(\varepsilon+2) = (4\pi/3)\,\Sigma\, N_i\alpha_i \tag{6}$$

where α_i and N_i are respectively the polarizability and the number density of the i-th species.
It is fair to recall that the MGT leads to infrared divergences, i.e. when

$$\varepsilon_i(\omega) + 2\varepsilon_M = 0 \tag{7}$$

at the Fröhlich modes[21] of independent particles of type i, which have a uniform polarization

$$\vec{\mu}_i = R_i^3 \left(\frac{\varepsilon_i\ \varepsilon_M}{\varepsilon_i+2\varepsilon_M}\right)\vec{E}_o \tag{8}$$

but the theory does not present any "percolation threshold".[12] For example, when only one type of inclusion (i=1) is present with concentration f,

$$\text{Max}\,[\text{Im}\varepsilon_{av}(\omega)] \simeq 8f(2+f)^{-2}\text{Im}\varepsilon_1(\omega) \tag{9}$$

Such a maximum occurs when $\text{Im}\varepsilon_1(\omega)$ is infinite, i.e, at $\omega=0$ when the inclusion is of metallic type, (or at $\omega=\omega_T$, where ω_T is the transverse optical phonon frequency,if the inclusion is a dielectric).
In the EMT, the distinction between matrix and inclusion disappears. All constituents are assumed to be spheres embedded in a continuous effective medium with probability f and (1-f). It suffices to replace ε_o by ε_{eff} and E_o by E_{eff} in the previous equation, and to consider two equations similar to eq.(4), one for the i-type of material, the other for the M-type. This leads to one of Bruggeman relations, e.g.

$$f\,\frac{\varepsilon_i-\varepsilon_{eff}}{\varepsilon_i+2\varepsilon_{eff}}+(1-f)\;\frac{\varepsilon_M-\varepsilon_{eff}}{\varepsilon_M+2\varepsilon_{eff}}=0 \tag{10}$$

which could be generalized to many constituents if one neglects fields variation in the sample.

At small concentrations, both theories give the same EDF which is only slightly different from that of the matrix. At larger concentrations ε_{eff} varies continuously and linearly with f in MGT, but quadratically with f in EMT. In the latter, one observes that ε_{eff} has a large variation at $f=1/3$ (in particular if $\varepsilon_i \gg \varepsilon_M$). This is interpreted as a percolation feature, -absent in MGT indeed.

Extensions of MGT and EMT have on one hand taken into account the shape of particles. Ellipsoidal particles are easily handled in both cases, -depolarisation factors being considered as free parameters in absence of better information on shape and size distribution.[23] Effects due to surface limited mean free path in metallic particles have also been considered in order to give an analytical form to ε different from the usual Drude formula.[24]

Describing as such the MGT and EMT apparently apply to different constituents. MGT is appropriate to systems conforming to the cermet geometry, where inclusions are 'well defined'. However, it is clear that roles of matrix and inclusions can be interchanged in EMT, -the EDF being symmetrical around $f=1/2$ indeed. It is also clear that predictions based on either theory will differ when the ratio $\varepsilon_i/\varepsilon_M$ becomes larger. As emphasized by Aschcroft,[22] 'a real composite system must ultimately be described statistically through the specifications of correlation functions that specify the microscopic arrangement of the constituents'. One expects clustering effects in many cases. A generalization of the above theories is to consider ε_M and ε_i as functions depending on the 'local fractional occupancy', and continuous redefinition of ε_{eff} at larger scales. This renormalisation group-like approach for continuous media does not seem to have been attempted yet. An approach taking a similar point of view has been recently given in ref.15.

Another effect susceptible to modify optical parameters is the existence of coating on particles. This introduces in fact two new parameters (thickness and nature of the coating). These effects have been mainly investigated by Kerker and coworkers.[25] Similarly electrical and magnetic higher order polar effects have been recognized important when particles are in contact.[6,26] Therefore in view of the above it is quite natural that much work has been devoted to study cases in which particles are not too far from each other and form chains or globular clusters.[27,28] In several of these cases quantitative improvements are even obtained.[29]

It is obvious that more realistic theories must be developed. We have proposed a continuum approach in which fields are obtained from the solution of Maxwell's equations and calculated in a multiple scattering like technique by matching boundary conditions on scattering objects and summing up all external fields to the sample in order to give measurable quantities.

III. MAXWELL'S EQUATIONS SOLUTION FOR AGGREGATED SPHERES

Mie light diffusion theory, as explicitly presented by Kerker,[30] allows one to calculate the light intensity scattered by an isolated particle. Mie has shown that this intensity is proportional to the square of an amplitude function $S(\vartheta,\varphi)$ which is a diagonal matrix for spherical particles with both elements $S_1(\vartheta)$ and $S_2(\vartheta)$ given by

$$S_1(\vartheta)= \sum_{n=1}^{\infty} \frac{2n+1}{n(n+1)} \{A_n\alpha_n(\cos\vartheta) + B_n\beta_n(\cos\vartheta)\} \tag{11}$$

$$S_2(\vartheta)= \sum_{n=1}^{\infty} \frac{2n+1}{n(n+1)} \{B_n\alpha_n(\cos\vartheta) + A_n\beta_n(\cos\vartheta)\} \tag{12}$$

with $\alpha_n(\cos\vartheta)=P^1{}_n(\cos\vartheta)/\sin\vartheta$, and $\beta_n(\cos\vartheta)=(d/d\vartheta)P^1{}_n(\cos\vartheta)$, where $P^1{}_n(\cos\vartheta)$ is the Legendre polynomial of order n. The diffusion extinction coefficient is

$$K_{diff} = N\pi(a/\alpha_o)^2 \sum_{n=1}^{\infty} (2n+1)\{|A_n|^2+|B_n|^2\} \tag{13}$$

while the total extinction coefficient is

$$K_{abs} = N\pi(a/\alpha_o)^2 \sum_{n=1}^{\infty} (2n+1)\mathrm{Re}(A_n+B_n) \tag{14}$$

for N identical particles of radius a, with $\alpha_o=2\pi n_M a/\lambda_o$ where λ_o is the wavelength of the light in vacuum and n_M the matrix refraction coefficient. A_n and B_n are obtained by writing the boundary conditions on the surface of the particle. The case of metallic particles characterized by a dielectric constant including plasmon effects can be found in ref.31 which contains more detailed calculation.

Rayleigh and Rayleigh-Gans theories [32] are more outdated and particular cases of Mie theory. Rayleigh theory applies to homogeneous spherical particles when $a<<\lambda$, but also if $|\hat{n}|a<\lambda$, which implies that the polarization field responds instantaneously. R-G theory differs from the preceding one by the introduction of a phase factor to describe that elements of the particle diffuse independently. Furthermore one supposes that particles are weakly absorbing, i.e., $|\hat{n}-1|<<1$, and $|\hat{n}-1|a<<\lambda$. When the dispersion of particles in the medium is large, such theories are useful for defining an EDF. In the following paragraphs, we will describe a method to obtain aggregate spectra in absence of strong theoretical requirements, e.g. large dispersion.

The magnetic field $\vec{H}$ and the electric field $\vec{E}$ are expanded in terms of spherical wave

vector functions,[32,33] inside each spherical particle and in the matrix. The incident field is also expanded in the same way. Such expansions contain Bessel functions which have been chosen with the required asymptotic behavior at infinity or at 'zero', i.e. at the center of each particle.
The complex wave number k is given by the dispersion relation

$$k^2 = \frac{\omega^2}{c^2} \; \varepsilon^T(k,\omega)\mu(k,\omega) \tag{15}$$

for transversal waves, where $\varepsilon^T(k,\omega)$ is the transverse dielectric constant and $\mu(k,\omega)$ the magnetic permeability of the appropriate medium. For longitudinal waves, we have

$$\varepsilon^L(k,\omega) = 0 \tag{16}$$

where $\varepsilon^L(k,\omega)$ is the longitudinal dielectric constant. Solutions of the above equations are chosen such that k has a positive imaginary part. Furthermore, we consider that there is no longitudinal plasmon in the matrix (this supposition leads to a divergenceless electric field).
Usual boundary conditions are then applied on each particle surface.[31] In the matrix, the solution of Maxwell equations is taken as the sum of the diffracted fields and the incident wave ($\vec{E}_o$, $\vec{H}_o$). Stratton[32] has shown how to relate spherical vector functions definedin different reference frames, e.g. with origin at the center of sphere i and j, by using a projection technique.
Finally, the field expansion coefficients are given as solution of a system of linear equations where the unknown $a^M_{\ell m}(j)$, $b^M_{\ell m}(j)$ are mere generalizations of A_n and B_n in Eq.(11). The ℓ, m, p, q indices pertain to usual spherical harmonics notations.
The suffix I refers to the incident wave coefficients while the following susceptibilities Δ_q and Γ_q are introduced in terms of material characteristics for a particular sphere.

$$\Delta_q = - \frac{j_q(kR)}{h^1_q(kR)} \cdot \frac{\varepsilon[j^+_q(k^T_j R)-f_q(k^L_j R)]-\varepsilon^T_j[j^+_q(kR)-f_q(k^L_j R)]}{\varepsilon[j^+_q(k^T_j R)-f_q(k^L_j R)]-\varepsilon^T_j[h^{1+}_q(kR)-f_q(k^L_j R)]} \tag{17}$$

$$\Gamma_q = - \frac{j_q(kR)}{h^1_q(kR)} \cdot \frac{\mu j^+_q(k^T_j R) - \mu_j \, j^+_q(kR)}{\mu j^+_q(k^T_j R) - \mu_j \, h^{1+}_q(kR)} \tag{18}$$

where j_q is the appropriate Bessel function and each function $g^+(z)$ is defined by

$$g^+(z) = \frac{1}{g(z)} \frac{d}{dz} [zg(z)] \tag{19}$$

while the function $f_q(z)$ is defined by

$$f_q(z) = q(q+1)j_q(z) / \left[z \frac{d}{dz} j_q(z)\right] \tag{20}$$

where R is the radius of the spheres, k_j^L, ε_j^L, ε_j^T are respectively the longitudinal wave number, longitudinal dielectric constant and transverse dielectric constant inside the sphere j.

When there is no longitudinal plasmon inside the sphere, the imaginary part of k_j^L tends to infinity, and $f_q(k_j^L R)$ vanishes. In this case, the electrical susceptibilities Δ_q are similar to the magnetic susceptibilities Γ_q with the interchange $\varepsilon \leftrightarrow \mu$.

Notice the important result that the magnetic susceptibilities are not influenced by the longitudinal plasmons. One has thus to solve the system :

$$b_{\ell m}^M(j) - \Gamma_\ell(j) \sum_{q,p} \sum_{i \neq j} [b_{qp}^M(i)\, \mathcal{J}_{qp\ell m}(i,j) + a_{qp}^M(i)\, \mathcal{C}_{qp\ell m}(i,j)] = \Gamma_\ell(j) b_{\ell m}^I(j) \tag{21a}$$

$$a_{\ell m}^M(j) - \Delta_\ell(j) \sum_{q,p} \sum_{i \neq j} [b_{qp}^M(i)\, \mathcal{C}_{qp\ell m}(i,j) + a_{qp}^M(i)\, \mathcal{J}_{qp\ell m}(i,j)] = \Delta_\ell(j) a_{\ell m}^I(j) \tag{21b}$$

In those equations, the summations over $q=[1,\infty]$ and over $p=[-q,+q]$ may be truncated in order to obtain some approximation to the fields. The above system explicitly includes all electromagnetic interactions $\mathcal{C}_{qp\ell m}(i,j)$ between particles. If these coupling terms do not exist, one obtains two distinct sets describing the electrical modes, and independently the magnetic modes.

It is possible to reformulate the solution of the system (21) into the inversion of two matrices with a dimension reduced by a factor two. Let $\underline{\underline{\Delta}}^{-1}$ and $\underline{\underline{\Gamma}}^{-1}$ be the diagonal matrices formed by the inverse of the susceptibilities, while $\underline{\underline{\mathcal{C}}}$ and $\underline{\underline{\mathcal{J}}}$ are formed from the $\mathcal{C}_{qp\ell m}(i,j)$ and $\mathcal{J}_{qp\ell m}(i,j)$. The $b_{\ell m}^M(j)$, $a_{\ell m}^M(j)$, $a_{\ell m}^I(j)$ and $b_{\ell m}^I(j)$ are defined as elements of the vectors b^M, a^M, a^I and b^I respectively. Then the system (21) may be concisely written as

$$(\underline{\underline{\Delta}}^{-1} - \underline{\underline{\mathcal{J}}}) \cdot \underline{a}^M - \underline{\underline{\mathcal{C}}} . \underline{b}^M = \underline{b}^I \tag{22a}$$

$$(\underline{\underline{\Gamma}}^{-1} - \underline{\underline{\mathcal{J}}}) \cdot \underline{b}^M - \underline{\underline{\mathcal{C}}} . \underline{a}^M = \underline{a}^I \tag{22b}$$

Limiting the sum in Eq.(21) to the term q=s, the solution of the problem then leads to invert two matrices of dimension N.s(s+2) and to make a few matrix products. Analytical expressions of the interaction terms for any kind of aggregate are not recalled here, but are presented in ref.4.

Numerical results have been presented in ref.4-7 for different types of clusters and of particles, sometimes taking the long wavelength limit in order to shorten the calculation, but including the quadrupolar, octupolar terms in most cases. Figures can be found in such publications. It is interesting to observe the necessity of including all effects in order to observe spectrum variations. Systematic trends are seen when a and are varied. Most importantly appears the conclusion that an 'isolated particle approximation' is rarely valid.[4,7] Furthermore, high order polar effects play an important role,[6] even in the long wavelength limit,[5] in particular (as expected) in dense clusters. Somewhat appealing but not used experimentally up to now is the information which can be obtained on cluster shape, size and distribution when polarized light is used. However more work has to be done in such direction to sort out specific aspects. It is known that metallic particles stick together to form needles. The observation[4,7] that depolarization effects are reduced for fields parallel to the axis of chain-like clusters is of experimental interest indeed. Of course some improvement would arise if ellipsoids rather than spheres were theoretically studied as above.

IV. EFFECTIVE DIELECTRIC FUNCTION

After having solved Maxwell's equations "exactly", one has to define an EDF in terms of an average over all field configurations. This requires the non trivial task of obtaining the probability function for the fields in a random medium. A second average must be made, but is generally omitted, i.e., because the particles inserted in the composite have irregular sizes, one has to average over their size distribution as well. This will also be done here; such a distribution is the log-normal distribution,[34]

$$f(R) = \frac{1}{\sqrt{2\pi}}\frac{1}{R\sigma}\exp[-\ \frac{1}{2}\ (\frac{\ell n(R/R_o)}{\sigma})^2]$$

where σ is a measure of the width of the distribution while R_o is a radius of particles (thus supposed to be spherical) such that 50% of them have a radius R less than R_o. Typical values of $\sigma_{LN} \equiv e^{\sigma}$ are between 1.1 and 1.5 and R_o is of the order of 500-1000 nm.

Beside the spherical hypothesis of particles (H0), three additional hypotheses are needed in order to render our treatment as analytical as possible : (H1). We suppose that due to the randomness in the distribution of particles the system can be described as an average medium in which the electrical field (hence the polarization field) is weakly varying from site to site ('site' refering to the position of a particle). The 'small parameters' are thus the gradient of material characteristics:(H2). The position of the particles in space is supposed tobe normally distributed, (as in the Gaussian broadening model of Fuchs)[35], i.e. a set of arbitrary samples from the system would show that the number of particles per volume obeys a gaussian law. This corresponds to assuming that the pair distribution function P(r) is a superposition of chi-square functions [36] (H3). The topology of each a priori large clusters of spheres is arbitrary but they are considered to be embedded in a large sphere (of radius a) outside of which the MGT applies, hence outside of which the EDF has the MGT form.

Therefore (H1) and (H3) indicate that our theory will interpolate between the EMT and the MGT by taking into account clustering effects, while (H2) is primarily needed below to obtain an analytical form for the field distribution function. Let us also discuss only the case of dipolar fields in absence of retardation effects.

As discussed above, the various fields outside and inside each spherical particle and outside the large sphere are expanded in terms of spherical wave vector functions. After applying the usual boundary conditions at the surface of each sphere the coefficients $b^M_{nm}(j)$ describing the fields outside the particle (j) are solutions of a single linear matricial equation (21b) rewritten here

$$a_{nm}(j) - \Delta_n(j) \sum_{i\neq j} \sum_{q,p} a_{qp}(i) < X_{qp}(i) | H_{nm}(j) > = d_{nm}\, \Delta_n(j) \tag{23}$$

where for a homogeneous spherical particle of radius R and in this long wavelength approximation,

$$\Delta_n(j) = \frac{n(\varepsilon_M - \varepsilon_j)}{n\varepsilon_j + (n+1)\varepsilon_M} R^{2n+1} \tag{24}$$

The coefficient d_{nm} describes the external field, while to obtain the interaction matrix elements $\langle X(i)|H(j)\rangle$ between the N spheres we refer to ref.4.

Here we neglect all terms with n or q>1. This can be done if the density of particles is not too high (hence when MGT usually applies). Mathematically, (H1) means that we can write $a_{1m}(j)/\Delta_1(j)=A_m$, i.e. A_m is a constant for all j ; this is obviously true in the absence of interactions between particles (or for spheres regularly distributed on a lattice). Henceforth, (H1) implies that no density gradient exists in the medium. Defining

$$G_{1m}(j) \equiv \sum_{i \neq j} \Delta_1(i) \langle X_{1p}(i) | H_{1m}(j) \rangle \tag{25}$$

and supposing that the external field is uniformly applied in the z direction between the parallel plates of an infinite condensor we are led to conserve only the m=0 term in the above equations. Observe that $d_{10}=\sqrt{4\pi/3}$. The solution of Eq.(23) is obviously (dropping here the index j)

$$a_{10} = -d_{10}\,\Delta_1(1+G_{oo})^{-1} \tag{26}$$

since $G_{pm}=0$ if $p \neq m$.

An algebraic calculation of G_{oo} allows one to recover the MGT and, hence, the Clausius-Mossoti relation for the EDF. We define the "field fluctuation" G as $G=G_{oo}-G_{MG}$, and search for the field distribution function of G in order to average a_{10} in Eq.(26). Because G is in general a complex function, one must look for a two variable distribution function W(G',G") with G≡G'+iG", in which the terms G' and G" are not statistically independent. By the method of characteristic function and after some algebra described in ref.3 we have found

$$W(G',G'') = \frac{1}{2\pi\sigma'\sigma''}\,\frac{1}{\sqrt{1-\rho^2}}\,\cdot$$

$$\cdot\exp\left[-\frac{1}{2}\,\frac{(G'/\sigma')^2+(G''/\sigma'')\ -2\ G'G''/\sigma'\sigma''}{(1-\rho^2)}\right] \tag{27}$$

with $(\sigma')^2=(4/5)(4\pi/3)\lambda\langle\Delta'^2_1/r^3_m\rangle_R$ and similarly for σ", while $\rho=\langle\Delta'_1\Delta''_1\rangle_R/(\langle\Delta'^2_1\rangle_R\langle\Delta''^2_1\rangle_R)$ From Eq.(11) one can obtain $\langle\langle a_{10}\rangle_R\rangle$. A typical simplification occurs when $\rho=\pm 1$, which corresponds (without proof here) in fact to the case of homogeneous spheres. (Δ'_1 is then proportional to Δ''_1). In such a case, our solution is formally similar to that of Fuchs[35] when f(R) is a delta function.

We define an EDF from the expression of the electrostatic energy stored (or lost) U rather than the fields[37]. The variation of U due to the presence of spheres is from a 'microscopic' view point

$$\Delta U = \frac{1}{2}\varepsilon_H \sum_{j=1}^{N} \int_{v_j} (1-\varepsilon^*_j/\varepsilon_M)\ \vec{E}_o\cdot\vec{E}^*_j\ dv \tag{28}$$

but is also given by

$$\Delta U = \frac{1}{2}\,\varepsilon_M(1-\varepsilon_M/\varepsilon_{eff})|\vec{E}_o|^2 \tag{29}$$

which thus defines ϵ_{eff}. The latter equation is so written because it is obviously imposed that the sources of the $\vec{D}$ field remain constant both in the 'effective' and 'microscopic' cases.(We also suppose that ϵ_M is real). Hence,

$$\epsilon_{eff} = \epsilon_M (1+ \sqrt{12\pi}\ \lambda \langle\langle a_{10} \rangle_R \rangle)^{-1} \tag{30}$$

in which $\lambda=(3/4\pi) f r_m^{-3}$, and where r_m is a cut-off radius.[35]
We have already discussed elsewhere [3] physical quantities as the reflectivity for metallic particles in a dielectric (vacuum) matrix. Taking into account field correlation even in the case of equal size particles has been seen to modify $R(\omega,f)$ for $f\neq 0$. It is also obvious that our theory leads to physical quantities bounded by EMT and MGT results. The reflectivity as a function of frequency for a (large) filling factor, just below the close packed limit, is shown on Fig.1, together with MGT and EMT results. The Mie peak at $\omega=\omega_p$ is of course absent in MGT and in our case. More important is the shift of the reflectivity peak toward lower frequency. This is apparently in better agreement with experimental results as those of Kreibig et al.[38]
On Fig.2, we show the real imaginary part of the EDF as a function of frequency for two filling factor cases and compare them to the MGT result. The dash line at $\omega=\omega_s$ correspond to the single particle Fröhlich mode.[21]

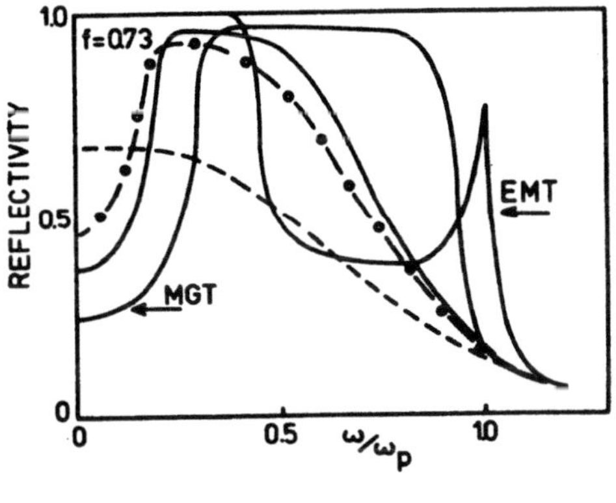

Fig.1 : Frequency dependence of the reflectivity for a metal insulator composite following usual approximations MGT, EMT and this theory at a metal volume fraction 73%. ——: $\sigma_{LN}=1.0$; $r_m=83.2$; — o — : $\sigma_{LN}=1.2$, $r_m=87.4$; ---: $\sigma_{LN}=1.4$, $r_m=98.6$.

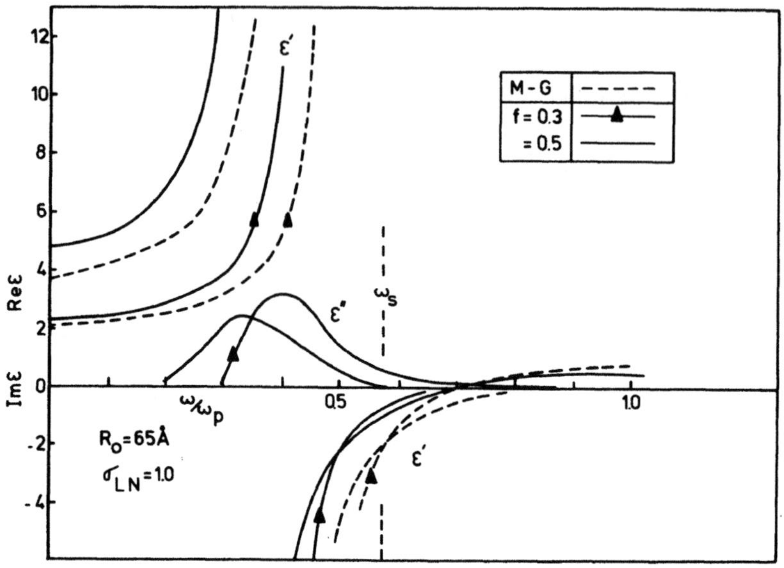

Fig.2: Real and imaginary part of the effective dielectric constant of a composite medium of metallic spherical particles in a dielectric as a function of the reduced frequency $\omega/\omega_{p,0}$ and for different filling factors f, when all the particles have a radius R_0=65 Å, as compared to MGT results.

References

1. A.Coniglio, in Magnetic Phase Transitions, Springer Series in Solid State Sc. vol.48, ed. by M.Ausloos and R.J.Elliott (Springer, Heidelberg, 1983)p.195
2. D.L.Johnson in Excitations in Disordered Systems, ed. by M.F.Thorpe (Plenum Press, New York, 1982) p.661.
3. J.M.Gérardy and M.Ausloos, Phys. Rev. B 26, 4703 (1982).
4. J.M.Gérardy and M.Ausloos, Phys. Rev B 25, 4204 (1982)
5. J.M.Gérardy and M.Ausloos, Phys. Rev. B 22, 4950 (1980)
6. J.M.Gérardy and M.Ausloos, Surf. Sc. 106, 319 (1981)
7. J.M.Gérardy and M.Ausloos, Phys. Rev. B 27, 6446 (1983)
8. J.C.Maxwell-Garnett, Phil. Tr. R.Soc.Lond. 203, 385 (1904)
9. D.A.G.Bruggeman, Ann.Phys.(Leipzig) 24, 636 (1935)
10. D.M.Wood and N.W.Aschcroft, Phil. Mag. 35, 269 (1977)
11. J.E.Gubernatis, in ref.12, p.84
12. Electrical Transport and Optical Properties of Inhomogeneous Media, ed. by J.C. Garland and D.B.Tanner, A.I.P. Proceedings 40, (1978); see e.g. R.Landauer, p.2
13. Macroscopic Properties of Disordered Media, Lectures notes in Physics vol.154 ed. by R.Burridge, S.Childress and G.Papanicolaou , (Springer, Heidelberg, 1982).
14. P.N.Sen and D.L.Johnson, Phys. Rev. B 27, 3133 (1983)
15. P.Sheng, Phys. Rev. Lett. 45, 60 (1980)
16. D.Bergman, Phys. Rev. Lett. 44, 1285 (1980)
17. G.W.Milton, Appl. Sc. A 26, 125 (1981)
18. J.J.McCoy in Mechanics Today, vol.6, ed. by Nemat-Nasser (Pergamon, Oxford, 1981) p.1
19. M.Beran in Composite Materials, vol.2, ed. by G.Sendeckyj (Academic Press, New York, 1974) p.209
20. C.J.F. Böttcher, Theory of Electric Polarisation (Elsevier, Amsterdam 1952)
21. H.Fröhlich, Theory of Dielectrics (Oxford Univ. Press, Oxford, 1948)
22. N.W.Aschcroft, in ref.13, p.1
23. C.G.Granqvist and O.Hunderi, Phys. Rev. B 18, 2897 (1978)
24. P.O'Neill and A.Ignatiev, Phys. Rev. B 18, 6540 (1978)
25. D.S.Wang and M.Kerker, Phys. Rev. B 25, 2433 (1982) and ref. therein
26 G.Bosi, F.E. Girouard and V.V.Truong, J.Appl. Phys. 53, 648 (1981)
27. P.Clippe, R.Evrard and A.A.Lucas, Phys. Rev. B 14, 1715 (1976)
28. C.G.Granqvist and O.Hunderi, Phys. Rev. B 16, 1353 (1977)
29. P.Chylek, D.Borie and R.G.Pinnick, Phys. Rev. B 27, 5107 (1983)
30. M.Kerker, The scattering of light (Ac. Press, New York, 1969)
31. R.Ruppin, Phys. Rev. B 11, 2871 (1975)
32. J.Stratton, Electromagnetic Theory (Mc Graw Hill, New York, 1941)
33. P.M.Morse and H.Feshbach, Methods of Theoretical Physics (McGraw Hill, New York, 1953)
34. C.G.Granqvist and R.A.Buhrman, J.Appl. Phys. 47, 2200 (1976)
35. R.Fuchs, in ref.12, p.276
36. M.G.Kendall and P.A.P. Moran, Geometrical Probability (Ch. Griffin, London, 1963)
37. see D.Bergman, in ref.12, p.46
38. U.Kreibig, A.Althoff and H.Pressmann, Surf. Sc. 106, 308 (1981)

V. LOCALISATION

THE ANDERSON TRANSITION AND THE ELECTRON-ELECTRON INTERACTION

C.Castellani
Istituto di Fisica, Università dell'Aquila - Italy
Gruppo Nazionale di Struttura della Materia del C.N.R.

C.Di Castro
Dipartimento di Fisica, Università "La Sapienza"
Roma,Italy
Gruppo Nazionale di Struttura della Materia del C.N.R.

It is a well established fact that disordered electronic systems undergo a metal-insulator transition by increasing the amount of disorder. However no theory is still available which accounts for all the experimental features.[1,2]

An historical survey of the main ideas involved in the choice of the "effective couplings" for this problem is presented. One obtains in this way a good account of the conceptual and technical difficulties behind a field theoretic treatment of the localization of the electrons in the presence of interaction. Since most of the content of these lectures has been published as one of the contributions to the LXXXIX Varenna Summer School on "Highlights of Condensed Matter Theory", we present here only a short summary of the most recent achievements.

In the absence of interaction among the electrons the Thouless argument based on the sensitivity to the boundary conditions distinguishes between the insulating localized phase and the metallic phase. The relevant energy scales lead to select "the conductance" g as "the invariant charge" of the renormalization group for this problem. The phenomenological scaling theory[5] of the Anderson localization transition has been built by interpolating between the behaviour of complitely localized and complitely delocalized electrons. The perturbative (in 1/g) quantum corrections to the metallic weakly disordered system with logarithmic singularities in 2-dimension were obtained by resuming the so called maximally crossed diagrams.[16,17] This correspond to a ladder summation of impurity scattering line in the "particle-particle" channel, thus leading to a scattering amplitude with a diffusion pole named Cooperon. The corresponding summation in the particle-hole channel has the same diffusion pole and is named diffuson. If intervines in the density-density response function and at higher order in 1/g in the conductivity.

A renormalization procedure to take care of these singularities was obtained by mapping the Anderson problem into an effective hamiltonian corresponding to a non-linear σ model.[8-14] Alternatively[20] gauge invariance has been used to reduce the multiplicative renormalization scheme of the singularities of the conductivity and of the density-density response function due to Cooperons and diffusons to a one renormalization parameter theory, thus leading to the well known one parameter scaling behaviour. The $2+\varepsilon$ expansion for the critical indices leads[8,9,11,12,13] to the "critical exponent $\mu = 1+0(\varepsilon^3)$ for the conductivity $\sigma \sim (n-n_c)^{\mu}$ (orthogonal case). When time reversal invariance is broken due to the presence either of magnetic impurities or of the magnetic field, the Cooperons acquire a mass and remain finite in the limit of small frequency and small wave vector. The diffusion localization is obtained in this case (unitary case) with a critical index $\mu = 1/2$. In the presence of spin-orbit coupling (symplectic case) an antilocalization term appears in the expression for the conductivity. In all these cases the single particle density of states remains finite.[15]

Despite these achievements, the theory based merely on disorder cannot account for the experimental values of the critical indices. The experiments seem to indicate two different sets of critical indices for doped semiconductors[1] (like Si:P) and for amorphous system[2] (like Nb_x Si_{1-x}). The critical index for the conductivity has a value near to 0.5 and unity respectively. In the last case the index value is the same as the one for pure localization but the single particle density of states goes to zero at the Fermi surface. There are also additional strong evidences[1] of the importance of the interaction among the electrons like the change in sign of the coefficient of the temperature dependence of the conductivity.

The interaction among the electrons has been considered[27] to introduce in addition to the "conductance" only one effective coupling. This is obtained by rescaling out the high value of the wave vector q in the screened Coulomb interaction made dimensioless by means of the density of states. A phenomenological two-parameter scaling theory then follows, which is associated to a new fixed point, the old Anderson fixed point becoming unstable.

Short-comings of this approach have been discussed in several papers.[30,31,32] We follow here the procedure of ref.(31). Gauge invariance is a good guiding line for checking selections of diagrams at any order in perturbation theory and provides useful global conditions which have to be satisfied by various physical quantities. A direct evaluation in 2-dimension and at T=0 of the density-density response function at first order in 1/g and in the interaction shows that, contrary to the assumption of the phenomenological scaling theory, the density of states appearing in the screening and in the Einstein relation (connecting the conductivity to the diffusion constant) remains finite. It is only the single particle density of states which shows a singular behaviour as it was originally shown in ref.(22)(23)(24). Moreover calculations of the single particle density of states at second order in 1/g show that at least two additional independent effective couplings (and not only one) are generated by the interaction among the electrons, one being associated to small values of the transfer momentum q and the other one to high values of q. The way they appear, indicates that the system flows towards a strong coupling regime. This invalidates all the previously proposed scaling theories based on perturbative analysis.

At a given order in 1/g, infinite resummations in the interactions have to be carried out. Finkelstein[32] semplifies the problem by suppressing the singularities due to the Cooperons via the presence of a small magnetic field, leaving the diffusons as the only effective propagator of the theory. He was then able to provide an effective hamiltonian in terms of three effective couplings as suggested above. Its renormalization leads to a strong coupling behaviour in terms of the scattering amplitudes with no metal-insulator transition. No transition is reached because after an initial decrease of the conductivity along the flow lines, it starts to increase because the Hartree term dominate, changing the sign of the β-function before the fixed point is reached. Since the Zeeman splitting on the electron spin states caused by the magnetic field is not taken into account, this behaviour is valid only in a narrow region of temperature and has never been observed experimentally.

The Finkel'stein theory has been reexamined and extended within the context of perturbation theory.(39) As Altshuler and Aronov(34) pointed out, Finkel'stein approach is better described in terms of singlet and triplet scattering amplitudes. If we suppress part of the effect due to the Hartree terms by suppressing the triplet channel via a spin flip scattering term, then a metal-insulator transition is recovered in $2+\varepsilon$ dimension. In the group equation for the conductivity only a universal term due to the long range part of the interaction is in fact present. This would lead to a critical index μ equal to unity (to order ε) as for the case of pure disorder. The single particle density of states goes to zero in a universal although complicated way.(38,39) Pure diffusion localization which is obtained in the presence of spin flip scattering would lead to an index μ equal to 1/2. By a proper choice of the coefficients of these two terms due to the interaction and to the diffusion localization respectively when combined in the group equation for the conductivity, an explanation of the Si:P experiments has been attempted.(37)

When the interaction among the electrons is short range,(39) once again no metal-insulator transition appears if both singlet and triplet scattering amplitudes are considered, whereas the diffusion localization with an index $\mu=1/2$ is stable against the introduction of the singlet interaction term. In this last case, the single particle density of states remains finite, in contrast with the results for long range forces.

Taking into account the full effect of the magnetic field by suppressing the $S_z = \pm 1$ channel of the diffusons one obtains(38,39) the exponent $\mu=1$ for short and long range forces. The single particle density of states vanishes in both cases although in a non universal way for short range forces.

The above theoretical picture, which is arising since the last few months, shows that the interaction among the electrons cannot be simply considered to introduce corrections to the Anderson transition. It actually introduces new features and many different universality classes have already been discussed when the localization due to pure disorder is suppressed. Crossed experiments on various physical quantities and further theoretical investigations are required before a clear picture of the metal-insulator transition in disordered medium can be obtained.

The following list of references is given only as a help to the members of the school to follow our presentation of the topics and is not intended to be complete.

References

I. Experiments

1. G.A.Thomas, M.Paalanen and T.F.Rosenbaum, Phys.Rev.B27 (1983) 3897

T.F.Rosenbaum, R.F.Milligan, P.A.Paalanen, G.A.Thomas and R.N.Bhatt, Phys.Rev.B27 (1983) 7509

2. G.Hertel, D.J.Bishop, E.G.Spencer, J.M.Rowell and R.C.Dynes, Phys.Rev.Lett. 50 (1983) 743.

II. Localization

a) Anderson model

3. P.W.Anderson, Phys.Rev. 109 (1958) 1492

b) Phenomenological scaling theory of the localization transition

4. F.J.Wegner, Z.Phys. B25 (1976) 327
5. E.Abrahams, P.W.Anderson, D.C.Licciardello and T.V.Ramakrishnan, Phys.Rev.Lett. 42 (1979) 673

c) Disordered system with n orbitals per site

6. F.J.Wegner, Phys.Rev. B19 (1979) 783
7. R.Oppermann and F.J.Wegner, Z.Phys. B34 (1979) 327

d) Localization and non-linear σ model

8. F.J.Wegner, Z.Phys. B35 (1979) 207
9. S.Hikami, A.I.Larkin and Y.Nagaoka, Prog.Theor.Phys. 63 (1980) 707
10. L.Schafer and F.J.Wegner, Z.Phys. B38 (1980) 113
11. K.B.Efetov, A.I.Larkin and D.E.Kheml'nitskii, Zh. Eksp. Teor. Fiz. 79 (1980) 1120 (Sov.Phys.JETP 52 (3) (1980) 568).
12. S.Hikami, Prog.Theor.Phys. 64 (1980) 1466; Phys.Rev. B24 (1981) 2671
13. E.Brézin, S.Hikami and J.Zinn-Justin, Nucl.Phys. B165 (1980) 528
14. A.M.Pruisken and L.Schafer, Phys.Rev.Lett. 46 (1981) 490; Nucl.Phys. B200 FS4 (1982) 20.
15. F.J.Wegner, Z.Phys. B44 (1981) 9

e) Perturbative quantum corrections and diagrammatic theory

16. D.S.Langer and T.Neal, Phys.Rev.Lett. 16 (1966) 984
17. L.P.Gorkov, A.I.Larkin and D.Khmelnitskii, Pisma Zh. Eksp.Theor.Fiz. 30 (1979) 248 (JETP Lett. 30 (1979) 228).
18. E.Abrahams and T.V. Ramakrishnan, J. Non-Cryst Sol. 35 (1980) 15.
19. D.Vollhart and P.Wolfle, Phys.Rev.Lett. 45 (1980) 842.
20. C.Castellani, C.Di Castro, G.Forgacs and E.Tabet, J.Phys. C16 (1983) 159

f) General reference on perturbation theory

21. A.A.Abrikosov, L.P.Gorkov and I.E.Dzyaloshinskii, Quantum field theoretical methods in statistical physics (Pergamon, Oxford, 1965)

III. Localization and interaction

A) Perturbative quantum corrections at d=2

22. B.L.Altshuler, A.G.Aronov and P.A.Lee, Phys.Rev.Lett. 44 (1980) 1288
23. H.Fukuyama, J.Phys.Soc.Japan 48 (1980) 2169
24. B.L.Altshuler, D.Khmelnitskii, A.I.Larkin and P.A.Lee, Phys.Rev. 22 (1980) 5142

... at d=3

25. B.L.Altshuler and A.G.Aronov, Solid State Commun. 30 (1979) 115
26. B.L.Altshuler and A.G.Aronov, Zh.Eksp.Fiz. 77 (1979) 2028

B) Tentative scaling theories

27. W.L.McMillan, Phys.Rev. B24 (1981) 2739
28. Y.Gefen and Y.Imry, preprint
29. G.S.Grest and P.A.Lee, Phys.Rev.Lett. 50 (1983) 693

C) Density of states, screening and metal insulator transition

30. P.A.Lee, Phys.Rev. B26 (1982) 5882
31. C.Castellani, C.Di Castro, G.Forgacs and E.Tabet, Nucl.Phys. 225, FS9, 441 (1983)
32. A.M.Finkel'stein, Zh.Eksp.Teor.Fiz. 84 (1983)
33. C.Castellani, C.Di Castro and G.Forgacs preprint (1984)

D) Singlet and triplet quantum corrections

34. B.L.Altshuler and A.G.Aronov, Solid State Commun. 46 (1983) 429
35. B.L.Altshuler and A.G.Aronov, Pis'ma Zh.Eksp. Teor.Fiz. 37 (1983)
36. P.A.Lee, J.Non Crys.Sol. 35 (1980) 21
    A.Kawabata, J.Phys.Soc. Japan 50 (1981) 2461
37. A.Kawabata, preprint
38. A.M.Finkel'stein, Pis'ma Zh.Eksp.Teor.Fiz. 9 (1983) 436
    A.M.Finkel'stein, preprint (1983)
39. C.Castellani, C.Di Castro, P.A.Lee, M.Ma, Phys.Rev. B (1984)

General references

40. Ill-condensed Matter eds. R.Balian, R.Maynard and G.Toulouse (North-Holland, New York, 1979)
41. Disordered systems and localization, Proc.Lecture Notes in Physics, 149, eds. C.Castellani, C.Di Castro and L.Peliti (Springer-Verlag, Berlin, 1981)
42. Anderson localization eds. Y.Nagaoka and H.Fukuyama (Springer-Verlag, Berlin 1982)
43. B.L.Altshuler, A.G.Aronov, D.E.Khmelnitskii and A.I.Larkin in "Quantum Theory of Solids" edited by I.M.Lifshits Mir Publishers Moscow (1982).

LOCALIZATION OF ELECTRONS IN INFINITELY LONG RANDOM SYSTEMS - DIAGRAMMATIC APPROACH

W. Weller
Sektion Physik
Karl-Marx-Universität
DDR-7010 Leipzig

Abstract

We describe the diagram analysis for the one-dimensional localization problem developed by Berezinskii in a simplified and modified form. Then we discuss the generalization to N coupled chains (a disordered wire). It is shown that all eigenstates are localized and that the localization length is approximately equal to Nl_2 (l_2 scattering length for backward scattering at one chain).

1. Introduction

As is well known, in a one-dimensional disordered system all electronic states are localized even for arbitrarily small random potential /1/. At present time the investigation of systems closer to reality like quasi-one-dimensional systems and wires is of special interest.

In these lectures devoted to the diagrammatical approach to the localization problem we consider infinitely long systems: a one-dimensional disordered chain and N coupled disordered chains (a disordered wire). For the one-dimensional problem a special diagram technique was developed by Berezinskii /2/. The generalization of the one-dimensional problem to N transverse states corresponding to the quantized transverse motion in the cross section of the wire essentially complicates the diagram analysis (already for N=2), and it was necessary to modify the technique.

In Section 2 we consider as a simple localization criterion the probability for returnal of the electron after infinitely long time. The restriction to returnal simplifies the diagram anlysis (the so called "central part" /2/ is avoided).

In Part I we explain in detail the diagram analysis for the one-dimensional chain already in the simplified (no central part) and modified (no ordering of the pairs of lines) version.

In Part II we consider N coupled chains. The localization problem in a wire was firstly considered by Thouless /3/ who showed by means of simple scaling arguments the localization of all states and the proportionality of the localization length to the area of the cross section. Anderson et al. /4/ and Azbel /5/ developed scaling theories for the one-dimensional case and for the case of N transverse channels. Part II of these lectures is based on a paper /6/, where the localization problem in a thin disordered wire was firstly considered in the framework of a microscopic model, and on further investigations /7/ done in collaboration with Yu. A. Firsov and V. N. Prigodin. Dorokhov treated the case of N=2 /8/ and the case of arbitrary N /9/ by means of the transfer matrix method. Efetov and Larkin /10/ considered a sufficiently thick wire on the basis of their supermatrix formulation. Numerical results for a thin wire were presented by MacKinnon and Kramer /11/; we refer to the upper branch in their Fig. 1(b) corresponding to weak disorder. All mentioned authors confirm the results of the first simple scaling theory by Thouless.

2. The Localization Criterion

Following Anderson /12/ and Economou and Cohen /13/ we consider the transition amplitude for an electron going from lattice point n at time t=0 to n′ at time t:

$$(n'|e^{-iHt}|n) = G^{ret}_{n'n}(t) \ . \tag{1}$$

G^{ret} is the retarded Green function. The configurationally averaged probability for returnal (n=n′) after infinitely long time is

$$P_{nn}(t\to\infty) = \lim_{s'\to+0} s'\int_0^\infty dt\, e^{-s't} \langle |G^{ret}_{nn}(t)|^2\rangle \tag{2a}$$

$$= \lim_{s'\to+0} \frac{s'}{\pi}\int_{-\infty}^{\infty} d\omega \langle G_{nn}(\omega+is')\, G_{nn}(\omega-is')\rangle \tag{2b}$$

If the integrand in (2b) is different from zero at an ω for an infinite system (at least infinite in one direction) the electronic states are localized at this ω. Thus, we have to calculate diagrammatically the averaged product of the retarded and the advanced Green functions and to look for a $1/s'$-singularity.

Part I. One-Dimensional Chain

I.1. Elements of the Diagram Analysis

We consider a one-dimensional infinite chain with lattice points n ($-\infty < n < \infty$, lattice constant a). The Hamiltonian of the non-interacting electrons is

$$H = t'' \sum_{\langle n,n' \rangle} a_n^+ a_{n'} + \sum_n v_n a_n^+ a_n \tag{3}$$

with unperturbed (tight-binding) energy

$$\varepsilon_k = 2t'' \cos ka , \tag{4}$$

and with the random impurity potential (site-diagonal randomness) v_n satisfying

$$\langle v_n \rangle = 0 ,$$

$$\langle v_n v_{n'} \rangle = \langle v_n^2 \rangle \delta_{nn'} . \tag{5}$$

The unperturbed (retarded) Green function is

$$G^{ret\,(0)}(n_1, n_2, \omega) = \frac{1}{\mathcal{N}} \sum_k \frac{e^{ika(n_1 - n_2)}}{\omega + is' - 2t'' \cos ka}$$

$$= - \frac{i}{2|t''| \sin \varphi_0} e^{(i\varphi_0 - s)|n_1 - n_2|} ; \tag{6}$$

$\mathcal{N}$ ($\to \infty$) is the number of lattice points, the energy of the electron defines the angle φ_0 ,

$$\omega = 2t'' \cos \varphi_0 ,$$

$$s = \frac{s'}{2|t''| \sin \varphi_0} .$$

With v_n as a perturbation we have to calculate diagrammatically the integrand in (2b) (we choose n=0):

$$K \equiv \langle G_{00}(\omega + is') \, G_{00}(\omega - is') \rangle$$

$$= \tag{7}$$

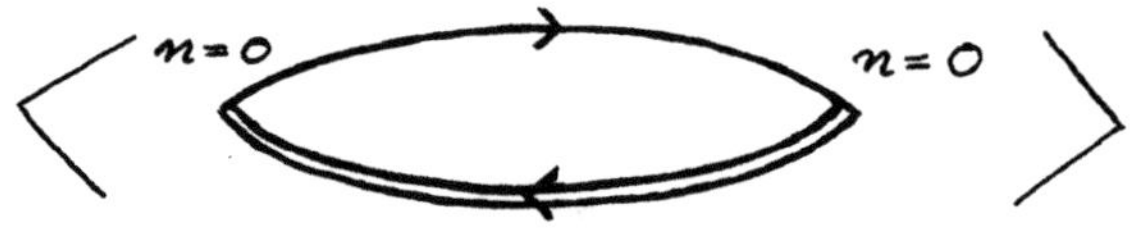

The perturbed Green functions in (7) (thick lines, double line for advanced function) are given by the Dyson equation

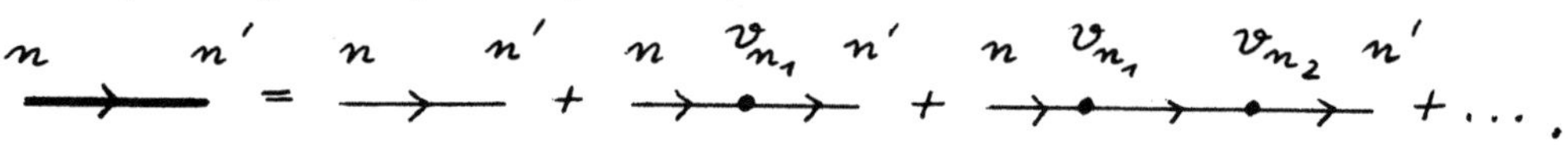

(8)

After insertion of (8) into (7) we have to perform the configurational average. We restrict to weak scattering (Born approximation), that means to the second order in v_n for every point n, and denote the average by a wavy line (impurity vertex): 

$$= \langle v_n^2 \rangle \qquad (9)$$

In order to obtain localization for a one-dimensional system it is necessary to sum up <u>all</u> diagrams (in the Born approximation). This includes the ladder diagrams (leading to elementary transport theory, to diffusion), but besides them the larger class of so called interference diagrams (diagrams with intersecting wavy lines in momentum space technique; an example is shown in Fig. 1a). As remarked by Berezinskii the analysis becomes much simpler in real space, because there are <u>no</u> intersections (Fig. 1b shows in real space the diagram of Fig. 1a). In this real space technique the n_i are ordered.

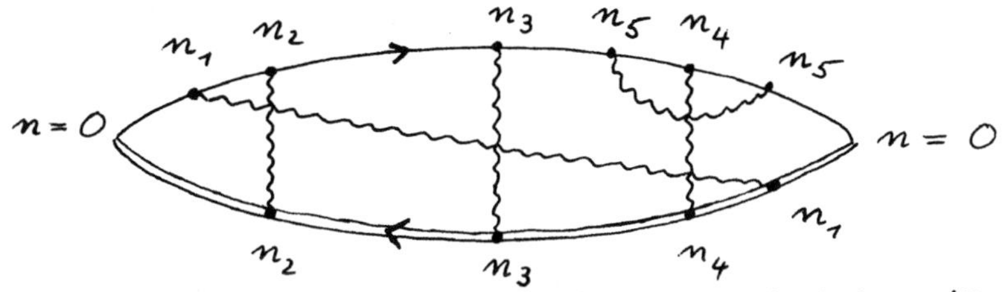

Fig. 1a. Interference diagram in momentum space technique (the n_i are summed over).

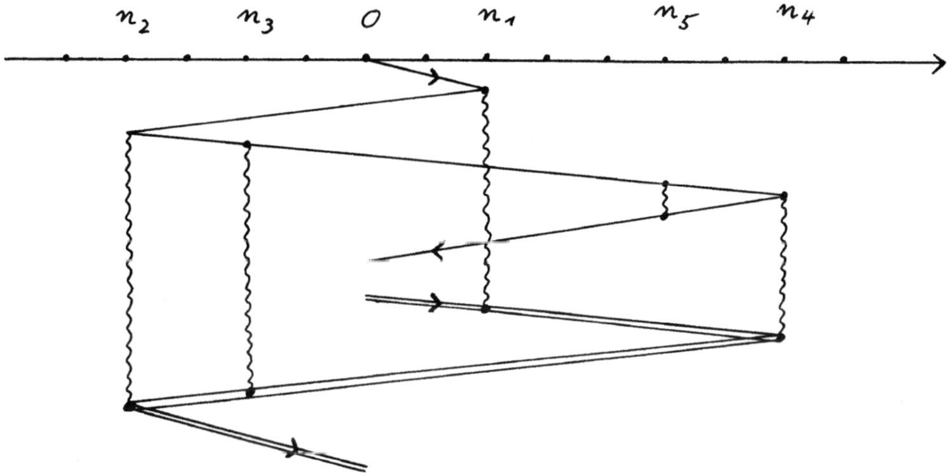

Fig. 1b. Diagram of Fig. 1a in real space. The order of the n_i is chosen as $n_2 < n_3 < 0 < n_1 < n_5 < n_4$ (the n_i are summed over conserving this order).

Then the modulus $|n_i - n_{i'}|$ in the Green function has a definite value, the Green function factorizes, and these factors are joined with the impurity vertices (4 factors at each vertex). Thus, each vertex contains the factor

$$\frac{\langle v_n^2 \rangle}{(2t'' \sin \varphi_0)^2} \equiv \frac{1}{l} . \tag{10}$$

l is the (dimensionless) scattering length; it is of the order of the localization length $L^{(1)}_{loc}$. Figs. 2a,b show the so called non-oscillating vertices. In Fig. 2c an example of an oscillating vertex is shown. The oscillating exponentials of the type $\exp(2i\varphi_0 n)$ from these vertices lead to negligible contributions in the sums over n (the essential domain for the summation over n is l, $l\varphi_0 \gg 1$ for weak scattering).[1)]

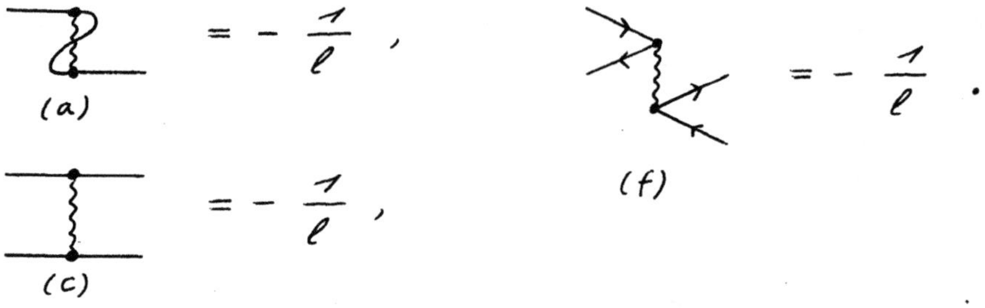

Fig. 2a. Non-oscillating vertices with retarded Green functions. The vertices with advanced Green functions are analogous and have the same analytical expressions.

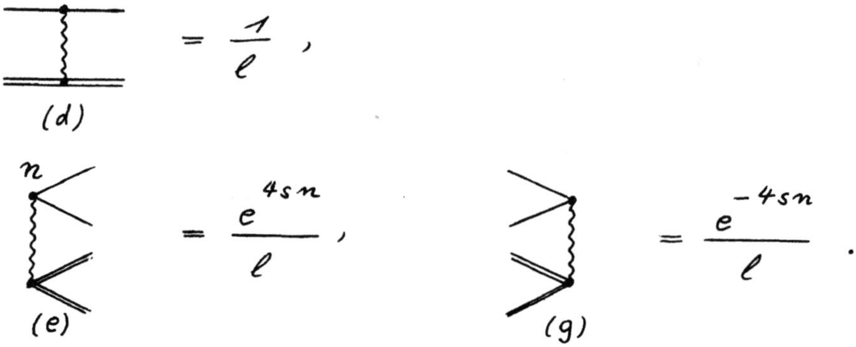

Fig. 2b. Non-oscillating vertices connecting retarded and advanced (═══) Green functions.

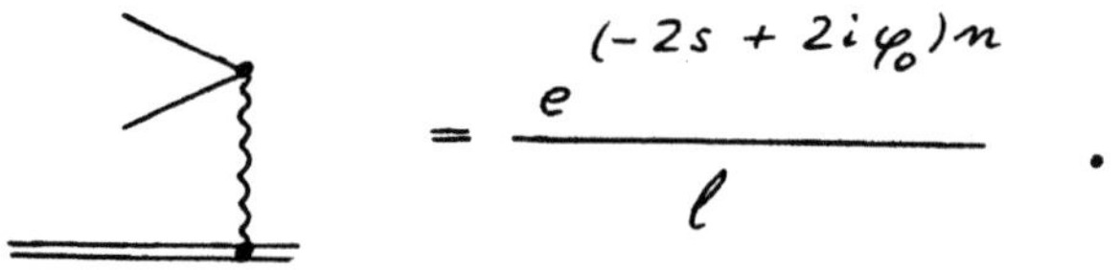

Fig. 2c. Example of an oscillating vertex.

---

1) The inclusion of oscillating vertices is however essential if the potential v_n contains a constant (non-random) part /14/.

I.2. Summation of Diagrams

Following Berezinskii /2/ we construct the diagrams by going along the n-axis from $-\infty$ to the right or from $+\infty$ to the left and adjoining to the existing pairs of lines ($\rightleftarrows$) at any n the vertices of Fig. 2 in any possible way. In Fig. 1b, for example, the vertex (d) is adjoined at n_3 to the existing one retarded and one advanced pairs. In this procedure it is essential to avoid the creation of closed loops. In /2/ this was achieved by ordering the pairs of lines at any point. For $N > 1$ such an ordering of the indexed (see Part II) pairs would lead to serious complications; thus, following /6/ we change the definition of the diagram blocks and the rules for adjoining the backward scattering vertices f,g: (i) No order of the pairs (created by e or f) is introduced. (ii) The two lines of a pair are never coupled into one end of the impurity line. (iii) If one line of a pair is coupled (by f or g) with the line in opposite direction of a second pair, the two remaining lines represent a new pair (Fig. 3).

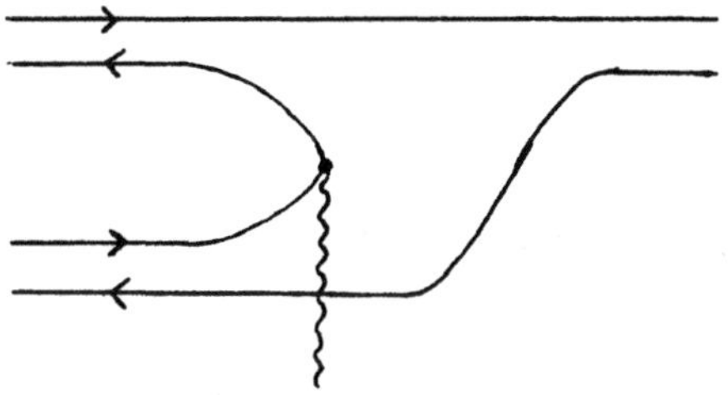

Fig. 3. Coupling of lines by backward scattering vertices.

We define the left block L(M, n-1/2) as the sum of all diagrams (from the left) with M retarded and M advanced pairs between n-1 and n. Adjoining the vertices of Fig. 2 in any possible way we obtain the difference equation

$$L(M, n+\tfrac{1}{2}) = L(M, n-\tfrac{1}{2}) + \dots + \frac{e^{4sn}}{\ell} L(M-1, n-\tfrac{1}{2}) + \dots + \frac{e^{-4sn}}{\ell} [(M+1)M]^2 L(M+1, n-\tfrac{1}{2}). \tag{11}$$

Not all terms are written down on the right-hand side. The first term corresponds to diagrams, where all lines go through n without scattering. The second term comes from the (e)-vertex: a new retarded pair and a new advanced pair are created by backward scattering (Fig. 4a). The third term comes from (g) (Fig. 4b); note the combinatorial factor counting the number of possible couplings of lines.

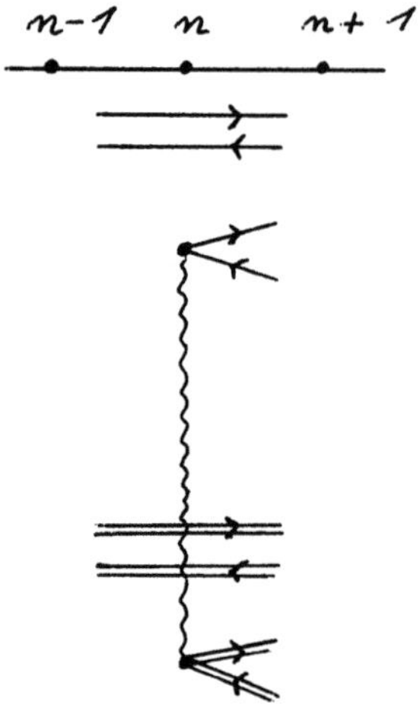

Fig. 4a. Insertion of vertex (e) (M=2)

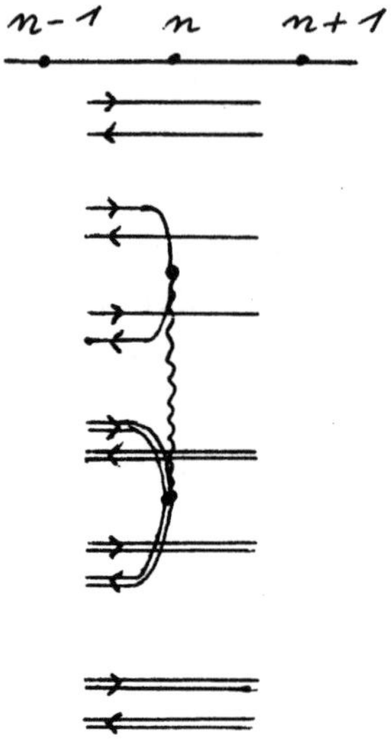

Fig. 4b. Insertion of vertex (g) (M=2)

By inspection of special diagrams it is obvious that all diagrams show the same exponential n-dependence:

$$L(M, n-\tfrac{1}{2}) = e^{4sMn}\,\tilde{L}(M) \tag{12}$$

With the further transformation

$$\tilde{\tilde{L}}(M) = (M!)^2\,\tilde{L}(M) \tag{13}$$

(11) goes over into

$$\{e^{4sM} - 1\}\,\tilde{\tilde{L}}(M) = \frac{M^2}{\ell}\{\tilde{\tilde{L}}(M+1) + \tilde{\tilde{L}}(M-1) - 2\tilde{\tilde{L}}(M)\}. \tag{14}$$

Only diagrams with $4slM \sim 1$ are essential (see below); for weak scattering $l \gg 1$, thus $4sM \ll 1$ and $\exp(4sM) - 1 \approx 4sM$ (nevertheless $M \to \infty$ for $s \to 0$). The block $\tilde{\tilde{L}}(M)$ is then a function of the variable

$$p = 4sM\ell\,; \tag{15}$$

for $s \to 0$ the curly bracket on the right-hand side becomes the second derivative and (14) goes over into the Bessel-type equation

$$\tilde{\tilde{L}}(p) = p \frac{d^2 \tilde{\tilde{L}}(p)}{d p^2} . \tag{16}$$

The <u>quadratic</u> combinatorial factor M^2 in (14) is due to the inclusion of all diagrams (all combinations of lines by adjoining the vertices). A selection of special classes of diagrams (like ladder or maximally crossed diagrams) would lead to a linear combinatorial factor M, as a consequence $\tilde{\tilde{L}}$ would be a function of the variable $p'=2\sqrt{sl}\,M$; whereas a dependence on p gives the 1/s-singularity necessary for localization, a dependence on p′ gives only a $1/\sqrt{s}$-singularity characteristic for diffusion. Obviously, l ($1/l \to 0$ for weak scattering) only scales the argument of the block $\tilde{\tilde{L}}$, so that there is no small parameter in the equations.

I.3. Fit of the Left Block to the Right Block

In order to obtain the sum of all diagrams for the correlator K we have (i) to connect (for example between n=0 and n=1) the lines of a left block $\tilde{L}(M^L)$ with the lines of a right block $\tilde{L}(M^R)$ (sum of all diagrams from the right, analytically equal to the left block) in any possible way (Fig.5), (ii) to sum over all possible pair numbers M^L, M^R of the left and the right blocks:

$$K = \frac{1}{(2t'' \sin\varphi_0)^2} \sum_{M^L=0}^{\infty} \tilde{L}(M^L)\, \tilde{L}(M^R=M^L+1)(M^L!)^2(M^R!)^2 + \tag{17}$$

+ 3 terms for other initial vertices.

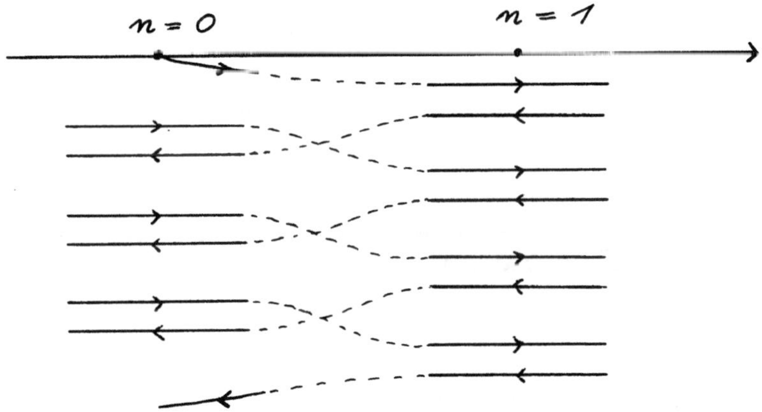

Fig. 5. Fit of a left block $\tilde{L}(M^L=3)$ to a right block $\tilde{L}(M^R=4)$. The initial vertex is specially chosen so that the electron goes to the right and returns from the right corresponding to the first term in (17). Furthermore, a special combination of the lines of the left block with the lines of the right block is drawn; the total number of combinations is $M^L!M^R!$. The fit among the advanced lines is anlogous.

The first factor on the r.h.s. is due to the two retarded and two advanced lines starting or ending at n=0. The combinatorial factors are used in the transformation (13). Because $\tilde{\tilde{L}}$ depends on p, for $s \to 0$ the sum over M can be converted into an integral:

$$\bar{K} = \frac{1}{4sl} \frac{1}{(2t'' \sin\varphi_0)^2} \int_0^\infty dp\, \tilde{\tilde{L}}^2(p) \quad + \tag{18}$$

+ 3 further terms for other initial vertices.

(18) shows the 1/s-singularity necessary for localization. Furthermore there appears 1/l. Assuming for the probability p_{on} that an electron initially at n=0 is found after infinitely long time at n,

$$p_{on} \sim \frac{1}{L^{(1)}_{loc}} e^{-|n|/L^{(1)}_{loc}} , \tag{19}$$

it is plausible that the localization length $L^{(1)}_{loc} \sim l$. We will not enter into the details.

Part II. N Coupled Chains

II.1. Elements of the Diagram Analysis

We consider N continuous chains (kinetic energy ε'') numbered by ν. The Hamiltonian without impurities is given by

$$H^{(0)} = \int dz \Big[\sum_\nu \tilde{a}^+_\nu(z)\, \varepsilon''\{-i\tfrac{d}{dz}\}\, \tilde{a}_\nu(z) \; + \tag{20}$$

$$+ \sum_{\nu \neq \nu'} t_{\nu\nu'}\, \tilde{a}^+_\nu(z)\, \tilde{a}_{\nu'}(z) \Big] .$$

We diagonalize with respect to the transverse motion,

$$\tilde{a}_\nu(z) = \sum_j U_{\nu j}\, a_j(z) , \tag{21}$$

energy eigenvalues $\delta\varepsilon_j'$. The unperturbed retarded Green function is (for weakly coupled chains, $|\delta\varepsilon_j'| \ll \varepsilon_F$)

$$G^{(0)}_{jj}(z, \varepsilon_F + is') = -\frac{i}{v_F}\, e^{i(k_F - \delta\varepsilon_j + is)|z|} , \tag{22}$$

where $\delta\varepsilon_j = \delta\varepsilon_j'/v_F$, $s = s'/v_F$. For the interaction with the impurities we add to $H^{(o)}$

$$H^{imp} = \int dz \sum_\nu V_\nu(z)\, \tilde{a}^+_\nu(z)\, \tilde{a}_\nu(z) , \tag{23}$$

with short ranged correlator for the random potential V:

$$\langle V_\nu(z)\, V_{\nu'}(z') \rangle = \delta_{\nu\nu'}\, W(|z-z'|) . \tag{24}$$

Generalizing /6,7/ the technique developed for one chain we now have to consider diagrams in space (z) with indexed lines. We again consider weak scattering (Born approximation) and obtain e.g. for a forward scattering vertex connecting 4 retarded lines

$\gamma =$

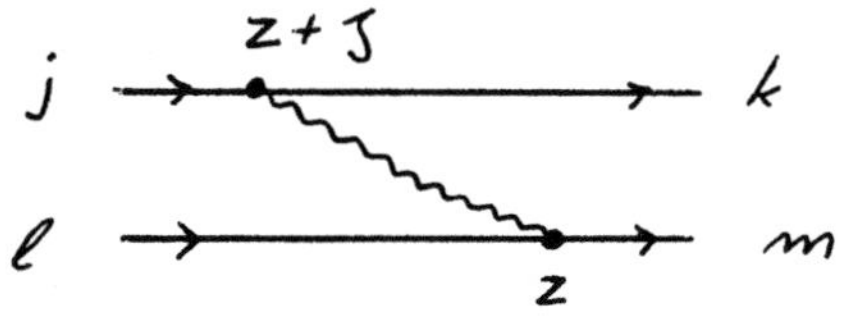

$$= -\frac{1}{v_F^2}\int d\zeta\, W(|\zeta|)\, e^{i(\delta\varepsilon_j-\delta\varepsilon_k)\zeta}\, e^{i(\delta\varepsilon_j-\delta\varepsilon_k+\delta\varepsilon_l-\delta\varepsilon_m)z} \,\cdot$$

$$\cdot \sum_\nu (\mathcal{U}^+)_{j\nu}\, \mathcal{U}_{\nu k}\, (\mathcal{U}^+)_{l\nu}\, \mathcal{U}_{\nu m} . \tag{25}$$

The exponent containing ζ can be neglected because of the short ranged W. We now simplify the model: (i) We take into account only terms for which the indices j,k,l,m coincide pairwise (including also the time reversed states, j=k, j=m or $j=\hat{l}$, $\hat{l}$ time reversed state l; in the following we shall omit such details). (ii) We neglect all further vertices where in the exponent the factor in front of z does not vanish (oscillating vertices). The characteristic integration range for z is the localization length $L^{(N)}_{loc} \sim N l_2$ (l_2 scattering length for backward scattering, see (28)). Thus, the necessary condition for strong oscillations is

$$\underset{j\neq k}{\mathrm{Min}}\, |\delta\varepsilon_j - \delta\varepsilon_k|\, N l_2 \gg 1 ; \tag{26a}$$

for the three-dimensional case (two-dimensional cross section with regular lattice)

$$8\pi^2 \frac{t^\perp}{\varepsilon_F} k_F l \gg 1 . \tag{26b}$$

For weak scattering ($k_F l_2 \gg 1$) this condition is satisfied already for small nearest neighbour transfer amplitudes $t^\perp$ ($t^\perp \ll \varepsilon_F$). Condition (26b) means physically that the time between two backward scatterings along a chain is large compared with the time necessary for the transfer from a chain to a neighbouring one.

In Fig. 6 two examples for vertices corresponding to (c,e) of Figs. 2a,b are shown. l_1 (scattering length for forward scattering) and l_2 are given by

$$\frac{1}{l_1} = \frac{2}{v_F^2} \int_0^\infty W(\zeta)\, d\zeta \,, \tag{27}$$

$$\frac{1}{l_2} = \frac{2}{v_F^2} \int_0^\infty W(\zeta) \cos(2k_F \zeta)\, d\zeta \,. \tag{28}$$

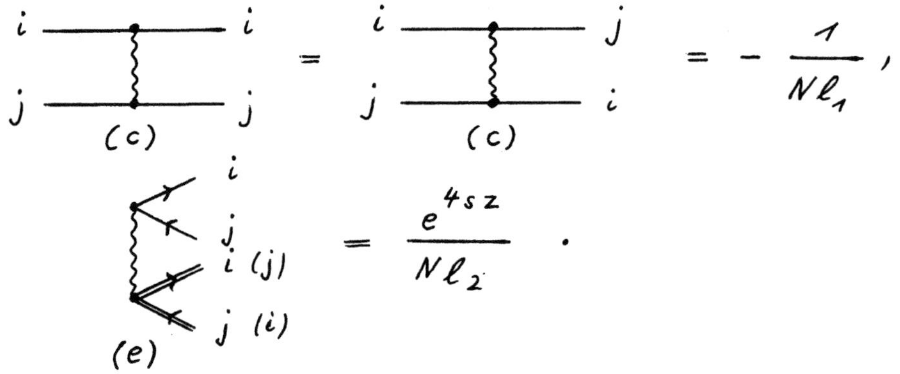

Fig. 6. The vertices (c) and (e)

Any pair of retarded (advanced) lines is labelled by i, j ($\bar{i}, \bar{j}$), first index for the line going from the left to the right, second index for the line in opposite direction. The left block now depends on the numbers m_{ij} ($m_{\bar{i}\bar{j}}$) of retarded (advanced) pairs at z forming a vector $m = \{m_{ij}, m_{\bar{i}\bar{j}}\}$. We apply transformations similar to (12,13):

$$L(z, \vec{m}) = e^{4sMz}\, \tilde{L}(\vec{m}) \,, \tag{29}$$

$$\tilde{\tilde{L}}(\vec{m}) = N^{2M} \prod_{ij} \left(m_{ij}!\, m_{\bar{i}\bar{j}}! \right) \tilde{L}(\vec{m}) \,. \tag{30}$$

Going along the z-axis from z to z+dz and adjoining in between the vertices in any possible way we again obtain for $\tilde{\tilde{L}}$ a difference equation:

$$4sM\, \tilde{\tilde{L}}(\vec{m}) = \{A + B + C + D + E + F + G\}\, \tilde{\tilde{L}} \,. \tag{31}$$

The terms are written down symbolically. Referring for all details to /6,7/ we give here only as an example the analytical expression for the term E corresponding to the insertion of the vertex (e) (compare Fig. 4a):

$$E\tilde{\tilde{L}} = \frac{N}{l_2} \sum_{\substack{il \\ i'l'}} \Delta_{\substack{il \\ i'l'}} m_{il}\, m_{\bar{i}'\bar{l}'}\, \tilde{\tilde{L}}(\vec{m} + \vec{\eta}_E), \qquad (32)$$

$$\vec{\eta}_E = \{-il, -\bar{i}'\bar{l}'\},$$

$$\Delta_{\substack{il \\ i'l'}} \equiv \delta_{il}\,\delta_{ii'}\,\delta_{il'} + (1-\delta_{il})(\delta_{ii'}\,\delta_{ll'} + \delta_{il'}\,\delta_{li'}).$$

The special deviations in the pair numbers due to the vertices are symbolically denoted by $\vec{\eta}$; for example $\vec{\eta} = \{il, -\bar{i}'\bar{l}'\}$ means that m_{il} is increased by 1 and $m_{\bar{i}'\bar{l}'}$ is decreased by 1.

II.2. Approximative Reduction to an Effective One-Dimensional Problem

The simplest approximation to (31) consists in the neglection of the special deviations $\vec{\eta}$, what is equivalent to the assumption that $\tilde{\tilde{L}}(\vec{m})$ only depends on the total number M of retarded (or advanced) pairs. Then (31) is reduced to

$$4sM\tilde{\tilde{L}}(M) = \frac{1}{Nl_2}\left(2-\frac{1}{N}\right)M^2\{\tilde{\tilde{L}}(M+1) + \tilde{\tilde{L}}(M-1) - 2\tilde{\tilde{L}}(M)\}. \qquad (33)$$

Comparison with (14) shows that (33) is the equation for the left block of an effective one-dimensional problem. In contrast to the real one-dimensional problem there appears Nl_2 in the effective equation (33) instead of l_2. As a consequence (33) leads to the relation between the localization lengths

$$L^{(N)}_{loc} \approx N L^{(1)}_{loc}. \qquad (34)$$

In (34) one has to expect a numerical factor of the order of 1. However, the determination of this factor needs an improved treatment of (31).

References

/1/ N.F. Mott, and D.W. Twose, Adv.Phys. 10, 107(1961).
/2/ V.L. Berezinskii, Zh.eksper.teor.Fiz. 65, 1251(1973).
/3/ D.J. Thouless, Phys.Rev.Letters 39, 1167(1977).
/4/ P.W. Anderson, D.J. Thouless, E. Abrahams, and D.S. Fisher, Phys.Rev. B22, 3519(1980).
P.W. Anderson, Phys.Rev. B23, 4828(1981).
/5/ M.Ya. Azbel, Phys.Rev.Letters 46, 675(1981).
/6/ W. Weller, V.N. Prigodin, and Yu.A. Firsov, phys.stat.sol.(b) 110, 143(1982).
/7/ W. Weller, Yu.A. Firsov, and V.N. Prigodin, Proceedings of the International Seminar on "Localization in Disordered Systems", Johnsbach 1983, ed. by W. Weller and P. Ziesche, B.G. Teubner Leipzig 1984.

/8/ O.N. Dorokhov, Solid state Commun.44,915(1982).
/9/ O.N. Dorokhov, Zh.eksper.teor.Fiz.85,1040(1983).
/10/ K.B. Efetov, and A.I. Larkin, ZH.eksper.teor.Fiz.85,764(1983).
K.B. Efetov, Adv.Phys.32,53(1983).
/11/ A. MacKinnon, and B. Kramer, Phys.Rev.Letters 47,1546(1981).
/12/ P.W. Anderson, Phys.Rev.109,1942(1958).
/13/ E.N. Economou, and M.H. Cohen, Phys.Rev.B5,2931(1972).
/14/ W. Weller, and M. Kasner, Contribution to MECO-Seminar 1984.

VI. LATTICE GAS

CRITICAL AND MULTICRITICAL POINTS IN FLUIDS AND MAGNETS

Jean SIVARDIERE
Département de Recherche Fondamentale
Centre d'Etudes Nucléaires de Grenoble
85 X - 38041 - GRENOBLE CEDEX - FRANCE

ABSTRACT :

In this article we review the geometrical properties of the phase diagrams of multicomponent fluids. Lattice gas models used to simulate the thermodynamical behaviour of these fluids are presented. We focus on critical and multicritical points, and on the analogies between fluids and magnets.

Spin one-half models are shown to describe systems with a single critical point. Spin one models may describe systems with a symmetrical tricritical point. Spin three half models may describe systems with a symmetrical tetracritical point and non-symmetrical tricritical points. The topology of phase diagrams is correctly reproduced within the molecular field approximation, and classes of systems having isomorphic phase diagrams are defined.

I. THE CRITICAL POINT

I.1. Critical points in fluids and magnets

A large variety of physical systems exhibit a critical point. As discussed some time ago by TISZA [1], two types of critical points may be distinguished :

a) Curie critical points

We consider here isotropic or anisotropic ferro and antiferromagnets, ferroelectrics, He4, superconductors, alloys, molecular crystals,...
In these systems, a second order phase transition (of the order disorder or displacive type) may be found as temperature is lowered below T_c :

- This transition corresponds to a lowering of the high temperature symmetry. An order parameter η may be defined. η is zero above T_c, its value below T_c describes the amplitude of the symmetry lowering.
- If a field H, conjugated to η, is applied to the system, $\eta \neq 0$ at any temperature so that the transition is suppressed : indeed the symmetry lowering is "already" induced by the application of H (in some cases, this experiment may be fictitious since H is unphysical ; for instance for an antiferromagnet, H is an unphysical staggered field).
- In the molecular field approximation, the specific heat anomaly at the transition is a discontinuity ; however in practice the singularity is a λ-shape divergence such as in He4, whence the name λ-critical point also given to Curie points.
- Below T_c, several equivalent domains may coexist. They are connected by the symmetry elements which are lost by the system at T_c. They are characterized by different phases of the order parameter η and by the same values of the true thermodynamic variables.

We concentrate from now on the particular case of a strongly anisotropic ferromagnet. Below T_c, two domains of opposite magnetizations $\pm$ M coexist in a zero external field H. As T is increased to T_c, the "difference" 2M between the two domains tends to zero, and at $T=T_c$ a single critical phase is found.

b) Andrews critical points

The Andrews critical point of a simple fluid is very similar to a Curie critical point. Consider a point of the liquid-gas (L -G) coexistence curve. As T is increased to T_c, the point moves to the critical point and the two phases at equilibrium become identical. We may introduce an order parameter $\eta = \rho - \rho_c$ (ρ being the density) which

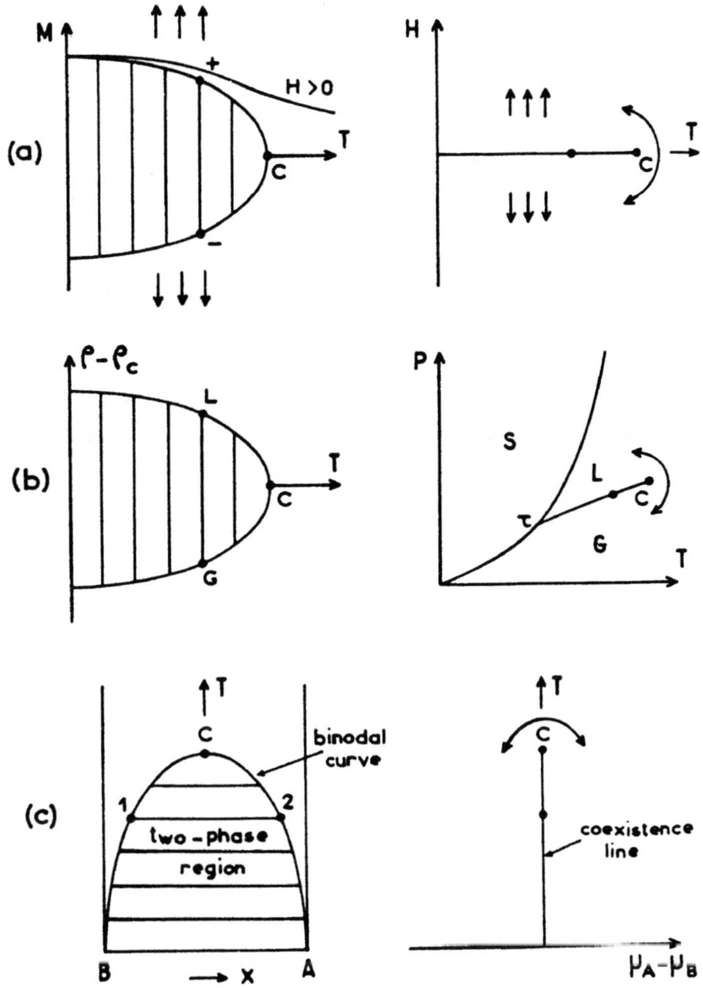

Figure 1 : Phase diagrams of an anisotropic magnet (a), a simple fluid (b), and a binary fluid (c). The curve ρ (T) of a simple fluid is nearly symmetrical with respect to $\rho = \rho_c$: this is the law of rectilinear diameters. Similarly the binodal curve of the binary fluid is often nearly symmetrical.

zero at T_c and takes two values η_L and η_G below T_c. η_L and η_G are approximately opposite according to the law of rectilinear diameters.

A similar situation is found in many non-ideal binary mixtures which present a phase separation below some critical temperature T_c. Two phases of different compositions x_1 and x_2 coexist ; as T is increased to T_c, $x_1 - x_2 \rightarrow 0$ and at T_c a single critical phase is found. The critical point is sometimes called the Alexeiev critical point. We may take here $\eta = \frac{1}{2} - x$ and we have approximately : $\eta_1 = -\eta_2$.

Two differences are found between systems with Curie and Andrews critical points.

1) In the first situation η is only a pseudo-thermodynamical variable, which is determined by the true thermodynamical variables such as temperature, pressure,... In the second situation, η is a true thermodynamical variable.

2) The first situation corresponds to a spontaneous symmetry breaking. In the second situation there is no symmetry breaking : the two phases at equilibrium are isotropic and are characterized by values of the order parameter which are only approximately opposite.

In spite of these two differences, there is a strong similarity between the two types of critical points, which is illustrated by Fig. 1 (see also Ref. [2]).

1) In all cases, the system is always homogeneous above T_c. Below T_c, two phases or domains may coexist.

2) The magnetization curve M(T) of the ferromagnet is similar to the binodal curves of the simple and binary fluids. Inside the binodal curve is a two-phase region. The lenght of a tie-line is proportionnal to η.

3) The critical point C is the top of the binodal curve and corresponds to a limit of stability of an homogeneous phase. In the vicinity of C, strong fluctuation phenomena are found, leading to critical diffusion of light or X-rays.

4) The extensive variables η used when considering the binodal curves (M, ρ , x) and generally called "densities" [3] may be replaced by the conjugated intensive variables or "fields" H : the magnetic field H, the pressure P or the chemical potential μ, the difference $\mu_A - \mu_B$ between the chemical potentials of the two chemical species A and B. The binodal curve is then replaced by a coexistence line terminating at the critical point C, since two coexisting phases at T are characterized by the same value of the field H.

5) The transformation of a phase η to the phase $-\eta$, which may coexist with it, is possible in two ways : either abruptly by going through the two phase region ; or continuously by "turning around" the coexistence line in the (H, T) plane.

I.2. The spin one-half Ising model

It is remarkable that the three systems discussed above - the anisotropic ferromagnet, the simple fluid, the binary fluid - can be described qualitatively using the same spin one-half Ising hamiltonian :

$$\mathcal{H} = -H \sum_i S_{iz} - J \sum_{i,j} S_{iz} S_{jz} \quad (1)$$

We drop the index z in the following. $\mathcal{H}$ obviously describes a strongly anisotropic ferromagnet, J is the positive exchange integral between spins and H the external applied field. In the Weiss molecular approximation, the magnetization is given by :

$$M=\frac{1}{2} \tanh \frac{\beta}{2} (H + 2 JM) \quad (2)$$

whence the critical temperature :

$$kT_c = \frac{J}{2} \quad (3)$$

The binodal curve M(T) is found for H = 0. For $H \neq 0$ no phase transition is found, the system is ferromagnetic at any T since the time reversal symmetry is broken by H. For $T \gg T_c$, the susceptibility is given by the Curie Law : $\chi \sim \frac{1}{T-T_c}$. (4)

The same hamiltonian (1) may be used to reproduce the behaviour of a binary fluid AB. This is the lattice-gas model of Bragg and Williams [4]. We suppose that the macroscopic volume of the fluid is cut in small cells i, j,... forming a fixed lattice and having roughly the size of a molecule A or B, in such a way that each cell is occupied, either by a molecule A or by a molecule B. A state variable $S_i = \pm \frac{1}{2}$ describes the state of each cell, and the composition x is given by :

$$x = \frac{1}{2} + \langle S_i \rangle = \frac{1}{2} + M \quad (5)$$

Introducing interactions J_{AA}, J_{AB}, J_{BB} and the chemical potentials μ_A and μ_B, the fluid is described by a hamiltonian of the form (1) with :

$$H = \mu_A - \mu_B + 4(J_{AA} - J_{BB})$$
$$J = J_{AA} + J_{BB} - 2 J_{AB} \quad (6)$$

Bragg and Williams used a mean field approximation to treat $\mathcal{H}$; they found that if J is positive, a critical point is found at $kT_c = \frac{J}{2}$. The equation of the binodal curve is (2) with H = 0 : below T_c, two phases M and - M coexist for H = 0.

	Anisotropic magnet	Simple fluid	Binary mixture
Critical point	Curie	Andrews	Alexeiev
Critical diffusion	Neutrons	Light	Light
Density	M	$\rho - \rho_c$	$x - \frac{1}{2}$
Field H	H	P	$\mu_A - \mu_B$
$T \gg T_c$	Curie law	Boyle-Mariotte law	Henry law
Model	Ising	Lee-Yang	Bragg-Williams
Mean field approximation	Weiss	Van der Waals	"

TABLE I : Analogies between the anisotropic magnet, the simple fluid and the binary fluid.

Similarly, the hamiltonian (1) describes the liquid-gas equilibrium in a simple fluid A. Indeed in the lattice-gas model introduced by Lee and Yang [5], the volume of the fluid is cut in small cells, each of which is either occupied by a molecule A ($S_{iz} = +\frac{1}{2}$) or empty ($S_{iz} = -\frac{1}{2}$). The density of the fluid is given by :

$$\rho = \frac{1}{2} + \langle S_{iz} \rangle = \frac{1}{2} + M \qquad (\rho_c = \frac{1}{2}) \tag{7}$$

In (1) one must take :

$$H = \mu_A + J_{AA}$$
$$J = J_{AA} \tag{8}$$

When the mean-field approximation is used, the state equation of the Boltzmann-Van der Waals gas is recovered ($v = \frac{1}{\rho}$ is the specific volume) :

$$P = -\frac{J}{v^2} + kT \,\mathrm{Log}\, \frac{v}{v-1} \tag{9}$$

Table 1 summarizes the comparison of the three systems, characterized by two states in each cell, a single effective interaction J and a single order parameter M.

II. THE SYMMETRICAL TRICRITICAL POINT

II.1. Critical lines and tricritical points

Let us consider a system with a critical point : a second order phase transition is found as T is lowered below T_c for a given value (H = 0 in the Ising model) of the field H conjugated to the order parameter η ; if $H \neq 0$ the transition is suppressed. Suppose now we vary an intensive variable or field H' *not conjugated to* η, for instance the pressure of a solid magnet, or He4, or a binary fluid. The critical temperature is then modified, and the critical point generates a ***critical line*** in the (T, H') plane. This line is the border of a ***coexistence surface*** generated by the coexistence lines found for each value of H'. (As discussed by Tisza [1], an erroneous comparison between these (second-order) critical lines and the (first-order) coexistence lines lead Keesom and Ehrenfest to an unphysical classification of phase transitions).

It often happens that the transition becomes first order as the field H' is increased beyond some value. The critical line then merges into a line of coexistence of several phases, for instance a ***triple line*** in the case of an anisotropic magnet : a paramagnetic phase coexists with two phases of opposite magnetisations at the first

System	η	H	H'	η'
He3–He4	Suprafluid order parameter	Unphysical	Chemical potential $\mu_3 - \mu_4$	Composition X_3
Antiferromagnet	Sublattice magnetization	Staggered field (unphysical)	Homogeneous field	net magnetization
NH_4 Cl, KDP	Order-disorder parameter	Unphysical	Pressure	Density
Superconducting film	Superconducting order parameter	Unphysical	Magnetic field	magnetization
Monomolecular layer	nematic order parameter	Unphysical	Surface Pressure	Density
Compressible ferromagnet	magnetization	magnetic field	Pressure	Density
Ferromagnetic alloy	magnetization	magnetic field	Chemical potential $\mu_A - \mu_B$	Composition
$KTaO_3$ – $KNbO_3$	polarization	electric field	$\mu_A - \mu_B$	Composition

TABLE II

order transition point. The limit of the critical line is called the ***tricritical point*** C', the existence of which was predicted by Landau [6] from a purely phenomenological point of view.

Many systems exhibit such a tricritical point : He3-He4 mixtures [7], anisotropic antiferromagnets such as $FeCl_2$ or DAG [8] in a magnetic field, NH_4Cl or KDP under pressure [9], thin superconducting films [10], monomolecular layers of elongated molecules [11], compressible ferromagnets [12], ferromagnetic alloys AB (B non magnetic) or mixtures of ferroelectric $KNbO_3$ and non-ferroelectric $KTaO_3$ [13], nematic-smectic liquid crystals [14]. For these systems, Table II indicates the physical nature of η, H, H' and η', the (non-ordering) density conjugated to H'. Fig. 2 shows typical phase diagrams in the (T, η'), (T, H') and (η', H') planes, found for H = 0. The (H', η') diagram shows that, for $T < T_c'$, increasing H' at constant T induces a first order transition, for instance a metamagnetic transition in the case of an antiferromagnet.

II.2. <u>The spin-one Ising model</u>

Blume [15] and Capel [16] developped a very simple model of a magnetic system exhibiting a tricritical point. We consider a lattice of S = 1 magnetic ions, submitted to an anisotropic crystalline field. J being the exchange integral, the hamiltonian is ($Q_i = S_i^2$) :

$$\mathcal{H} = - D \sum_i Q_i - H \sum_i S_i - J \sum_{i,j} S_i S_j \qquad (10)$$

If $D > 0$, the magnetic transition is second order. Blume and Capel showed that it becomes first order for large enough negative values of D. Here $M = \langle S_i \rangle$ is the ordering density η conjugated to the field H, and $Q = \langle Q_i \rangle$ is the non-ordering density η', conjugated to the field D = H'.

The phases diagrams presented in Fig. 2 are typical of a large variety of systems, similar to the Blume-Capel magnet, characterized by two order parameters or densities : the ordering density η and the non-ordering density η'. However as we shall see it is quite possible that both η and η' are ordering densities in the same system if two types of interactions, J and K respectively, are present. For instance if we consider He3-He4 mixtures or a magnetic AB alloy, an interaction J leads to suprafluid or magnetic ordering but a second interaction K may lead to phase separation and possibly compete with J.

Of course systems with two order parameters and two interactions cannot be represented by a spin one half Ising model, but we may try a spin one model since if

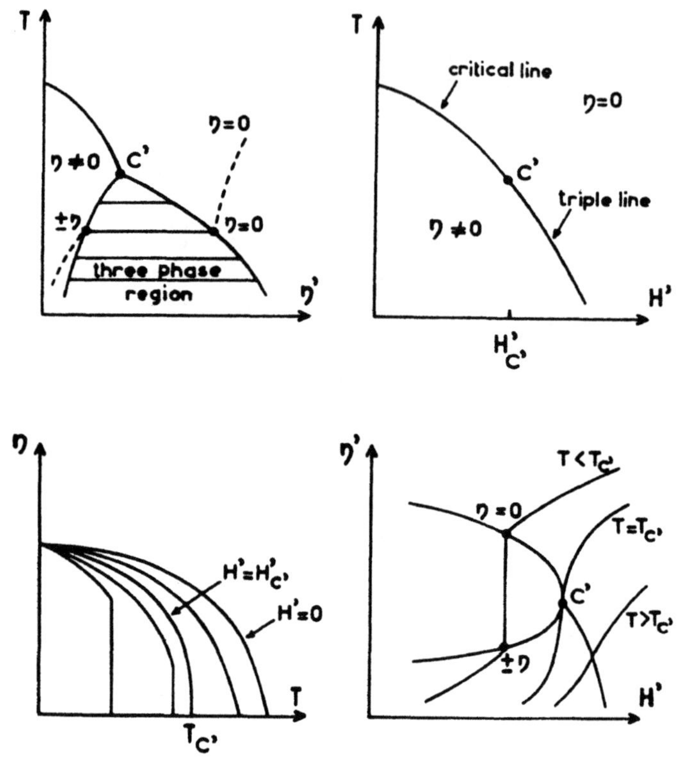

Figure 2
The tricritical point C', H' and η' are conjugated field and density H=0. The surface below the critical and triple lines in the (T,H') plane is a coexistence surface (±η). The dashed curve in the (T,η') plane is a constant H' line.

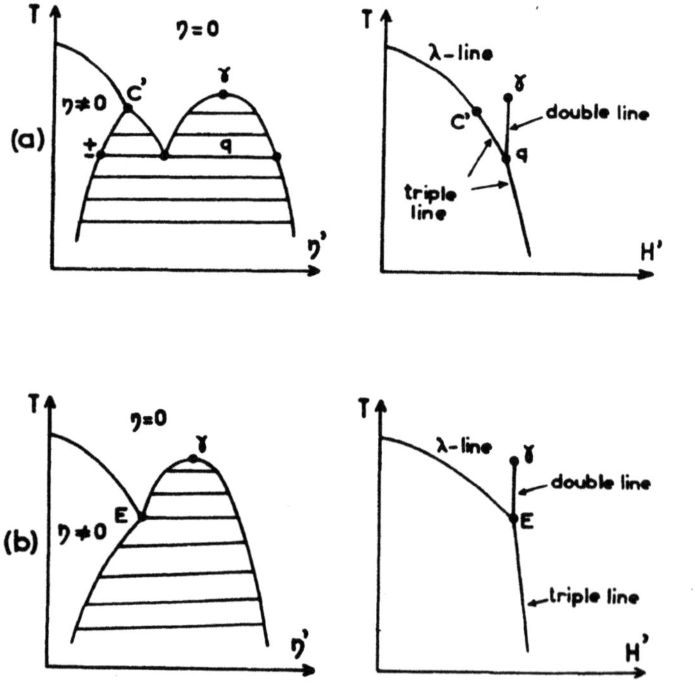

Figure 3
The phase diagrams described by the BEG model: mixtures kept at constant pressure (a) 0.365 > J/K > 0.26; (b) J/K < 0.26. The case J/K > 0.365 corresponds to fig.2 .

$S = 1$, two order parameters may be defined : the dipolar one M representing say, η and the quadrupolar one Q representing η'. Then the following generalization of Blume-Capel hamiltonian may be considered :

$$\mathcal{H} = - H \sum_i S_i - D \sum_i Q_i - J \sum_{i,j} S_i S_j - K \sum_{i,j} Q_i Q_j \qquad (11)$$

$\mathcal{H}$ can be treated in the mean field approximation and the following equations of state are found :

$$Q = \frac{2\, e^{\beta(D + 2KQ)} \operatorname{Cosh} \beta (H + 2JM)}{1 + 2e^{\beta(D + 2KQ)} \operatorname{Cosh} \beta (H + 2JM)} \qquad (12a)$$

$$M = Q \tanh \beta (H + 2JM) \qquad (12b)$$

II.3. The BEG model for constant pressure systems

The spin one Ising model has been used in a very elegant way by Blume, Emery and Griffiths [17] to represent He3-He4 mixtures. They consider a lattice gas, each cell is occupied either by a He4 atom or by a He3 atom : in the first case $Q_i = 1$, in the second case $Q_i = S_i = 0$ so that $1 - Q$ is the concentration in He3. The internal degree of freedom of He4 atoms ($S_i = \pm 1$) is used to simulate suprafluid ordering driven by J. K drives phase separation. Various possible phase diagrams in the $H = 0$ plane are described by the BEG model :

- If $\frac{J}{K} > 0.365$, phase separation driven by the intermolecular interactions K is not observed. In He3-He4 for instance, phase separation is driven by suprafluid ordering since $K \simeq 0$ (see fig. 2).
- If $0.365 > \frac{J}{K} > 0.26$, phase separation is driven either by K or by the suprafluid interaction J (fig. 3 a). A tricritical point C' and a critical point γ are observed ; the triple line and the normal liquid-normal liquid coexistence line (double line) meet at a quadruple point q.
- If $\frac{J}{K} < 0.26$, the critical line (λ-line) terminates at a critical end point E : E represents a critical phase in equilibrium with a non critical phase ; no tricritical point is found (fig. 3 b).

A model equivalent to the BEG model has been used to describe AB binary alloys (A magnetic, non magnetic) [18]-[20]. At a site (or in the cell) occupied by an A atom, $S_i = \pm 1$; at a site occupied by a B atom, $S_i = 0$.
J drives magnetic ordering, K drives phase separation. If $Q = 1$ (no B atoms) the model

	$S_i = 1$	$S_i = 0$	$S_i = -1$	System
Constant Pressure	$He4^+$	He3	$He4^-$	He3–He4 mixture
	A ↑	B	A ↓	Magnetic alloy or magnetic fluid mixture
	(A)	(B)	(A)	Two-dimensional nematic mixture
	A	B	C	Ternary mixture ($L \neq 0$)
Variable Pressure	$He4^+$	□	$He4^-$	He4
	A ↑	□	A ↓	Paramagnetic gas
	(A)	□	(A)	Two-dimensional simple nematic
	□ A	□□	A □	Simple fluid
	A	□	B	Binary mixture ($L \neq 0$)

TABLE III : Schematic description of the systems represented by the $\mathcal{S} = 1$ Ising model.

reduces to a spin one-half Ising model, so that magnetic ordering is second order. As the concentration in B is increased, it becomes first order and induces phase separation.

The BEG model might describe as well a mixture of elongated molecules A and spherical molecules B. Increasing the concentration in B is expected to lower the nematic ordering temperature [21]. However nematic ordering in pure A is "already" a first-order trantition, so that no tricritical point can be found. The BEG model applies only to a two dimensional AB mixture : in monomolecular layer, an elongated molecule may be thought as having only two possible orientations ($S_i = \pm 1$), and nematic ordering induced by anisotropic molecular interactions J is indeed second-order [11].

II.4. The BEG model for variable pressure systems

In the preceding systems - He3-He4 mixtures, magnetic or nematic mixtures -, all cells of the lattice gas are occupied, and the pressure is then fixed : the model describes the influence of dilution on some cooperative ordering and possibly the interplay between this ordering and phase separation. Suppose we replace the non-active atom or molecule (He3, B) by a vacancy (state $S_i = 0$). We are now considering a pure fluid at variable pressure : He4 [22], a paramagnetic gas [19], a monomolecular layer of elongated molecules at variable surface pressure (table III). Another possibility has been considered [22]. Consider a pure fluid and suppose that each cell of the lattice gas is divided into two subcells : preferential occupation of the right (or left) subcell simulates solidification. A gas or liquid phase is such that $Q \neq 0$, $M = 0$; $M \neq 0$ for a solid phase.

Whatever the physical nature of the internal ordering of the fluid ($M \neq 0$), it can be shown that the pressure P is given by :

$$P = - kT \operatorname{Log} (1 - Q) - KQ^2 - JM^2 \qquad (13)$$

where Q is the density. It is then easy to get the possible phase diagrams of the fluid [22]. The interaction K drives condensation and determines the liquid-gas critical point γ. J drives the suprafluid, nematic, magnetic or positional ordering. D is the chemical potential of the fluid.

- If $\frac{J}{K} > 0.365$, no liquid-gas equilibrium is found.
- If $0.365 > \frac{J}{K} > 0.26$, the ordinary phase diagram of a fluid is found. q is the triple point, in fact a quadruple point since the solid phase or ordered liquid is found as two twins or domains $\pm$ M.
- If $\frac{J}{K} < 0.26$,the liquid-ordered liquid transition is always second-order as in

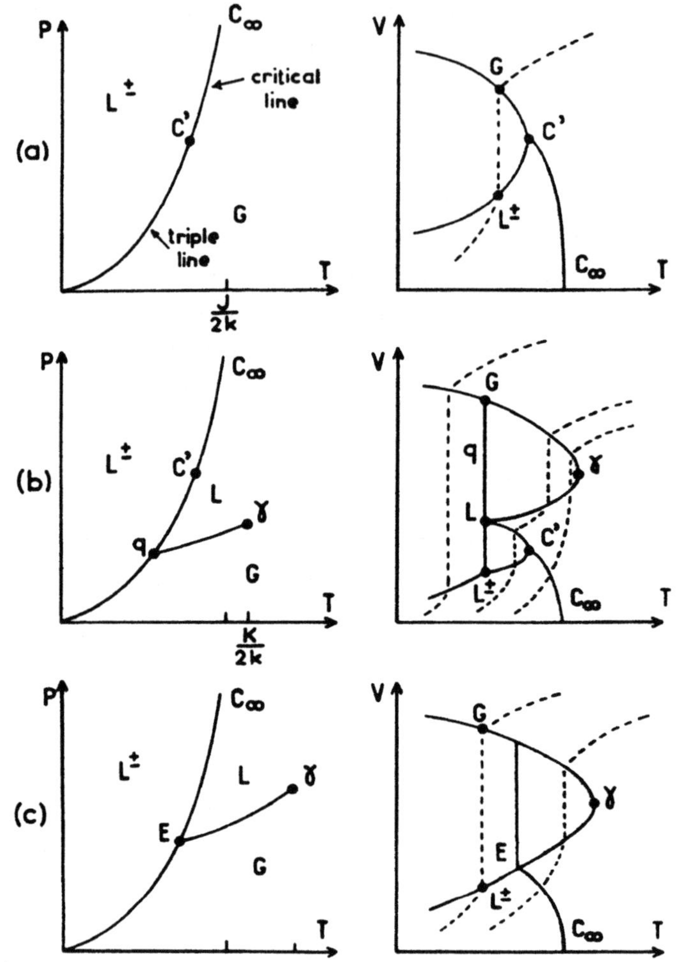

Figure 4 : The phase diagrams described by the BEG model : pure fluids at variable pressure (a) $\frac{J}{K} > 0.365$; (b) $0.365 > \frac{J}{K} > 0.26$; (c) $\frac{J}{K} < 0.26$

P is the pressure, V the specific volume. G is a gas phase, L a disordered liquid, $L^{\pm}$ an ordered liquid. C_{∞} is the critical point at infinite pressure. Dashed lines are isobars.

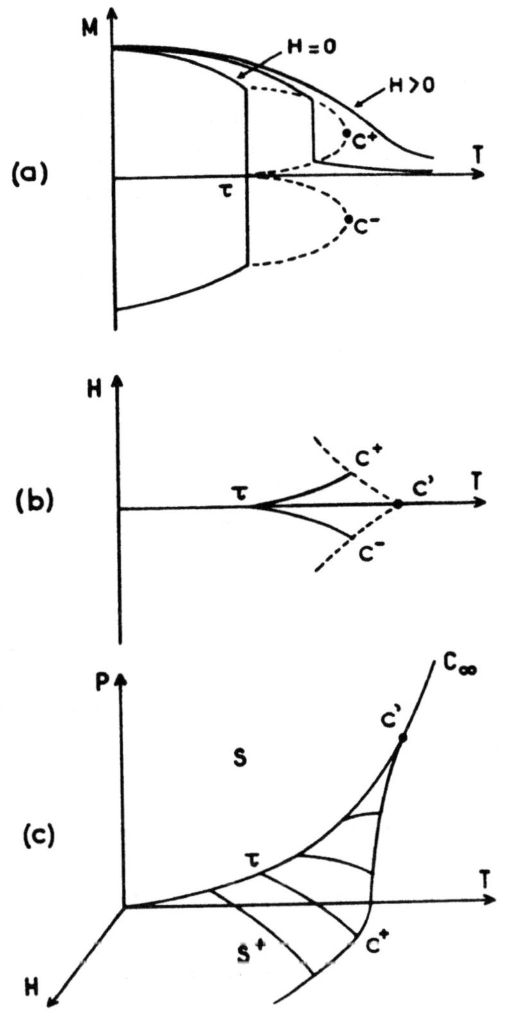

<u>Figure 5</u> : Phase diagram of the paramagnetic gas.(a) influence of an external field on the first order magnetic transition. C^+ and C^- are non-symmetrical critical points. The dashed lines are binodal lines.(b) Projections of the coexistence lines τC^+ and τC^- in the (H, T) plane. The dashed lines are the projections of the non-symmetrical critical lines. (c) Three dimensional (P, H, T) phase diagram. The coexistence surface S^- is not represented.

He4 (fig. 4).

The two limiting situations are interesting. If $\frac{J}{K} = 0.365$, a point on the triple line is a quadruple point representing a critical liquid-gas phase γ (M = 0) in equilibrium with two condensed ordered phases $\pm$ M. If $\frac{J}{K} = 0.26$, the quadruple point is in fact a tricritical point in equilibrium with a gaseous phase.

Suppose the model describes a simple fluid or simple nematic :

- At high enough pressure, the model exhibits critical melting or critical nematic ordering, which is never found experimentally in agreement withgeneral symmetry arguments given by Landau [6].
- This difficulty can be removed by introducing at least three states for each cell (three subcells, or three orientations) : the internal transition - solidification or nematic ordering - is then always first order (as the transition in the Potts model) and of course there is no longer a tricritical point [23].

The model however applies to a special type of fluid : monomolecular layer adsorbed on the surface of a liquid. Landau arguments do not apply to such a two-dimensional system, and indeed the low density - high density transition observed experimentally may be second order [11].

Consider now, as a second example, a paramagnetic gas with a large exchange integral J. Of course the model ignores the possibility of solidification of the gas. As pressure is decreased, the critical temperature is reduced and a critical line is found : it is the limit of a coexistence surface S, each point of which represent two domains of opposite magnetizations. When P is sufficiently low, the magnetic transition becomes first order. The triple line C'q and the liquid gas coexistence line γq meet at the quadruple q where four phases coexist : two paramagnetic phases of different densities and two condensed magnetic domains. At lower pressures the triple line describes sublimation of a magnetic condensed phase.

II.5. The Griffiths picture of a tricritical point

The tricritical point C' found for instance in the spin one Ising model for a paramagnetic gas is said to be symmetrical since it is found in the H = 0 plane. Similarly the critical line C_∞ C' and the triple line C'q are said to be symmetrical : two of the three phases which coexist at each triple point τ are related by time-reversal symmetry.

If H $>$ 0 is applied to the magnet, the critical line C_∞ C' disappears, the magnetization gap at each triple point τ is reduced (fig. 5) and T_τ it self is modified.

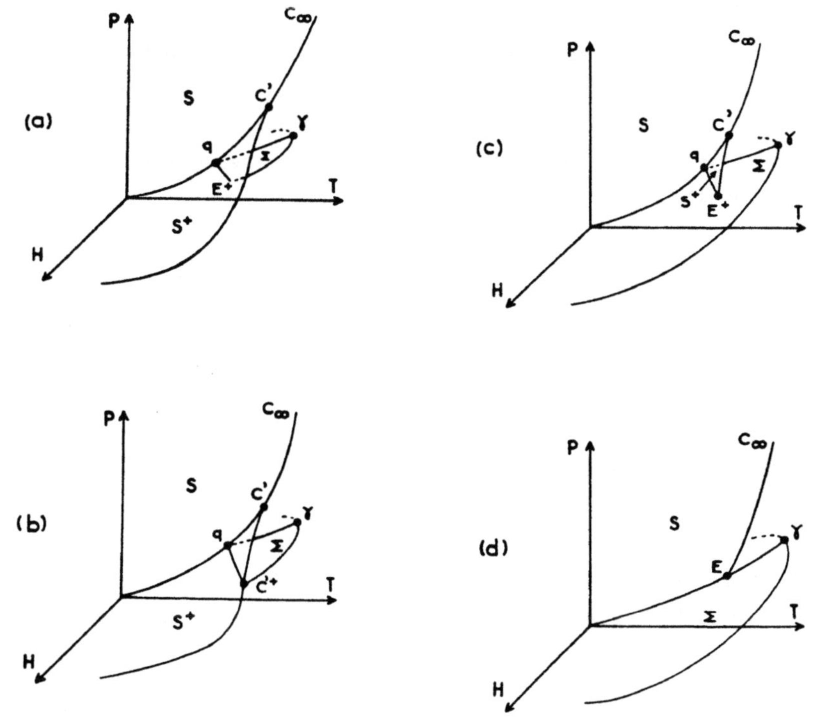

Figure 6 : Three dimensional phase diagram of the paramagnetic gas. Only the $H > 0$ part of the diagram is represented. (a) $0.365 > \frac{J}{K} > \frac{1}{3}$; (b) $\frac{J}{K} = \frac{1}{3}$; (c) $\frac{1}{3} > \frac{J}{K} > 0.26$; (d) $\frac{J}{K} < 0.26$. Σ is the liquid-gas coexistence surface, qE^+ is a triple line.

If H is large enough, the transition is suppressed and a critical point C^+ is found ; for very large H, the system is magnetic at any T. Finally we see that two critical lines develop from C' in the $H > 0$ and $H < 0$ regions respectively ; each of them limits, together with the triple line, a wing-like coexistence surface S^+ or S^-. The critical point C^+_- or C^- may be called non-symmetrical since the two phases which coexist just below $C^{\pm}$ do not have opposite magnetizations.

In conclusion we have the Griffiths picture of C' [24]-[25] :

- Three critical lines, limiting three coexistence surfaces, meet at the tricritical point C'.
- The three coexistence surfaces S, S^+ and S^- intersect along the triple line
- As one goes from τ to C' along the triple line, the three phases in equilibrium become identical.

Moreover the coexistence line q , if found for H = 0, is modified by the application of H : q belongs to a coexistence surface Σ, symmetrical in H and limited by a critical line. Σ meets S^+ and S^- along triple lines.

- If $0.365 > \frac{J}{K} > \frac{1}{3}$, these triple lines terminate at critical end points, E^+ and E^- respectively, which belong to the critical line limiting Σ (fig. 6a).
- If $\frac{1}{3} > \frac{J}{K} > 0.26$, the critical end points belong to the non-symmetrical critical lines (fig. 6c).

As shown by Mukamel and Blume [26], the intermediate situation is quite interesting : the non-symmetrical critical lines meet the critical line limiting Σ, so that three tricritical points are found for $\frac{J}{K} = \frac{1}{3}$. This value introduces a new, accidental, symmetry in the system, so that the two tricritical points found for $H \neq 0$ may be considered as symmetrical, as it will become apparent below.

II.6. The generalized spin one Ising model

The spin one Ising model can be given other interpretations [26]-[29]. Consider first a binary mixture AB at variable pressure. It may be represented by the following lattice gas model. Each cell is characterized by a state variable S_i : $S_i = 0$ if the cell is empty, $S = + 1$ (- 1) if it is occupied by an A (B) molecule. Introducing binary interactions J_{AA}, J_{AB}, J_{BB} and the following linear combinations :

$$K = \frac{1}{4} (J_{AA} + J_{BB} + 2 J_{AB})$$

$$J = \frac{1}{4} (J_{AA} + J_{BB} - 2 J_{AB}) \qquad (14)$$

$$L = \frac{1}{4} (J_{AA} - J_{BB})$$

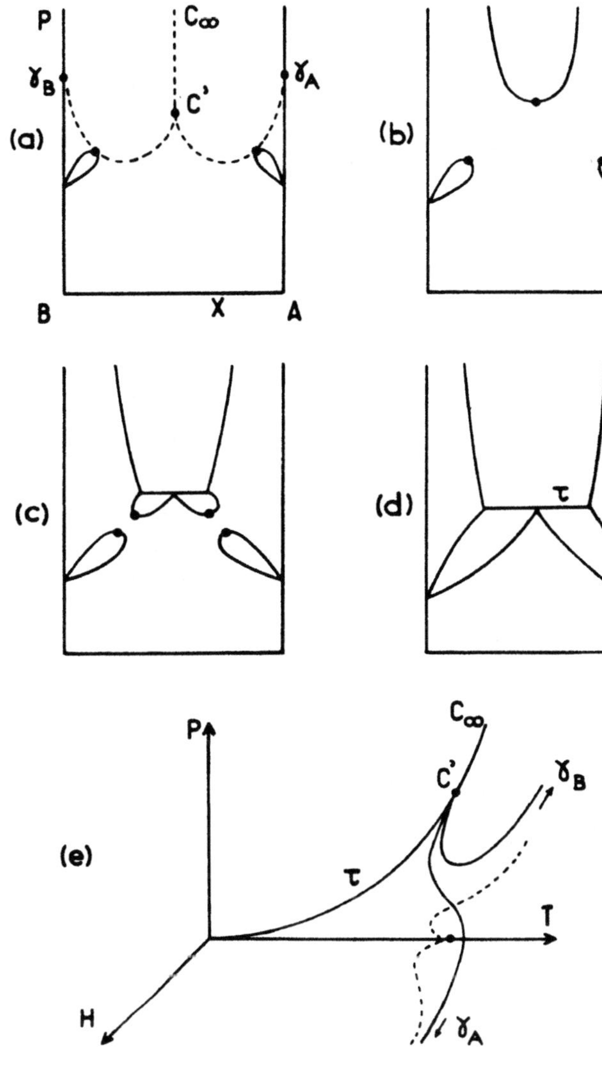

Figure 7 : Schematical phase diagrams of a symmetrical binary mixture for $\frac{J}{K} >$ 0.365. T is lowered from (a) to (d), τ is a triple point, the dashed line is the critical curve. (e) is the (P, H, T) phase diagram : the dashed line is the projection of the liquid gas critical curve in the (H, T) plane.

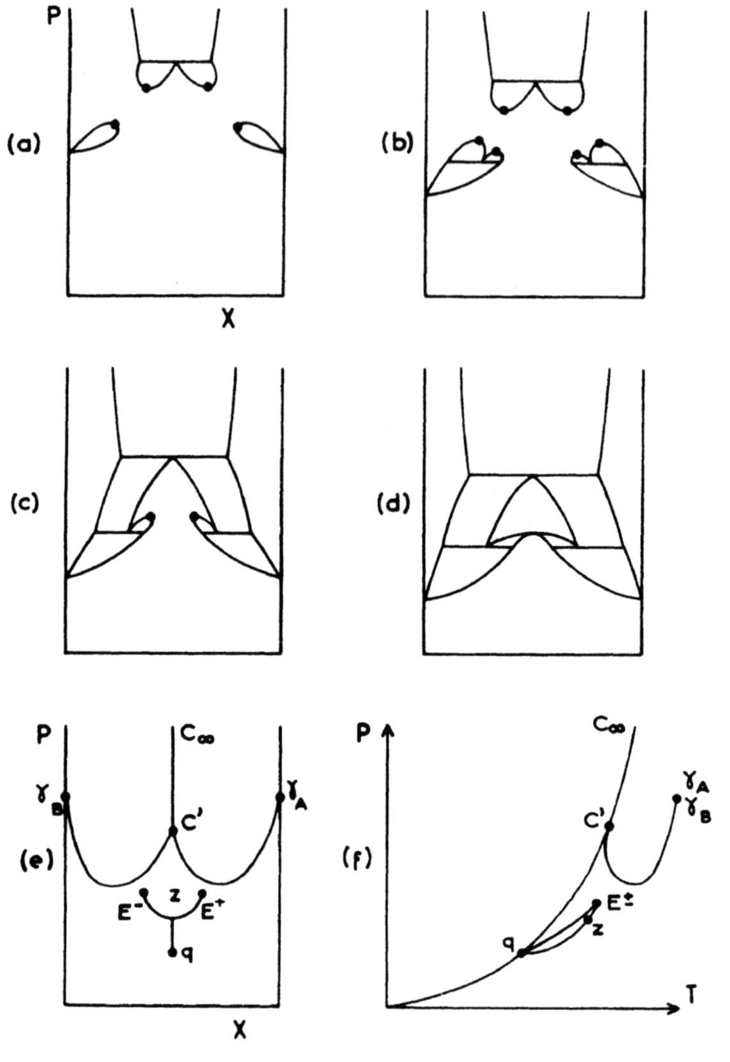

Figure 8 : Phase diagrams of a symmetrical binary mixture for $0.365 > \frac{J}{K} > \frac{1}{3}$. T is lowered from (a) to (d). (e) γ_A C' γ_B and E^+ Z E^- are critical curves. E^+ and E^- are critical end points, Z is the critical azeotrope, qZ is the azeotropic line.(f) qC', qE^+ and qE^- are triple lines projected on the (P, T) plane.

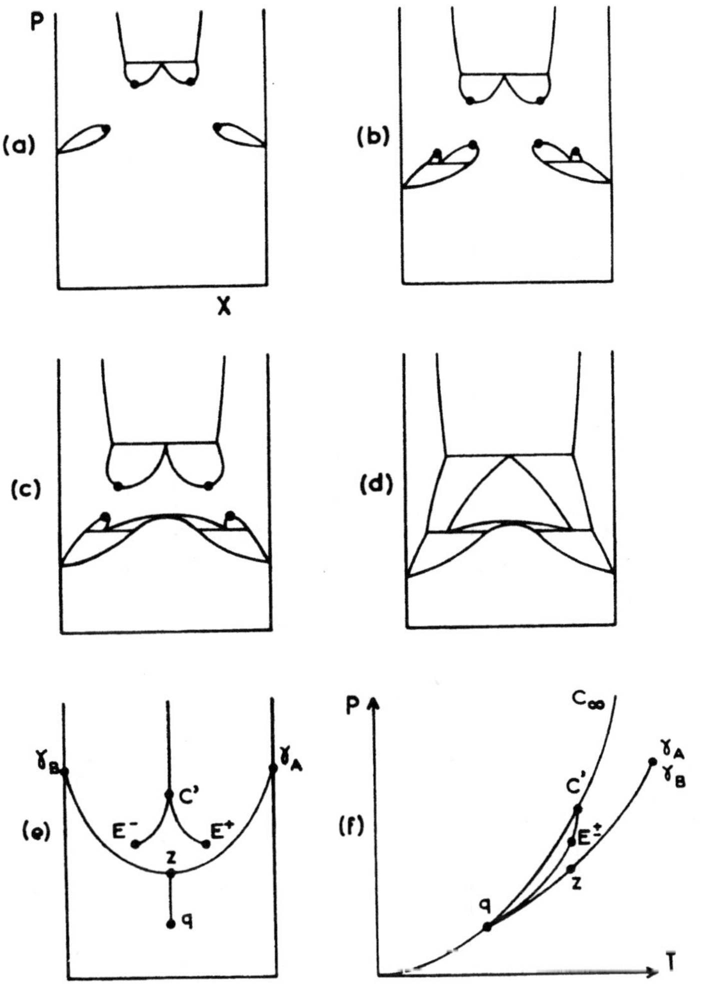

<u>Figure 9</u> : Phase diagrams of a symmetrical binary mixture for $\frac{1}{3} > \frac{J}{K} > 0.26$. T is lowered from (a) to (d).

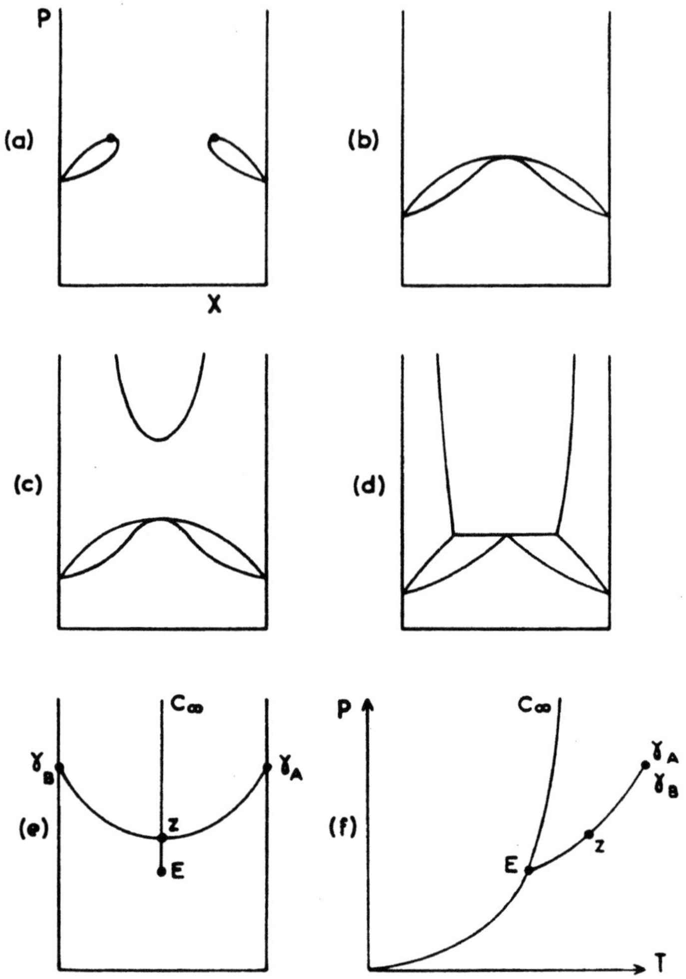

<u>Figure 10</u> : Phase diagrams of a symmetrical binary mixture for $\frac{J}{K} < 0.26$. T is lowered from (a) to (d). (e) γ_A Z γ_B is the Andrews critical curve which presents a minimum at Z, the critical azeotrope ; E is the critical end point, ZE is the azeotropic line.

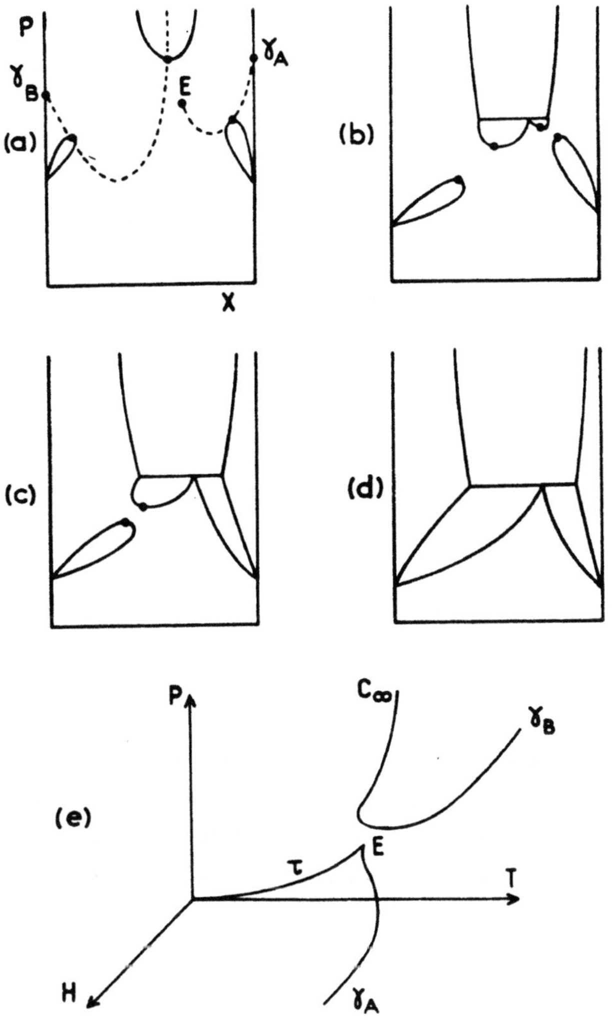

Figure 11 : Phase diagrams of a binary mixture for $\frac{J}{K} > 0.365$ and L slightly positive.

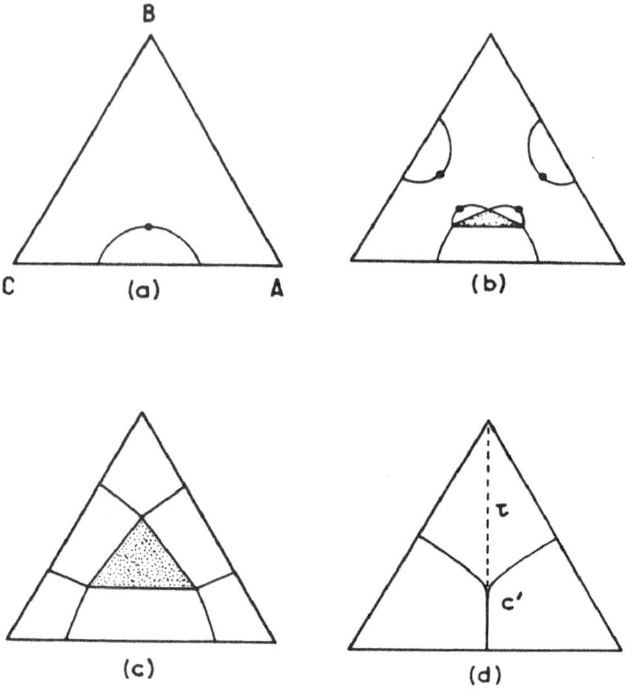

<u>Figure 12</u> : Phase diagrams of a symmetrical ternary mixture for $\frac{J}{K} > 0.365$. T is lowered from (a) to (c). Stippling denotes a three phase region. (d) is the activity triangle. The dashed line is a triple line, the full lines are critical lines.

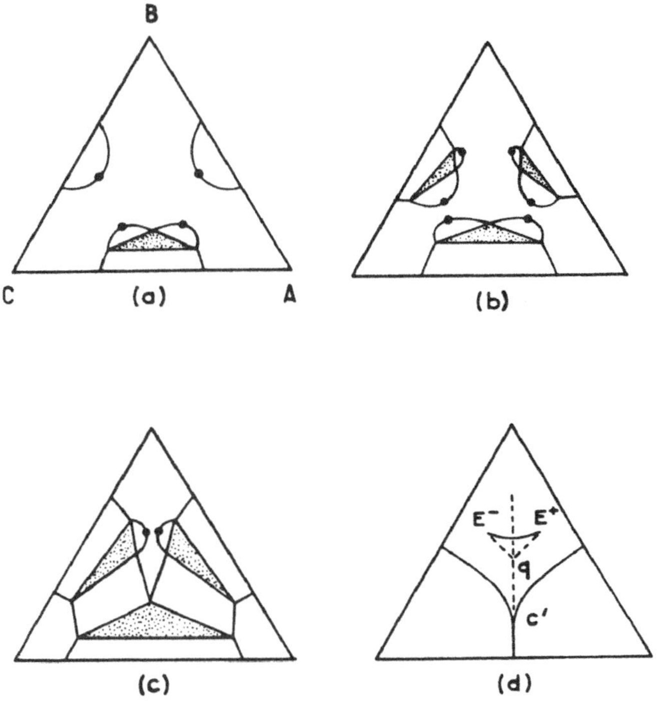

<u>Figure 13</u> : Phase diagrams of a symmetrical ternary mixture for $0.365 > \frac{J}{K} > \frac{1}{3}$. E^+ and E^- are critical end points, q is a quadruple point.

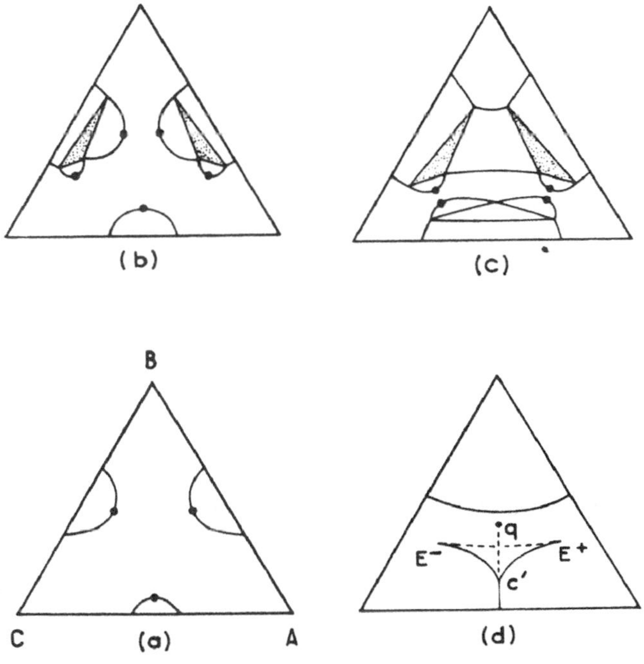

<u>Figure 14</u> : Phase diagrams of a symmetrical ternary mixture for $\frac{1}{3} > \frac{J}{K} > 0.365$.

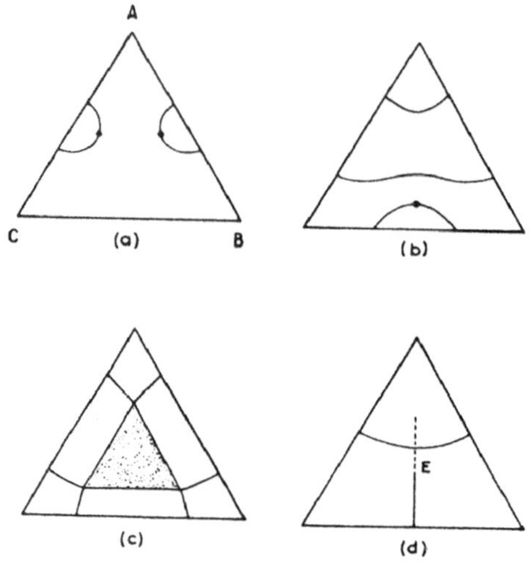

Figure 15 : Phase diagrams of a symmetrical ternary mixture for $\frac{J}{K} < 0.26$.

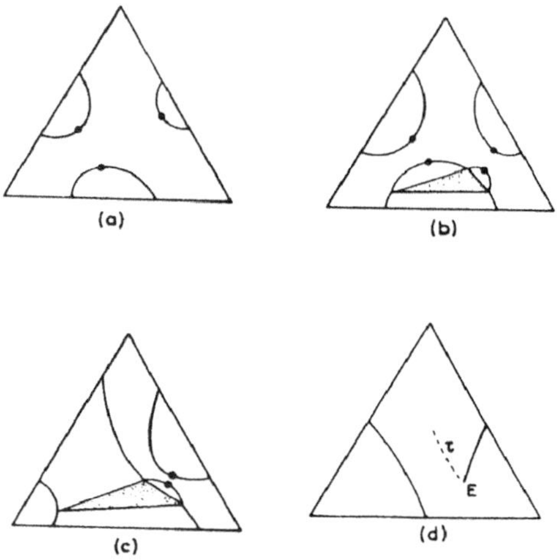

Figure 16 : Phase diagrams of a ternary mixture for $\frac{J}{K} > 0.365$ and L slightly positive.

the mixture is discribed by the generalized spin-one Ising hamiltonian :

$$\mathcal{H} = - H \sum_i S_i - D \sum_i Q_i - J \sum_{i,j} S_i S_j - K \sum_{i,j} Q_i Q_j - L \sum_{i,j} (S_i Q_j + S_j Q_i) \quad (15)$$

Of course the crossed dipolar-quadrupolar interaction L, which describes the "difference" between pure A and pure B, has no counterpart in magnetic systems because of time-reversal invariance. Q is the density of the mixture, $\frac{M}{Q}$ describes the deviation from equiconcentration. If μ_A and μ_B are the chemical potentials of A and B, one has :

$$D = \frac{1}{2} (\mu_A + \mu_B) \qquad H = \frac{1}{2} (\mu_A - \mu_B) \quad (16)$$

Let us now consider a ternary mixture A B C at constant pressure. It can be represented by a similar lattice-gas model : S_i = 0,1, − 1 if the cell i is occupied by a B, A, C molecule respectively. Then K, L, J, D and H are linear combinations of the intermolecular interactions and chemical potentials, and the composition of the mixture is given by :

$$X_A = \frac{Q + M}{2} \qquad X_B = 1 - Q \qquad X_C = \frac{Q - M}{2} \qquad (X_A + X_B + X_C = 1) \quad (17)$$

In a symmetrical AB mixture, all phase diagrams are symmetrical with respect to $x = \frac{1}{2}$. K induces condensation, and J induces phase separation in a liquid phase. Fig. 7 presents the phase diagrams found for $\frac{J}{K} > 0.365$. γ_A and γ_B are Andrews critical points of A and B respectively. No liquid-gas equilibrium is found for x = 0. The symmetrical curve $C_\infty C'$ limits the liquid-liquid coexistence surface. Fig. 8, 9 and 10 show the phase diagrams found for $0.365 > \frac{J}{K} > \frac{1}{3}$; $\frac{1}{3} > \frac{J}{K} > 0.26$; $\frac{J}{K} < 0.26$ respectively (the situation for $\frac{J}{K} = \frac{1}{3}$ is given in Ref. [28]). Details are given in the captions.

If L ≠ 0, no symmetrical tricritical point can be found, since phases with the same Q and opposite M do not exist. Fig. 11 shows how the phase diagrams of Fig. 7 ($\frac{J}{K} > 0.365$, L = 0) are modified when L is slightly positive. Fig. 12 and 13 of Ref. [27] show how the phase diagrams of Fig. 10 are modified.

Fig. 12 to 15 present the phase diagrams of a symmetrical ternary mixture : triangles a, b and c are composition triangles ; d is the activity triangle [30].

	$n_i = 0$	$S_i = 1$	$n_i = 1$, $S_i = 0$	$S_i = -1$	System
Constant Pressure	B	A_x	A_y	A_z	Binary nematic mixture
	C	A ↑	B	A ↓	Ternary mixture A magnetic
	D	A	B	C	Quaternary mixture
Variable Pressure		A_x	A_y	A_z	Three dimensional nematic fluid
		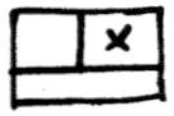	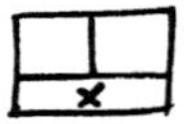	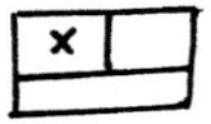	simple fluid
		$He4^+$	He3	$He4^-$	He3–He4 mixture
		A ↑	B	A ↓	paramagnetic binary fluid
		A	B	C	ternary mixture

TABLE IV : The singlet-triplet Ising model and its various interpretations.

Fig. 16 shows how the phase diagrams of Fig. 12 are modified when L is slightly positive. Fig. 4 of Ref. [28] shows how the phase diagrams of Fig. 15 are modified.

III. THE NON-SYMMETRICAL TRICRITICAL POINT

III.1. The spin 3/2 Ising model

When studying systems represented by the spin one Ising model, we found a tricritical point resulting from the coalescence of three phases, two of them being linked by a symmetry relation. We shall see that in ternary mixtures at variable pressure, or quaternary mixtures at constant pressure, non-symmetrical tricritical points are found which correspond to the coalescence of three phases without any symmetry relation. Consequently we consider now a lattice-gas model with four possible states in each cell, that is a spin $\frac{3}{2}$ Ising model [31]-[35] or equivalently a two spin one half model [36] or a singlet-triplet model [23] [37] .

In the singlet-triplet model, two state variables n_i and S_i are defined for each cell : $n_i = 0$ if the cell is empty and 1 if it is occupied ; $S_i = 1, 0, -1$ indicates how the cell is occupied. It is then possible to describe systems of variable pressure. In another interpretation, $n_i = 0$ if the cell is occupied by a D molecule and 1 if it is occupied by another type of molecule A, B or C (see table IV) : a quaternary mixture at constant pressure is then described : a D molecule has replaced a vacancy .

Four state systems are characterized by six different interactions and three order parameters. Consider for instance a ternary mixture A B C □ . The three order parameters are the density : $n = \langle n_i \rangle$, $Q = \langle n_i Q_i \rangle$ and $M = \langle n_i M_i \rangle$ whence the concentrations :

$$X_A = \frac{Q + M}{2n} \qquad X_B = \frac{n - Q}{n} \qquad X_C = \frac{Q - M}{2n} \tag{18}$$

The hamiltonian can be written :

$$\mathcal{H} = -\sum_i n_i - \sum_i n_i Q_i - \sum_i n_i S_i - k \sum_{i,j} n_i n_j - l \sum_{i,j} (n_i n_j Q_j + n_i Q_i n_j) - m \sum_{i,j} (n_i n_j S_j + n_i S_i n_j) - K \sum_{i,j} n_i Q_i n_j Q_j - J \sum_{i,j} n_i S_i n_j S_j - L \sum_{i,j} (n_i Q_i n_j S_j + n_i S_i n_j Q_j). \tag{19}$$

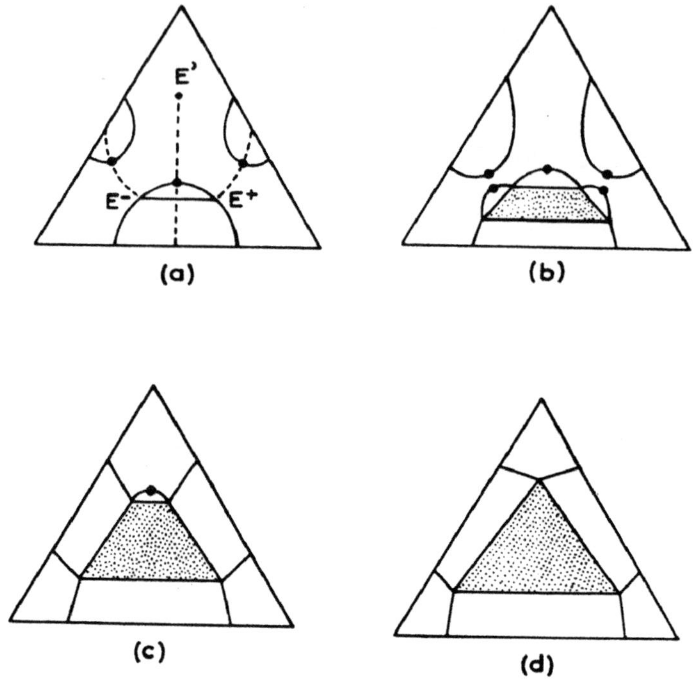

Figure 17 : Phase diagram of a ternary mixture : $L = 0, K < 0, \frac{J}{K} > \alpha$. T is reduced from (a) to (d). Dashed lines in (a) are critical lines. Stippling denotes four phase and three-phases regions.

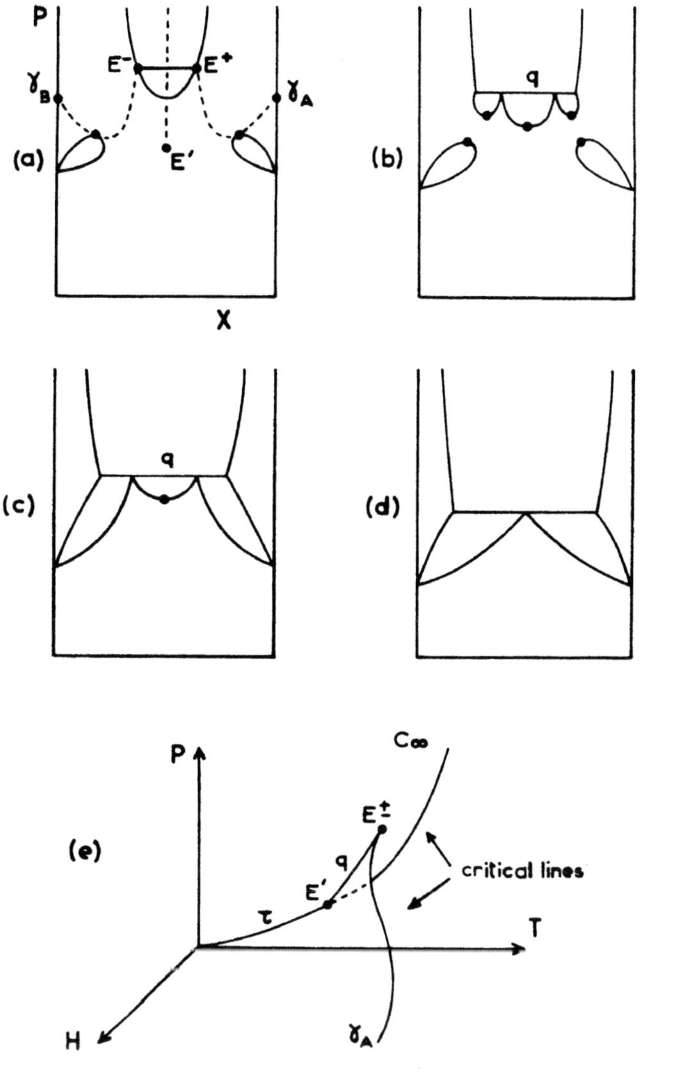

Figure 18 : Phase diagram of a binary mixture : L = 0, K < 0, $\frac{J}{K} > \alpha$. T is reduced from (a) to (d). Dashed lines in (a) are critical lines. (e) is the P, H, T phase diagram : $E^{\pm}$ E'is the quadruple line.

with :

$$\mu = \mu_B \qquad \nu = \frac{1}{2}(\mu_A + \mu_C - 2\mu_B) \qquad \lambda = \frac{1}{2}(\mu_A - \mu_C) \tag{20}$$

and :

$$k = J_{BB}$$

$$l = \frac{1}{2}(J_{AB} + J_{BC} - 2\, J_{BB})$$

$$m = \frac{1}{2}(J_{AB} - J_{BC}) \tag{21}$$

$$K = \frac{1}{4}(J_{AA} + J_{CC} + 2\, J_{AC}) + J_{BB} - J_{AB} - J_{BC}$$

$$L = \frac{1}{4}(J_{AA} - J_{CC}) + \frac{1}{2}(J_{BC} - J_{AB})$$

$$J = \frac{1}{4}(J_{AA} + J_{CC} - 2\, J_{AC})$$

Three equations of state [37] are found in the mean field approximation. The following phenomena are then reproduced :

- solidification or nematic transition in a simple fluid ;
- condensation, phase separation and suprafluid ordering in a He3-He4 mixture at variable pressure (or condensation, phase separation and magnetic ordering in a paramagnetic gas mixture) ;
- phase separation in a three dimensional binary nematic mixture ;
- phase separation in ternary and quaternary mixtures : ABC□ and ABC D.

III.2. The symmetrical tetracritical point

As shown in Ref. [26] and [30] a symmetrical tetracritical point found in the S = 1 Ising model for L =0, $K < 0$ and $J + K > 0$, more precisely for

$$\frac{J}{K} > \alpha = \frac{2(1 - Q)}{3\,Q - 1} \qquad \text{with :}$$

$$Q = \frac{1}{1 + \sqrt{10}}$$

Fig. 17 shows the phase diagram for a ternary mixture with $\frac{J}{K} > \alpha$. As T is lowered, one finds successively : a symmetrical critical line, a double critical point (two critical end points E^+ and E^- which coexist), a quadruple line which terminates at a critical end point E', and a triple line. Fig. 18 illustrates the analogous situations in a binary mixture at variable pressure.

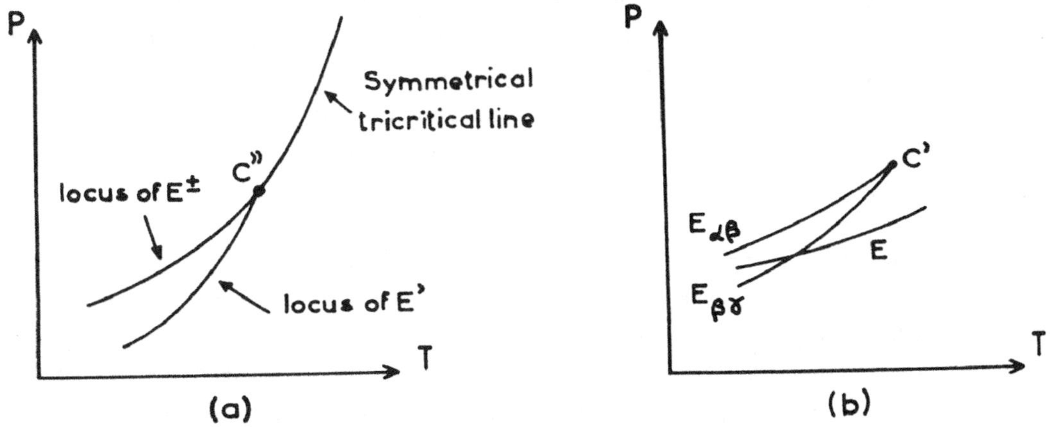

Figure 19 : (a) (P,T) phase diagram of a symmetrical ternary mixture A B C □. C" is the tetracritical point, $E^{\pm}$ a double critical point, and E' a critical end point coexisting with two symmetrical phases. (b) (P, T) phase diagram of a non-symmetrical ternary mixture A B C □. $E_{\alpha\beta}$, $E_{\beta\gamma}$ and E are critical end points.

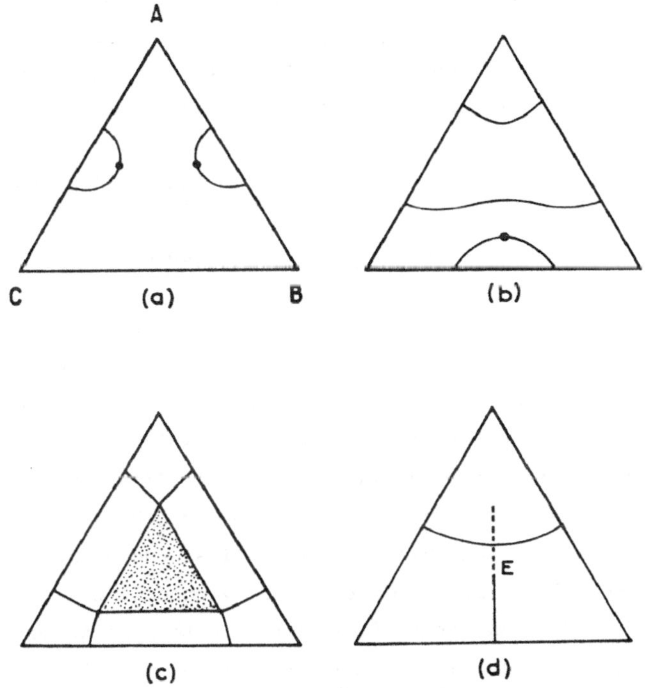

Figure 20 : Phase diagram of a non-symmetrical ternary mixture A B C near a tetracritical point for $P < P_{C''}$. T is reduced from (a) to (c). Dashed lines in (a) are critical lines. (d) is the activity triangle. $E_{\alpha\beta}$, $E_{\beta\gamma}$ and E are critical end points, dashed lines are triple lines.

Consider Fig. 18 e and suppose that $\frac{J}{K}\rightarrow\alpha$. The quadruple line $E^{\pm}$ E' shrinks and reduces to a single point C''. This point is a ***symmetrical tetracritical point*** : four phases, linked two by two by symmetry, are made to coalesce in C'' as $\frac{J}{K}\rightarrow\alpha$. However, the neighbourhood of C'' is similar to that of a tricritical point : Three critical lines C_{∞}-C'', $\gamma_{A(B)}$C'' meet at C'' (not four as the name given to C'' might suggest).

We consider now a ternary mixture A B C □ at variable pressure and temperature, for the special symmetrical case L = 0 and m = 0 (symmetry in A and C). Krinsky and Mukamel [32] have shown that a tetracritical point C'' is found for some range of variation of the interactions (as dicussed above, a point C'' is found in a mixture A B C at constant pressure only for a special set of the interactions : L = 0, $\frac{J}{K}=\alpha$). C'' is an invariant point. Its existence may be understood as follows. Suppose that the pressure of the gas is very large, almost infinite : the phase diagram in the composition triangle is determined by the ratio $\frac{J}{K}$. Suppose that $\frac{J}{K}<\alpha$: Phase diagrams are similar to those of Fig. 12 as T is lowered. When the pressure is lowered, the phase diagram is determined by all the interactions J, K, k and l. However we may consider "effective interactions" J' and K' which describe the influence of dilution of the gas by vacancies. Dilution is not expected to have a strong influence on J', but modifies K' and may even change its sign so that $\frac{J'}{K'}$ may become larger than α : phase diagrams are similar to those of Fig. 17 as T is lowered, with a double critical point and a quadruple line. For an intermediate pressure $P_{C''}$, a tetracritical point C'' is found at a given temperature $T_{C''}$. Fig. 19 a gives the qualitative phase diagram in the (P, T) plane.

III.3. The Kohnstamm picture of a tricritical point

We consider a ternary A B C □ or a quaternary mixture A B C D. Suppose we find an equilibrium of 3 phases, and vary T, P and chemical potentials μ_i. If one pair (α, ß) of phases becomes critical in the presence of the third, the corresponding point is the critical end point of a critical line which bounds a coexistence surface. In a binary system, such a point would be invariant. In a ternary mixture, it generates a line of critical end points. If phases ß and γ become critical when the physical conditions are changed, another line of critical end points may be considered. Finally if the three ***phases*** become critical simultaneouly the corresponding point *C'* may be called ***tricritical*** and it is ***invariant*** : in this point two lines of critical end points meet (in fact, as we shall see, they form a cusp). This is the Kohnstamm picture of a tricritical point [38] [39]. We shall see that only two, and not three, critical end point lines meet in C'.

Such a point C' can be found in the vicinity of a symmetrical tetracritical point C'' (in the same way as a non symmetrical critical point can be found

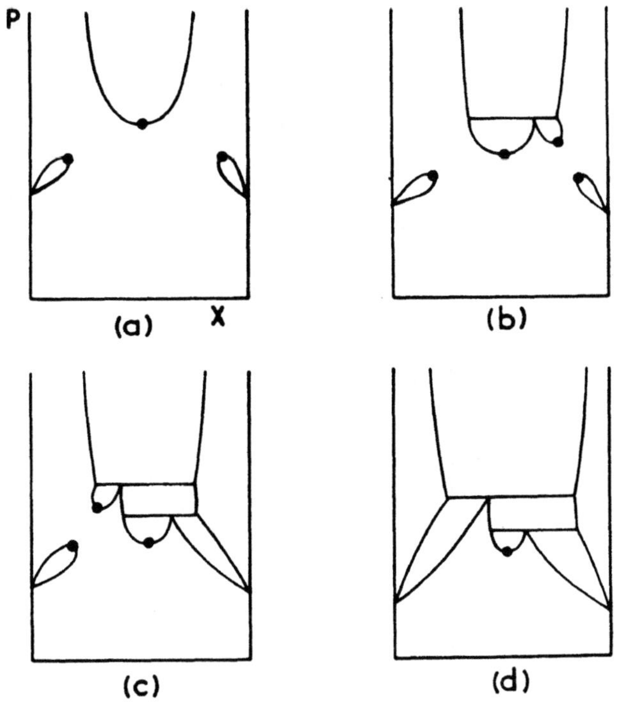

Figure 21 : Phase diagram of a non-symmetrical binary mixture A B □ near a tetracritical point. T is reduced from (a) to (d).

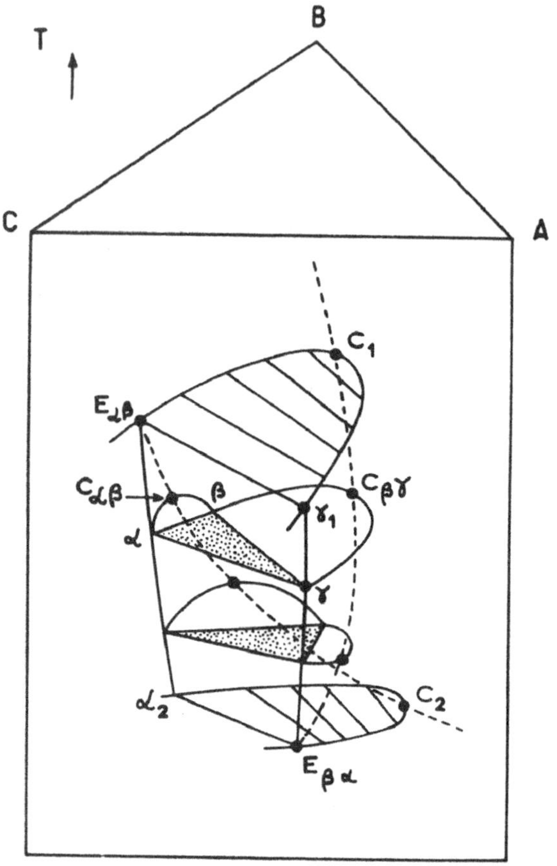

Figure 2l : Temperature composition phase diagram of a non-symmetrical ternary mixture for $P < P_{C'}$. The locus of β is not represented here, as well as the third critical line terminating at E and which plays no essential role here. γ_1 and $E_{\alpha\beta}$ are in equilibrium at $T = T_1$; α_2 and $E_{\beta\gamma}$ are in equilibrium at $T = T_2$.

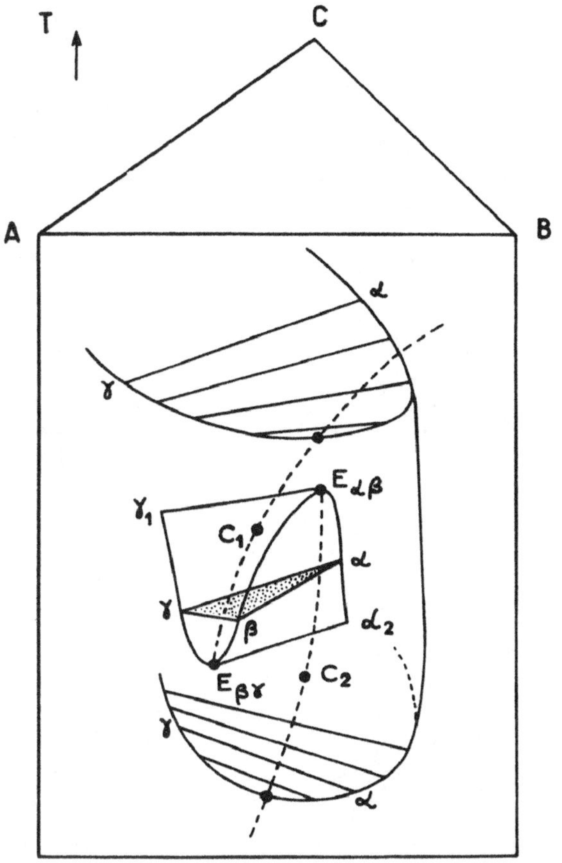

Figure 23 : Temperature-composition phase diagram of a non-symmetrical ternary mixture for $P < P_{C'}$. α, β and γ are in equilibrium for $T_1 > T > T_2$. The dashed lines are critical lines drawn on the binodal surface.

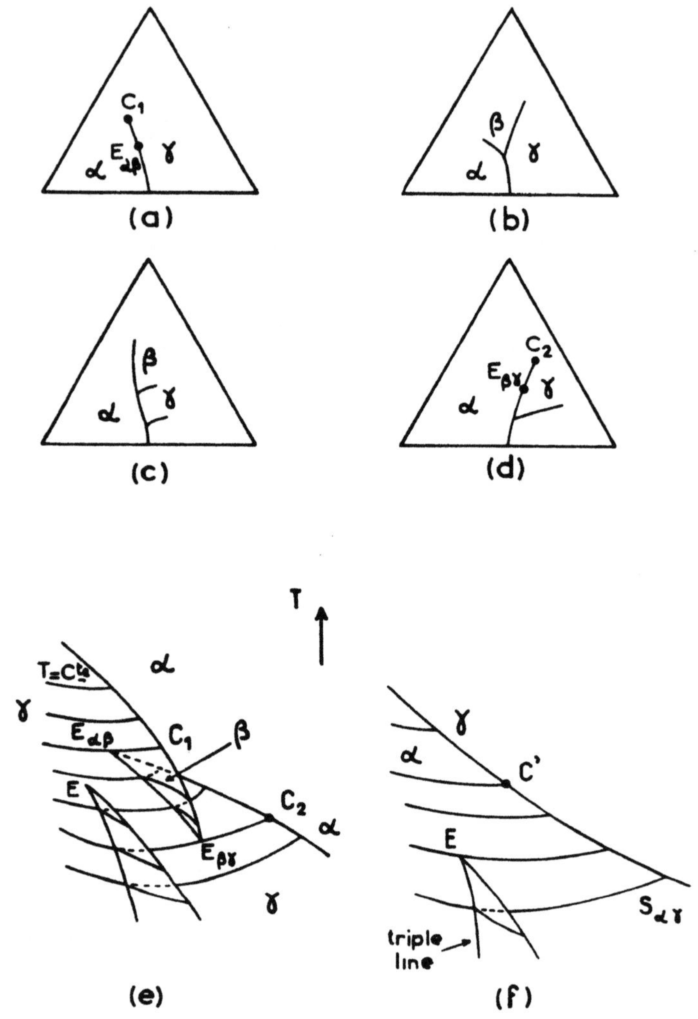

<u>Figure 24</u> : Phase diagram of a non-symmetrical ternary mixture for $P < P_{C'}$. (a) to (d) represent the activity triangle ; (a) $T = T_1$; (b) and (c) : $T_1 > T > T_2$; (d) $T = T_2$. (e) and (f) represent the three dimensional temperature activity phase diagram for $P < P_{C'}$ and $P = P_{C'}$ respectively.

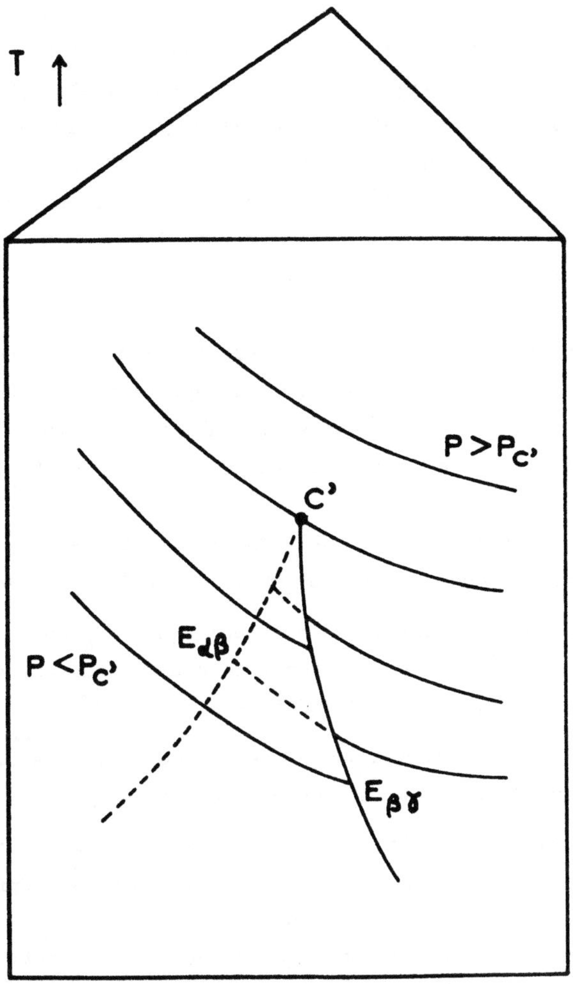

Figure 25 : Temperature composition diagram of a ternary mixture. The loci of $E_{\alpha\beta}$ and $E_{\beta\gamma}$ are shown as P is increased to $P_{C'}$. Isobar critical lines are shown.

in the vicinity of a symmetrical tricritical point). Let us consider a ternary mixture A B C □, and suppose that a point C'' is found if the interactions are symmetrical. If the interactions become slightly non-symmetrical, isobar phases diagrams for $P > P_{C''}$ are similar to those of Fig. 16. If $P < P_{C''}$, isobar phase diagrams are similar to those of Fig. 20 which derives from Fig. 17 (similarly Fig. 21 derives from Fig. 18 for a binary mixture A B □) : three critical lines are still found, each with a critical end point ; the quadruple line splits into two triple lines, and E' becomes an ordinary critical end point.

If now the pressure is increased, the length of the triple line $E_{\alpha\beta}\, E_{\beta\gamma}$ is reduced to a single point. This point, in agreement with Kohnstamm picture, is a tricritical point C' : the three phases α, β, γ in equilibrium along the triple line coalesce in C' ; two lines of critical end points (loci of $E_{\alpha\beta}$ and $E_{\beta\gamma}$) meet in C'.

Fig. 22 shows more precisely the three-dimensional temperature composition phase diagram corresponding to Fig. 20, that is for $P < P_{C'}$. For $T_{E_{\alpha\beta}} < T < T_{E_{\beta\gamma}}$, three phases α, β, and γ coexist. Two critical lines $C_{\alpha\beta}$ and $C_{\beta\beta}$ terminate in $E_{\alpha\beta}$ and $E_{\beta\gamma}$ respectively. As P is increased, the triple-phase region shrinks and disappears at $P_{C'}$. For $P < P_{C'}$, only two phases α and γ coexist for $T > T_{E_{\alpha\beta}}$ and $T < T_{E_{\beta\gamma}}$: the phase β is intermediate between α and γ, and found only in a limited range of temperature whereas α and γ are quite different. Only α and γ are found for $P > P_{C'}$.

Fig. 23 shows the same phase diagram viewed from the AB side of the composition triangle, with the two critical lines drawn on the binodal surface and the locus of β. Fox [34] has presented similar phase diagrams for quaternary mixtures at fixed pressure. Fig. 24 shows the evolution of the phase diagram in the activity triangle as T is decreased for $P < P_{C'}$ and the three dimensional temperature activity phase diagram for $P < P_{C'}$ and $P = P_{C'}$. C' is a point on the critical line limiting the coexistence surface $S_{\alpha\gamma'}$ whereas a symmetrical tricritical point is at the intersection of three critical lines.

Finally Fig. 19 b and 25 show that C' is found for a maximum of P and T, and that the two critical point lines $E_{\alpha\beta}$ and $E_{\beta\gamma}$ form the boundary of a single critical surface. This result is in agreement with experimental results of the Russian physicists quoted in Refs [38] and [39].

IV. Discussion : The validity of the lattice-gas model

A large variety of phase diagrams can be calculated using the lattice-gas model and the mean-field approximation, they reproduce the experimental phase diagrams at least qualitatively. Let us discuss the range of validity of the model more precisely.

a) The lattice-gas model is a good approximation only as long as it applies to regular solutions, in which the volume of the different molecular species are similar. When this condition is not satisfied, special features of the phase diagrams cannot be reproduced. For instance in a real simple fluid, the law of rectilinear diameters is satisfied only approximately ; in a real binary fluid, the minimum liquid-gas critical point is different from the critical azeotrope, and the liquid-liquid binodal curve is not symmetrical.

b) The mean-field approximation is generally correct as long as one does not consider the vicinity of the critical points and remains interested only in the topology of the phases diagrams. In very special cases however, this approximation might introduce spurious phase transitions or indicate a wrong order for a phase transition : this is the case for the three states Potts model [40].

c) The Ising approximation may be questionned too. The order parameter M has only a finite number of phases whereas in some systems the corresponding order parameter can be varied continuously in phase (Heisenberg magnet, suprafluid). This approximation affects the density of states and the low temperature spectrum, but not the topology of the phase diagram. Finally we find that the crucial point is the following : the order of an ordering process in M is to described correctly for the topology of the phase diagram to be reproduced. This order does not depend on the discrete or continuous character of the order parameter M but, as this parameter takes only discrete values, on the number of these values. If the transition is second order (magnetic or suprafluid transition, phase separation in a binary fluid) two values may be used ; if it is first order, three values at least are necessary (solidification, nematic ordering in three dimensions).

d) The validity of the spin $\frac{1}{2}$; 1 and $\frac{3}{2}$ models used to describe phase separation in multicomponent fluids is limited in two other ways. First solidification is ignored by the model : more complicated models have been considered to describe magnetic or nematic mixtures [41] and even simple fluids [42]. Second, the symmetrical models are irrealistic since no symmetry breaking is found in fluids.

e) In conclusion, the aim of lattice-gas model is not provide a theory for fitting experimental results. It is to simulate the behaviour of some systems (to exhibit also their analogies) so that the qualitative features - the topology - of the phase diagrams are reproduced at least in a qualitative way, in particular the interplay between the various ordering processes.

REFERENCES

(1) L. TISZA (1966), Generalized thermodynamics, The M.I.T. Press, Boston.

(2) H.E. STANLEY (1971), Introduction to phase transitions and critical phenomena, Clarendon Press, Oxford.

(3) R.B. GRIFFITHS and J.C. WHEELER (1970), Phys. Rev. A2 1047.

(4) K. HUANG (1963), Statistical Mechanics, Wiley, New-York.

(5) T.D. LEE and C.N. YANG (1952), Phys. Rev. 87, 410.

(6) L. LANDAU (1965), Collected papers of L. Landau, edited by D. ter Haar, Pergamon, London.

(7) E.H. GRAF, D.M. LEE and J.D. REPPY (1967), Phys. Rev. Letters 19, 417.

(8) I.S. JACOBS and P.E. LAWRENCE (1967), Phys. Rev. 164, 866.

(9) C.W. GARLAND and B.B. WIENER (1971), Phys. Rev. B3 1634.

(10) A.M. GOLDMAN (1973), Phys. Rev. Letters 30 1038.

(11) J.L. FIRPO et al. (1981), J. Chem Phys. 74 2569.

(12) C.P. BEAN and D. RODBELL (1962), Phys. Rev. 126, 104.

(13) J.L. MEIJERING (1963), Philips Res. Repts 18 318.

(14) J.W.DOANE (1972), Phys. Rev. Letters 28, 1694.

(15) M. BLUME (1966), Phys. Rev. 141, 517.

(16) H.W. CAPEL (1966), Physica 32, 966.

(17) M. BLUME, V.J. EMERY and R.B. GRIFFITHS (1971), Phys. Rev. A4, 1071.

(18) A. HINTERMANN and F. RYS (1969), Helv. Phys. Acta 42 608.

(19) J. BERNASCONI and F. RYS (1971), Phys. Rev. B4, 3045.

(20) M. WORTIS (1974), Phys. Letters 47 A, 445.

(21) M. PAPOULAR and J.P. LAHEURTE (1973), Solid State Commun. 12, 71.

(22) J. LAJZEROWICZ and J. SIVARDIERE (1975), Phys. Rev. A11, 2079.

(23) J. SIVARDIERE (1981), J. Phys. C. 14, 3829.

(24) R.B. GRIFFITHS (1970), Phys. Rev. Letters 24, 715.

(25) R.B. GRIFFITHS (1973), Phys. Rev. B7, 545.

(26) D. MUKAMEL and M. BLUME (1974), Phys. Rev. A10, 610.

(27) J. SIVARDIERE and J. LAJZEROWICZ (1975), Phys. Rev. A11, 2090.

(28) J. SIVARDIERE and J. LAJZEROWICZ (1975), Phys. Rev. A11, 2101.

(29) J.L. MEIJERING (1950), Philips Res. Repts **5**, 333.

(30) J.L. MEIJERING (1951), Philips Res. Repts **6**, 183.

(31) D. FURMAN, S. DATTAGUPTA and R.B. GRIFFITHS (1977), Phys. Rev. **B15**, 441.

(32) S. KRINSKY and D. MUKAMEL (1975), Phys. Rev. **B11**, 399.

(33) S. KRINSKY and D. MUKAMEL (1975), Phys. Rev. **B12**, 211.

(34) J.R. FOX (1978), J. Chem. Phys. **69**, 2231.

(35) H.K. HARDY and J.L. MEIJERING (1955), Philips Res. Repts **10**, 358.

(36) J. SIVARDIERE (1976), J. Physique **37**, 1267.

(37) J. SIVARDIERE (1981), J. Phys. **C14**, 3845 and (1982), **C15**, 1611 and 1619.

(38) B. WIDOM (1973), J. Phys. Chem. **77**, 2196.

(39) R.B. GRIFFITHS (1974), J. Chem. Phys. **60**, 195.

(40) J.P. STRALEY and M.E. FISHER (1973), J, Phys. **A6**, 1310.
See also : D. JASNOW, T. OHTA and J. RUDNICK (1979), Phys. Rev. **B20**, 2774.
And : M. KAUFMAN et al. (1981), Phys. Rev. **B23**, 3448.

(41) J. SIVARDIERE (1980), J. Physique **41**, 1081.

(42) J. SIVARDIERE (1981), Phys. Stat. Sol. **b103**, 823.

EXISTENCE OF PHASE TRANSITIONS IN SEVERAL MOLECULAR LATTICE GASES

Dale A. Huckaby
Department of Chemistry, Texas Christian University
Fort Worth, Texas 76129, USA

Four different variations of the Peierls argument are used to prove the existence of phase transitions in a number of molecular lattice gases. The models treated include lattice gas models for water, for carbon tetrachloride, for methane adsorbed on graphite, and for the coadsorption of small and large atoms.

I. The Peierls Argument

A molecular lattice gas is a model for a system of molecules in which the constituent molecules in the model are restricted to reside on fixed sites in one or more possible orientations. The grand canonical partition function for the lattice gas is given as

$$\Xi = \sum_{\xi} e^{-H(\xi)/kT}, \tag{1}$$

where $H(\xi)$ is the generalized Hamiltonian for a configuration ξ. $H(\xi)$ contains the sum of the interaction energies of the molecules in configuration ξ minus the sum of their respective chemical potentials.

In 1936 Peierls[1] gave a proof of the existence of a phase transition in the ferromagnetic Ising model, a model which was later shown by Yang and Lee[2] to be equivalent to a lattice gas with intermolecular attractions. The original Peierls argument was later made rigorous by Griffiths[3] and by Dobrushin.[4,5]

One version of the argument goes as follows. If the border of a lattice Λ contains molecules in one ordered configuration, then regions within the border which contain molecules or vacant sites not consistent with the configuration at the border of Λ can be separated from the border by closed contours. Depending upon the model, a line segment, a segment of area, or an element of volume can be used as a contour segment.

If N_α is the number of sites of Λ which are in the same ordered configuration α as is on the border of Λ, then

$$|\Lambda| - \langle N_\alpha \rangle < \sum_{|\gamma|} m_{|\gamma|} n_{|\gamma|} P_\gamma , \tag{2}$$

where $n_{|\gamma|}$ is the number of contours of length $|\gamma|$, $m_{|\gamma|}$ is the maximum number of sites enclosed by a contour of length $|\gamma|$, and P_γ is an upper bound to the probability a given contour γ of length $|\gamma|$ occurs in a configuration. Here the thermal average is only over configurations with ordered structure α at the border of Λ.

As discussed by Dobrushin[6] and Griffiths,[7] if p is the number of ordered structures which can be superimposed by some symmetry operation, then there is an ordered phase in the model at sufficiently low temperature (or at sufficiently high density in the case of hard core repulsion) in the thermodynamic limit ($|\Lambda| \to \infty$) if $\langle N_\alpha\rangle/|\Lambda| > 1/p$, independent of $|\Lambda|$. Series expansions suffice to show that the model is disordered at high temperatures (or low densities for the hard core case), thereby proving the existence of an order-disorder transition in the model.

The number $n_{|\gamma|}$ of possible closed contours containing $|\gamma|$ segments in a d-dimensional lattice is bounded as $n_{|\gamma|} < a\, c^{|\gamma|}|\Lambda|/|\gamma|$, where a is a constant, c is the maximum number of ways to continue a contour, and $|\Lambda|$ is proportional to the number of places at which the contour can be begun. The division by $|\gamma|$ results since the choice of the first segment is arbitrary. The maximum number of sites $m_{|\gamma|}$ contained in a contour of length $|\gamma|$ is bounded from above by the number of sites within a d-dimensional spherical region having the contour as its border. Hence $m_{|\gamma|} < b\,|\gamma|^{d/(d-1)}$, where b is a constant. Substituting into Eq. (2) gives

$$\langle N_\alpha\rangle/|\Lambda| > 1 - ab \sum_{|\gamma|} c^{|\gamma|}|\gamma|^{1/(d-1)} P_\gamma . \qquad (3)$$

The difficult part in completing the proof involves obtaining an upper bound for P_γ such that $\langle N_\alpha\rangle/|\Lambda| > 1/p$ at sufficiently low temperature. If a 1-1 correspondence can be found between configurations $\{\xi_\gamma\}$ which contain the contour γ and configurations $\{\xi'_\gamma\}$ which do not contain the contour γ, such that $H(\xi_\gamma) - H(\xi'_\gamma) > \alpha|\gamma|$, where $\alpha > 0$, then $P_\gamma < \exp(-\alpha|\gamma|/kT)$, and Eq. (3) then gives $\langle N_\alpha\rangle/|\Lambda| > 1/p$ for sufficiently low T.

This correspondence was effected for the ferromagnetic Ising model (attractive lattice gas) by Griffiths.[3] In that case a configuration ξ'_γ can be formed from ξ_γ by reversing all spins in ξ_γ which are inside γ. The argument was extended by Dobrushin[6] to include the case of the repulsive lattice gas in which there are two ordered structures related by translation. The Dobrushin argument was extended by Heilmann[8] to include cases in which there are more than two ordered structures related by translation. Heilmann[9] further extended the Peierls argument to include lattice gases with ordered structures related by reflection or rotation.

We shall give examples of molecular lattice gases in which hole-particle interchange, translation, and reflection are used in order to obtain an upper bound to P_γ. Before discussing these examples, we first describe another, quite different, method for obtaining an upper bound to P_γ.

II. Reflection Positivity

Reflection positivity is an idea that originated in field theory. The first application in statistical mechanics of the use of reflection positivity combined with the Peierls argument was given by Fröhlich and Lieb.[10]

We shall now outline the basic ideas[11-14] for use with the square lattice having periodic boundary conditions; i.e., composed of squares $\Lambda = \{(a,b): a,b = 0,1,...,2M-1\}$ with coordinates (a,b) computed modulo 2M onto $0 \le x,y < 2M$. Reflection lines $R_a^- = \{(a,y): y \in \mathbb{R}\}$ and $R_a^+ = R_{a+M}^-$ divide Λ into three disjoint regions,

$$\Lambda_a^+ = \Lambda \cap \{(x,y): M+a < x < 2M+a,\ y \in \mathbb{R}\},$$

$$\Lambda_a^- = \Lambda \cap \{(x,y): a < x < M+a,\ y \in \mathbb{R}\}, \text{ and}$$

$$\Lambda_a^\circ = \Lambda \cap (R_a^- \cup R_a^+).$$

(For some models, the reflection lines are chosen to pass between lattice sites.) There is a natural involution Θ_a: $(x,y) \to (2a-x,y)$ which reflects the coordinates through the reflection lines $R_a^\pm$.

If C_a^+ is the set of allowed molecular configurations on Λ_a^+ (and similarly for C_a^- and C_a°), then suppose $\Theta_a C_a^\pm = C_a^\mp$ and $\Theta_a C_a^\circ = C_a^\circ$, the latter transformation being invariant. We then, for any function $f: C \to \mathbb{C}$, define $\Theta_a f$ as $(\Theta_a f)(\xi) = f(\Theta_a(\xi))\ \forall\ \xi \in C$. Denote $\xi = (\xi_a^-, \xi_a^\circ, \xi_a^+)$, where $\xi_a^\pm \in C_a^\pm$ and $\xi_a^\circ \in C_a^\circ$. Let $F_a^+ = \{f: f(\xi) = f(\xi_a^\circ, \xi_a^+)\ \forall\ \xi \in C\}$. Then $\Theta_a f(\xi) = f(\xi_a^\circ, \xi_a^-)$ if $f \in F_a^+$. Define F_a^- analogously. Then it can be shown for some models that[12]

$$\sum_{\xi \in C} \overline{f(\xi)}\ \Theta_a f(\xi) \ge 0 \text{ if } f \in F_a^+ \cup F_a^-. \tag{4}$$

If the Hamiltonian can be written as $H = H_a^+ + \Theta_a H_a^+ = H_a^- + \Theta_a H_a^-$, where $H_a^\pm \in F_a^\pm$, then Eq. (4) gives $\langle \bar{f}\Theta_a f\rangle \ge 0\ \forall\ f \in F_a^- \cup F_a^+$. A standard Cauchy-Schwarz proof then gives

$$|\langle fg\rangle|^2 \le \langle \bar{f}\Theta_a f\rangle \langle \bar{g}\Theta_a g\rangle\ \forall\ f \in F_a^+,\ g \in F_a^-. \tag{5}$$

Let L be a nonempty set of small, nonoverlapping regions, S_r, where S_r is associated with the point r. If the Hamiltonian restricted to a region $S_r(\xi)$ is minimal, the region is said to be a "ground state region;" otherwise, it is an "excited state region." Let

$$Q_r(\xi) = \begin{cases} 1 \text{ if } S_r(\xi) \text{ is an excited state region} \\ 0 \text{ otherwise.} \end{cases}$$

If $Q(L) = \prod_{r\in L} Q_r(\xi)$, then $\langle Q(L)\rangle$ is the probability that L is a set of excited state regions.

Suppose each contour γ is associated with a certain set L_γ of excited state regions, such that $|L_\gamma| \geq \delta|\gamma|$, where δ is a constant. Then $P_\gamma \leq g^{\delta|\gamma|}$, where $g = \max_L \langle Q(L)\rangle^{1/|L|}$.

If $Q = Q_a^+ Q_a^-$, where $Q_a^\pm(L) = Q(L \cap (\Lambda_a^\pm \cup \Lambda_a^\circ))$, then from Eq. (5), $\langle Q\rangle^2 \leq \langle Q_a^+ \Theta_a Q_a^+\rangle \langle Q_a^- \Theta_a Q_a^-\rangle$. Let

$$f(Q) = \begin{cases} \langle Q\rangle^{1/|L|} & \text{if } |L| \neq 0 \\ 1 & \text{if } |L| = 0. \end{cases}$$

Then $f(Q) \leq f(Q_a^+\Theta_a Q_a^+)^{|L_+|/|L|} f(Q_a^-\Theta_a Q_a^-)^{|L_-|/|L|}$. If L_m maximizes $f(Q)$, it also maximizes $f(Q_a^+\Theta_a Q_a^+)$. (Proof by contradiction.) Hence, if $r\in L_m$, then $\Theta_a r \in L_m$ as well. But this is then true for <u>any</u> pair of lines $R_a^\pm$, $R_b^\pm$. Hence L_m contains regions "all over" Λ, which facilitates the calculation of a bound to g and hence to P_γ.

An alternative to Eq. (2) is obtained as follows. If $S_r(\xi)$ is a ground state region, and $S_{r'}(\xi)$ is not in a configuration compatable with $S_r(\xi)$, then either r or r' is surrounded by a closed contour. The probability that $S_{r'}$ is not in a ground state configuration compatable with S_r is then bounded as

$$h \leq \sum_{|\gamma|} m_{|\gamma|} n_{|\gamma|} g^{\delta|\gamma|}. \tag{6}$$

Here $m_{|\gamma|}$ is the maximum number of places to begin γ and enclose r or r'; therefore, $m_{|\gamma|}$ is less than twice the number of contour places inside a circle of circumference $|\gamma|$. Here $n_{|\gamma|}$ is the number of contours of length $|\gamma|$ beginning at a fixed contour place. By the same argument which led to Eq. (3),

$$h \leq a' \sum_{|\gamma|} c^{|\gamma|} |\gamma|^{d/(d-1)} g^{\delta|\gamma|}, \tag{7}$$

where $a' > 0$ is a constant.

The probability that S_r and $S_{r'}$ are both ground state regions belonging to the same ground state is then bounded as $P > (1-g)(1-h)$. If $g \to 0$ as $T \to 0$, then $P > 1/2$ if T is sufficiently small, implying the existence of an ordered phase in the model.

We now turn to some examples of the use of different methods of obtaining upper bounds to P_γ.

III. The Attractive Lattice Gas

The Peierls argument was used by Griffiths[3] and Dobrushin[4,5] to prove the existence of a phase transition in the ferromagnetic Ising model in zero field. The attractive lattice gas (equivalent to the ferromagnetic Ising model in a field[2]) has a Hamiltonian $H(\xi) = N_{\bullet\bullet}\varepsilon - N_{\bullet}\mu$, where $N_{\bullet\bullet}$ is the number of pairs of first neighbor sites both of which are occupied by molecules in configuration ξ, $\varepsilon < 0$ is the energy of attraction, and μ is the chemical potential. If q is the coordination number of the lattice, then $N_{\bullet\bullet} = q\, N_{\bullet}/2 - N_{\bullet\circ}/2$, where $N_{\bullet\circ}$ is the number of pairs of first neighbor sites in configuration ξ which contain one molecule and one vacancy. If a contour segment is drawn between each vacant site and an occupied neighboring site, then $N_{\bullet\circ}$ is the number of contour segments in ξ.

For this model (as for the ferromagnetic Ising model[3]), associated with a configuration ξ containing a contour γ is a configuration ξ' obtained by interchanging molecules and vacancies at all sites inside γ (this is a 1-1 correspondence). If $\mu = \varepsilon q/2$ (which corresponds to zero field in the equivalent Ising model) then $H(\xi) - H(\xi') = -\varepsilon|\gamma|/2$. Hence $P_\gamma < \exp(-|\varepsilon||\gamma|/2)$ with $|\varepsilon| > 0$, and, as discussed above, the model has a phase transition.

Here the dense phase represents liquid, and the rarefied phase represents vapor. The transition has been proved to be first order.[15] The contours can be thought of as the interfaces between vapor and liquid, they being improbable as the vapor condenses.

IV. A Lattice Gas Model for Water

We next consider two- and three-dimensional versions of a lattice gas model for water.[16,17] The rotation of the molecules is restricted in the model so as to allow only two different orientations of a molecule on a site. The main idea in the model is to disregard the asymmetry (but not the direction) of the hydrogen bond; i.e., one does not distinguish between the hydrogen atom and the lone pair of electrons.

To be more precise, the two orientations allowed in the models may be pictured as orientations of trigonal-planar molecules on the triangular lattice or as orientations of tetrahedral molecules on the body-centered cubic (bcc) lattice, where the center of a molecule occupies a lattice site, the bonds pointing toward neighboring lattice sites. One such orientation for the two-dimensional version is illustrated in Fig. 1a, and one orientation for a molecule in the three-dimensional version is illustrated in Fig. 1b. This pictorial representation has the following meaning: if two molecules occupy neighboring sites such that the bonds of the pictorial representations overlap, a hydrogen bond with energy $\varepsilon < 0$ is said to form.

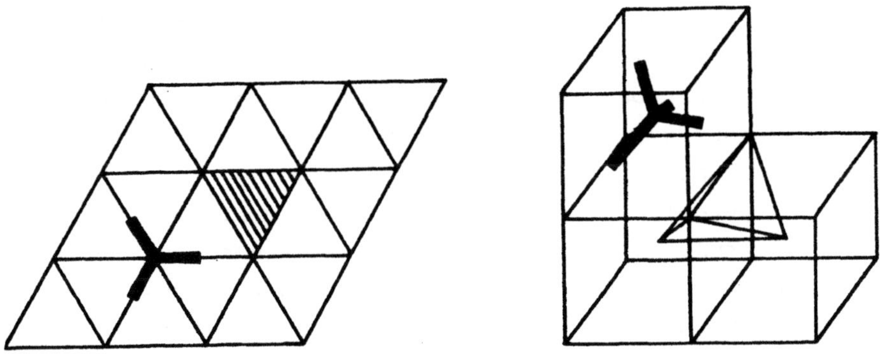

(a) (b)

Figure 1. (a) One orientation of a trigonal-planar molecule on the triangular lattice together with a triangular simplex. (b) One orientation of a tetrahedral molecule on the bcc lattice together with a tetrahedral simplex.

As such, $p = 3$ translationally related hydrogen-bonded structures can exist on the triangular lattice, each occupying an open honeycomb structure. This structure gives a model for hexagonal ice, Ih. Similarly, $p = 4$ translationally related hydrogen-bonded structures can exist on the bcc lattice, each occupying an open diamond structure. This structure models cubic ice, Ic. Two interlocking diamond structures can simultaneously fill the bcc lattice (giving $p = 2$). Such a structure provides a model for ice VIII, a high pressure form of ice. In order to show that under certain conditions an open, ice Ic structure can exist on the bcc lattice, a repulsion $\gamma \geq 0$ will be added between two molecules on neighboring sites if neither points a bond toward the other.

A Peierls argument, given by Heilmann and Huckaby,[17] proceeds as follows. Each lattice is composed of simplices (triangles or tetrahedra); a typical simplex for each lattice is illustrated in Fig. 1. If the Hamiltonian restricted to a simplex is minimal, the simplex is said to be a ground state simplex; otherwise it is said to be a contour segment. In two dimensions a ground state simplex is consistent with an ordered ice-like structure if $3\varepsilon/2 < \mu < 0$. In three dimensions, a ground state simplex is consistent with an open-ice structure if $2\varepsilon < \mu < \min\{(2\varepsilon+4\gamma),0\}$. (Hence open-ice is never a ground state structure unless $\gamma > 0$.) In three dimensions, a ground state simplex is consistent with a dense ice VIII structure if $\gamma < -\varepsilon$ and $\mu > \max\{(2\varepsilon+4\gamma),(4\varepsilon+8\gamma)\}$. Therefore, depending on μ, both an open-ice and a dense ice structure can be a ground state structure if $0 < \gamma < -\varepsilon$.

We fix the border of the lattice with an ordered ground state structure s. Then any simplex which is not of the same ground state structure as the boundary must be separated from the boundary by an outer contour. An outer contour is a contour which is not separated from the boundary by another contour.

With each configuration ξ containing an outer contour γ we associate a configuration ξ' (this is a 1-1 correspondence) generated in the following manner. Let T_r be a unit translation which converts ground state structure r into structure s. Translate by T_r all $r \neq s$ regions (together with their interiors) that border γ. Then replace all contour segments of γ with simplices having configurations belonging to structure s. Then $H(\xi) - H(\xi') \geq \alpha|\gamma|$, where $\alpha = H_{\star} - H_o > 0$. Here H_o is the value of the Hamiltonian restricted to a ground state simplex, and $H_{\star}$ is the next smallest value of the restricted Hamiltonian, i.e., the value of the Hamiltonian restricted to a "first excited state" simplex. Hence $P_{\gamma} < \exp(-\alpha|\gamma|/kT)$ and by Eq. (3), there is an ordered phase at sufficiently low temperature for the range of parameters for which the ground state corresponds to one of the ordered ice-like structures discussed above.

V. Lattice Gas Models for Carbon Tetrachloride and for Methane Adsorbed on Graphite

Here we consider a two dimensional lattice gas of trigonal-planar molecules on a triangular lattice and an analogous three dimensional lattice gas of tetrahedral molecules on a bcc lattice. In both models, a lattice site can be either vacant or occupied by a molecule in either of the two orientations in which the molecular bonds point toward neighboring lattice sites. One such orientation on each lattice is illustrated in Fig. 1. Two molecules on neighboring sites attract with an energy $\varepsilon \leq 0$ if one molecule points a bond toward the other and repel with a steric bond-bond repulsion $\gamma \geq 0$ if both molecules point bonds toward each other.

The two dimensional model may be used as a model for methane adsorbed on graphite, where a methane molecule is adsorbed over the center of a conjugated ring of the graphite structure, three hydrogens of the methane overlapping the π bonds of the graphite in either of two possible orientations. The ordered structure consisting of all sites occupied by molecules in a single orientation has been indicated by Monte Carlo studies to exist in a system of methane adsorbed on graphite.[18]

The analogously ordered structure in the three dimensional model is the same as the local orientational ordering proposed on the basis of X-ray diffraction studies to exist in liquid carbon tetrachloride.[19,20] The complete translational order which is present in the lattice gas is, of course, absent in the true liquid. Nonetheless, the model could be used to study the vaporization of liquids having orientational ordering.

For the special case of hard-core repulsion ($\gamma \to \infty$ and $\varepsilon = 0$), Huckaby[21] proved the existence of a phase transition using a quite different method[22] than that presented here. (The non-interacting case $\varepsilon = \gamma = 0$ has no transition and is not considered here.)

A Peierls argument for the present model, given by Huckaby, Dougherty, and Pękalski,[23] proceeds as follows. As discussed in the model for water given in Sec. IV, the triangular and bcc lattices are composed of simplices (triangles and tetrahedra). A typical simplex for each lattice is illustrated in Fig. 1.

As in the case for water, if the Hamiltonian restricted to a simplex is minimal, the simplex is said to be a ground state simplex; otherwise it is said to be a contour segment. For the present models, if $\mu > 3\varepsilon$ in two dimensions or $\mu > 4\varepsilon$ in three dimensions, then a ground state simplex is one which is consistent with an ordered structure in which every site of the lattice is occupied by molecules in a single orientation. There are $p = 2$ such structures in each model. Unlike the case in the model for water, the $p = 2$ ordered ground state structures in the present models are related by reflection rather than by translation.

We fix the border of the lattice to be consistent with an ordered ground state structure. Then any simplex which is not of the same ground state as the boundary must be separated from the boundary by an outer contour γ.

Heilmann[9] has developed an extension of the Peierls argument which provides estimates for P_γ for models containing two ground state structures related by reflection. Using an extremely complex set of transformations, he demonstrated how to associate with a configuration ξ, containing an outer contour γ, a configuration ξ' (this is a 1-1 correspondence) which provided a bound to P_γ at sufficiently low temperature. For the present models, this bound can be written as

$$P_\gamma \le D^{|\gamma|} e^{-|\gamma|(H_\star - H_o)/kT}, \qquad (8)$$

where $D > 0$ is a constant, and $H_\star - H_o$ is the difference in the value of the Hamiltonian restricted to a first excited state simplex and a ground state simplex. Eq. (8) combined with Eq. (3) then implies the existence of an ordered phase in the model.

VI. A Lattice Gas Model for Coadsorption

We now consider a lattice gas model for the coadsorption of small (A) and large (B) atoms on a square lattice Λ with periodic boundary conditions as defined in Sec. II. Two B atoms are excluded from occupying a pair of nearest neighbor sites and repel with energy ε_B if they occupy a pair of next nearest neighbor sites. An A atom on a lattice site repels a B atom on a neighboring site with energy ε_{AB} and repels an A atom on a neighboring site with energy ε_A. We denote by μ_A the sum of the binding energy and chemical potential of an adsorbed A atom (and similarly for a B atom).

Using reflection positivity and the Peierls argument, Huckaby and Kowalski[14] were able to prove the existence of several ordered phases in the model. We shall now outline this proof.

The lattice Λ consists of unit squares with lattice sites at each vertex. A configuration on a square which has a minimum value, H_o, of the Hamiltonian restricted to a square is called a ground state square. Otherwise, the square is said to be an excited state square. The ground state configurations on Λ can be readily constructed from these ground state squares. Regions of the $(\mu_A, \mu_B, T=0)$ plane corresponding to different ground state squares are illustrated in Fig. 2a for the case $0 < \varepsilon_A < \varepsilon_{AB} < \varepsilon_B$ and in Fig. 2b for the case $0 < \varepsilon_B < \varepsilon_{AB} < \varepsilon_A$.

If we define reflection lines $R_a^{\pm}$ as in Sec. II, then Λ is divided into the three regions Λ_a^+, Λ_a°, and Λ_a^-. If C_a^+ is the set of allowed configurations on Λ_a^+ (and similarly for C_a^- and C_a°), then $\Theta_a C_a^{\pm} = C_a^{\mp}$ and $\Theta_a C_a^\circ = C_a^\circ$. Since the Hamiltonian can be written as $H = H_a^+ + \Theta_a H_a^+ =$

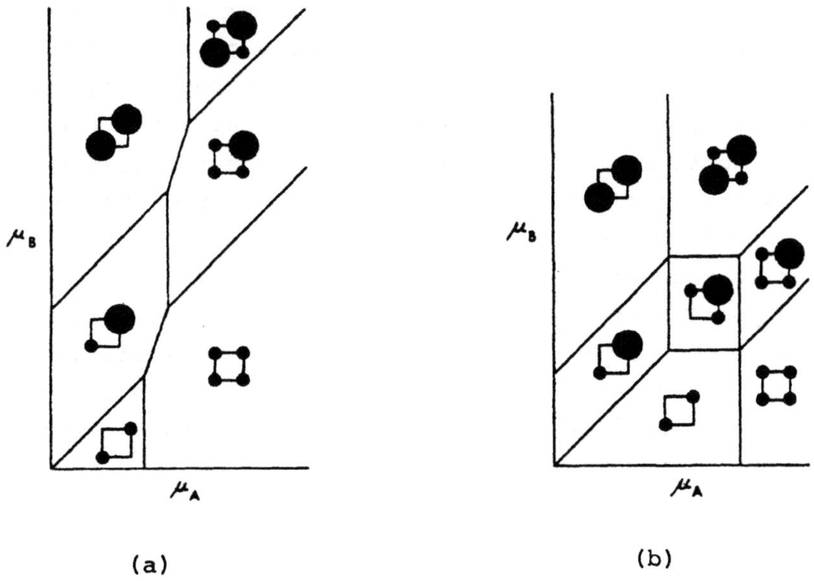

Figure 2. Zero-temperature phase diagrams for the cases (a) $0 < \varepsilon_A < \varepsilon_{AB} < \varepsilon_B$ and (b) $0 < \varepsilon_B < \varepsilon_{AB} < \varepsilon_A$.

$H_a^- + \Theta_a H_a^-$, then the inequality in Eq. (5) is satisfied.

Let Q_r be the projection onto configurations in which a square centered at r is an excited state square. Let $Q(L) = \prod_{r \in L} Q_r$, where L is a nonempty set of squares. The probability $\langle Q(L) \rangle$ that L is a set of excited state squares is bounded as $\langle Q(L) \rangle \leq g^{|L|}$, where $g = \max_L \langle Q(L) \rangle^{1/|L|}$.

Since $Q = Q_a^+ Q_a^-$, then from Eq. (5) we can deduce that a set L_m which maximizes $\langle Q(L) \rangle^{1/|L|}$ must contain all $4M^2$ squares in Λ. If $H_\star$ is the value of a "first excited state square," then

$$g \leq (\Xi^{-1} 3^{4M^2} e^{-4M^2 H_\star / kT})^{1/4M^2}, \tag{9}$$

where 3^{4M^2} is the maximum number of configurations on Λ. Since $4M^2 H_o$ is the Hamiltonian for a ground state configuration, then $\Xi > \exp(-4M^2 H_o/kT)$. Eq. (9) then gives $g < 3\, e^{-\alpha/kT}$, where $\alpha = H_\star - H_o > 0$.

If an excited state square and a ground state square share an edge, the edge is said to be a contour segment. Two contour segments are connected if they divide into two sets the four squares with which they share a vertex, such that one of the two sets contains only ground state squares. This prevents the branching of contours.

Suppose, in configuration ξ, that $S_r(\xi)$ is a ground state square, and $S_{r'}(\xi)$ is in a configuration that does not belong to the same ground state as $S_r(\xi)$. then either $S_r(\xi)$ or $S_{r'}(\xi)$ is surrounded by a closed contour γ composed of $|\gamma| \geq 4$ segments. Since each contour segment is bordered by a disordered square, and since no disordered square is bordered by more than four segments, then γ borders at least $|\gamma|/4$ disordered squares. Hence $P_\gamma \leq g^{|\gamma|/4}$, and Eq. (7) insures that an ordered phase exists in the model when the ground state is two- or four-fold degenerate. From Fig. 2 we can ascertain the existence of four or five ordered phases in the model.

Reflection positivity can also be used to prove the existence of ordered phases in some models which have highly degenerate ground states. This was done by Heilmann and Lieb[12] for a lattice gas model of a liquid crystal and by Heilmann[24] for the first neighbor exclusion lattice gas on the face-centered cubic lattice.

Acknowledgments

Support from The Robert A. Welch Foundation Grant P-446 and from the National Science Foundation is gratefully acknowledged. The author wishes to express his appreciation to the Université de Liège for their hospitality while this manuscript was being written.

References

1. R. Peierls, Proc. Cambridge Phil. Soc. 32, 477(1936).
2. C. N. Yang and T. D. Lee, Phys. Rev. 87, 404(1952).
3. R. B. Griffiths, Phys. Rev. 136A, 437(1964).
4. R. L. Dobrushin, Dokl. Akad. Nauk SSSR 160, 1046(1965). (English translation: Sov. Phys. Dokl. 10, 111(1965)).
5. R. L. Dobrushin, Teor. Veroyat. Ee Primen 10, 209(1965). (English translation: Theor. Prob. Appl. 10, 193(1965)).
6. R. L. Dobrushin, Funkts. Analiz Prilozh. 2, 44(1968). (English translation: Funct. Anal. Applic. 2, 302(1968)).
7. R. B. Griffiths, in Phase Transitions and Critical Phenomena, edited by C. Domb and M. Green(Academic: New York, 1972), Vol. I, p. 1.
8. O. J. Heilmann, Lett. Nuovo. Cimento 3, 95(1972).
9. O. J. Heilmann, Comm. Math. Phys. 36, 91(1974).
10. J. Fröhlich and E. H. Lieb, Phys. Rev. Lett. 38, 440(1977).
11. E. H. Lieb, in Lecture Notes in Physics 80, 59(Springer-Verlag, 1978).
12. O. J. Heilmann and E. H. Lieb, J. Stat. Phys. 20, 679(1979).
13. J. Fröhlich, R. B. Israel, E. H. Lieb, and B. Simon, J. Stat. Phys. 22, 297(1980).
14. D. A. Huckaby and J. M. Kowalski, J. Chem. Phys. 80, 2163(1984).
15. J. Ginibre, A. Grossmann, and D. Ruelle, Comm. Math. Phys. 3, 187 (1966).
16. G. M. Bell and D. A. Lavis, J. Phys. A 3, 568(1970).
17. O. J. Heilmann and D. A. Huckaby, J. Stat. Phys. 20, 371(1979).
18. S. F. O'Shea and M. L. Klein, J. Chem. Phys. 71, 2399(1979).
19. K. Nishikawa and Y. Murata, Bull. Chem. Soc. Jpn. 52, 293(1979).

20. K. Nishikawa, K. Tohji, M. Shima, and Y. Murata, Chem. Phys. Lett. 64, 154(1979).
21. D. A. Huckaby, J. Stat. Phys. 17, 371(1977).
22. L. K. Runnels, Comm. Math. Phys. 40, 37(1975).
23. D. A. Huckaby, A. M. Dougherty, and A. Pękalski, Phys. Rev. A 26, 3528(1982).
24. O. J. Heilmann, J. Phys. A 13, 1803(1980).

CONTRIBUTED PAPERS

SYSTEMS UNDER INFLUENCE OF DICHOTOMOUS NOISE

Ulrich Behn
Sektion Physik der Karl-Marx-Universität Leipzig
DDR-7010 Leipzig, GDR

1. Introduction

Stochastic differential equations describe a large variety of physical phenomena. The noise appearing in this equations models the influence of external fields or the interaction with internal subsystems. Quantities of interest are the stationary probability density, moments and correlation functions, and asymptotic properties, e.g. the decay of correlations. Qualitative changes in this quantities caused by the noise are called noise induced transitions /1/.
The noise is usually assumed to be a simple stochastic model process with definite properties which allow in some cases even exact solutions.
The simplest continuous stochastic process is the Gaussian white noise (GWN) with autocorrelation $W(t-t') = \langle \xi_t \, \xi_{t'} \rangle = \Gamma \delta(t-t')$, i.e. random events at different times are not correlated. The Ornstein-Uhlenbeck-process (OUP) has an exponential decay of correlations $W(t-t') = D/\tau \exp(-|t-t'|/\tau)$, i.e. a finite correlation time.
The simplest discrete stochastic process is the dichotomous Markovian process (DMP) jumping between two states (Fig. 1). The autocorrelation is exponential as for the previous case.

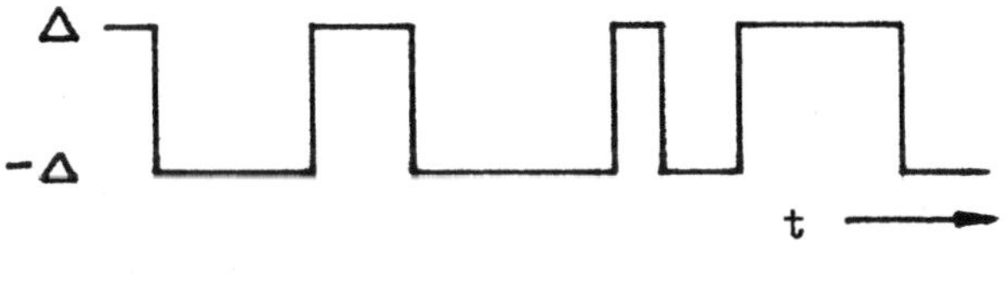

Fig. 1.
Typical realization of the dichotomous Markovian process I_t.

In this paper we consider two models, a microscopic and a phenomenological one, under influence of the dichotomous noise. In both exampels there appear qualitative new features compared with the case of the GWN and the OUP. The differences are due to the discrete nature of the DMP and the finite correlation time. The finite correlation time gives the possibility to change the driving process from slow to fast compared with the characteristic time of the system under consideration.

2. The Dichotomous Markovian Process

The DMP jumps with equal probability between the two states $\pm\Delta$. The process can be written as

$$I_t = I_K = \Delta(-1)^K \ , \qquad (2.1)$$

where K is the number of jumps in the time t, governed by the Poisson probability distribution

$$P(K,t) = (\alpha t)^K / K! \ \exp(-\alpha t). \qquad (2.2)$$

α denotes the mean number of jumps in unit time (jump frequency). The average sums up over all possible realizations of the process,

$$\langle f(I_t) \rangle = \sum_K P(K,t) f(I_K) = \int \mathcal{D} P(I) f(I). \qquad (2.3)$$

It can be considered as a functional integral with Poisson measure. The Markovian property is reflected by the following theorems /2,3/

$$\langle I_{t'} I_t \varphi[I] \rangle = \langle I_{t'} I_t \rangle \langle \varphi[I] \rangle \ , \qquad t' > t \ ; \qquad (2.4)$$

$$d/dt \langle I_t \varphi[I] \rangle = -2\alpha \langle I_t \varphi[I] \rangle + \langle I_t \dot{\varphi}[I] \rangle \ , \qquad (2.5)$$

where $\varphi[I]$ is a functional depending only on I_τ at times τ prior to t.
One readily calculates $\langle I_t \rangle = 0$ and the exponential autocorrelation

$$W(t-t') = \Delta^2 \exp(-2\alpha |t-t'|) \ , \qquad (2.6)$$

whose Fourier transform is a Lorentzian.
In the white noise limit ($\alpha, \Delta \longrightarrow \infty$; $\Delta^2/2\alpha = \Gamma = \text{const.}$) the results of the GWN are reproduced.

3. Quasiparticle Motion in a Discrete Stochastic Potential

We consider the motion of a tight binding quasiparticle in a local time dependent stochastic potential described by

$$H(t) = \sum_{<i,j>} T_{ij}\, a_i^+ a_j + \sum_i \varepsilon_i(t)\, n_i \,. \tag{3.1}$$

$\varepsilon_i(t)$ simulates the dynamics of that part of the system which is not of explicit interest. One can arrive at such a description, e.g., doing the partial trace over the noninteresting variables or using projection operator techniques. Since the probability distribution of the driving stochastic process $\varepsilon_i(t)$ is either very complicated or not explicitely known one considers simplified model processes.
The effect of phonons on excitonic systems, for instance, has been modeled by both the GWN /4,5/ and the OUP /6/.
Here we discuss the case of the dichotomous Markovian process /7/, where $<\varepsilon_i(t)\varepsilon_j(t')> = \delta_{ij} < I_t I_{t'}> = \delta_{ij}\,\Delta^2 \exp(-2\alpha|t-t'|)$.
The equation of motion of the site diagonal elements of the Green function defines a multiplicative stochastic process,

$$i\partial/\partial t\; G_i(t,t';\varepsilon) - \int d\tau\, \Delta_i(t,\tau;\varepsilon) G_i(\tau,t';\varepsilon) - \delta(t-t') = \varepsilon_i(t) G_i(t,t';\varepsilon). \tag{3.2}$$

The left hand side can be averaged neglecting intermediate returns to the initial (and final) site i so that the site self energy Δ_i does not depend on ε_i (single site approximation). This is the better the shorter the correlation time $1/2\alpha$. The right hand side can be averaged inserting the Dyson equation, $G = G^o + G^o \varepsilon G$, and applying (2.4) with the result

$$\int d\tau\;\; G_i^o(t,\tau) < \varepsilon_i(t)\; \varepsilon_i(t') > < G_i(\tau,t';\varepsilon) > = \int d\tau\, \Sigma(t,\tau) G(\tau,t'). \tag{3.3}$$

Thus, the self energy reads in energy representation

$$\Sigma(\omega) = \Delta^2\; G^o(\omega + 2i\alpha)\;, \tag{3.4}$$

or combined with $G^o(\omega) = (\omega - \Delta(\omega))^{-1}$ and $G(\omega) = (G^o(\omega)^{-1} - \Sigma(\omega))^{-1}$,

$$\Sigma(\omega) = \Delta^2\; (G(\omega + 2i\alpha)^{-1} + \Sigma(\omega + 2i\alpha))^{-1}\;. \tag{3.5}$$

This result has been first derived in /7/ and was later obtained by Inaba /8/ using a diagram technique. Inaba observed additionally that the vertex corrections for the diffusion constant vanish so that this quantity can be calculated knowing only the one-particle self energy. Here we discuss only the qualitative behaviour of the density of states (DOS) given by $\rho(\omega) = -\mathrm{Im}\, G(\omega)/\pi$.

In the fast motion limit ($\alpha \to \infty$) one obtains the unperturbed DOS since the particle cannot follow the very fast sequence of jumps and feels only the average potential which is zero.
In the static limit ($\alpha \to 0$) the potential becomes time independent and the results of the coherent potential approximation (CPA) for a static binary alloy are explicitely reproduced.
In the white noise limit (see above) one obtains the correct self energy /5/, $\Sigma(\omega) = -i\,\Delta^2/2\alpha$, causing simply a broadened DOS.
In the atomic limit ($T_{ij}=0$) we are left only with the second term of the Hamiltonian (3.1), a problem already treated by Kubo in the stochastic theory of line shape /9/. The DOS shows a transition from bimodal shape for small α/Δ to monomodal shape for large α/Δ (Fig. 2). In the former case the particle has time enough to feel both values of the potential $\pm\Delta$. In the latter case it feels only the zero average (motional narrowing).

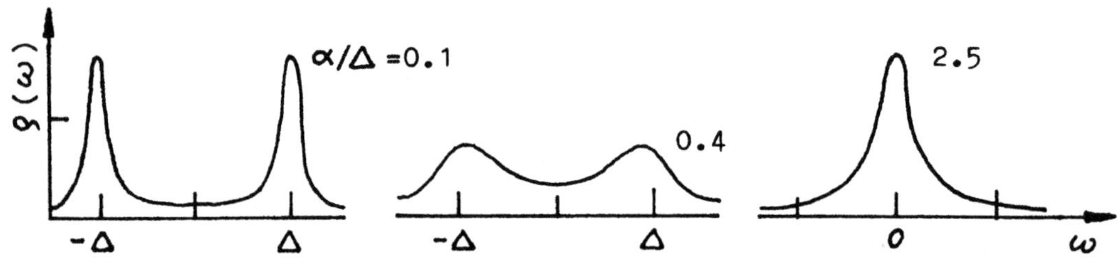

Fig. 2. Density of states in the atomic limit. The bifurcation occurs at $\alpha/\Delta = 1/\sqrt{2}$.

In the intermediate regime there is as in the atomic limit a competition between the two parameters α and Δ favourizing monomodal and bimodal shape, respectively (Fig. 3). In the bimodal case there is only a pseudogap between the two subbands due to the nonzero energy transfer for any finite α.
The scheme has been applied in a dynamic approximation /10/ for the Hubbard model which can be related to (3.1) with the help of a functional Fourier transform /11/. Describing electrons of a given spin direction the stochastic field simulates the occupation of the lattice sites with electrons of opposite spin. The jump amplitude 2Δ plays the role of the truncated Coulomb repulsion U, whereas the jump frequency α is self consistently determined by the mean kinetic electron energy.

In a different approach Paquet and Leroux-Hugon /12/ introduced a dynamic allowing jumps between the two minima of the free energy. To treat this problem they exploited the result of the dynamical CPA (3.5).

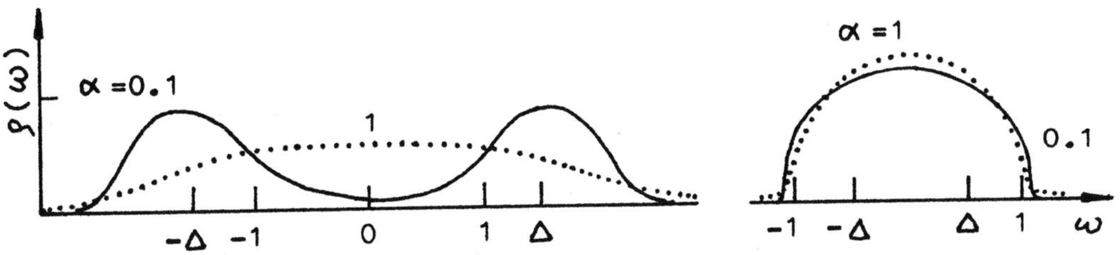

Fig. 3. Density of states in the intermediate regime for large (left), respectively small (right) jump amplitude. The energy is measured in units of the half unperturbed band width (semielliptic model density).

Further applications to superionic conductors, to the motion of quasiparticles in solitary lattice distortions, to X-ray emission and absorption in metals, or to adsorption on surfaces seem possible.

4. A Simple Nonlinear Model

We consider the Stratonovich model /13/,

$$\dot{x}(t)=ax(t)-x^3(t)+x(t)I_t \quad , \qquad (4.1)$$

which is frequently used to study nonequilibrium systems /13-16/ and experimentally realized for instance in parametric excited electric circuits /16/.

In the deterministic case a bifurcation of the stable stationary state occurs when the control parameter a changes the sign. The noise couples like a to the linear term describing stochastic changes of this parameter.

The stationary probability density $P_s(x)$ can be obtained exactly for the GWN /13,14/, and approximatively for the OUP /15/. In both cases one obtains a bifurcation of the maximum of P_s at values $a_c > 0$. In the latter case there appear additionally maximal values of P_s without deterministic equivalent (noise induced states).

Considering the influence of the DMP /17/ we start with the more general model,

$$\dot{x}(t)=f(x)+g(x)I_t=F(x,I_t) \quad , \tag{4.2}$$

f and g beeing arbitrary functions. For this model the stationary probability density is known in a semiexplicit form

$$P_s(x) \propto (1/F_+(x)-1/F_-(x)) \exp\left\{ -\alpha\int^{x} dy(1/F_+(y)+1/F_-(y)) \right\} . \tag{4.3}$$

$F_\pm(x)$ is the shorthand for $F(x,\pm\Delta)$ defined in (4.2). (4.3) has been derived by several methods /18,19/, e.g. exploiting theorem (2.5). In the important case that $F_\pm(x)$ is a polynomial of order n with simple zeros $a_\nu^\pm$ we can evaluate (4.3) explicitely /17/,

$$P_s(x) \propto g(x) \prod_{\nu=1}^{n} (x-a_\nu^+)^{\lambda_\nu^+} (x-a_\nu^-)^{\lambda_\nu^-} \quad , \tag{4.4}$$

with $\lambda_\nu^\pm = -\alpha/F'_\pm(a_\nu^\pm)-1$. (4.4) shows that the zeros of $\lambda_\nu^\pm$ separate the parameter space in regions characterized by different combinations of signs of the $\lambda_\nu^\pm$. For positive (negative) values of these exponents P_s vanishes (diverges) as the corresponding bases tend to zero. Analyzing this behaviour and the extreme values of P_s given by /19/

$$f-(\Delta^2/2\alpha)gg'+ff'/\alpha-f^2g'/(2\alpha g)=0 \quad , \tag{4.5}$$

one obtains the qualitative behaviour of $P_s(x)$ in the parameter space, i.e. the "phase diagram".
To determine the support of P_s for the specific model (4.1) we consider the realizations $I_t=\pm\Delta$ separately and perform a linear stability analysis of the stationary states $F_\pm(x_s)=0$. For $-\Delta<a$ the support consist of the hatched region in Fig. 4 since the system once catched there cannot escape.
Thus we obtain from (4.4)

$$P_s(x)= N\ |x|^{-2(\lambda+\mu)-1}(a+\Delta-x^2)^{\lambda-1}(x^2-(a-\Delta))^{\mu-1} \quad , \tag{4.6}$$

with $2\lambda=\alpha/(a+\Delta)$ and $2\mu=\alpha/(a-\Delta)$. P_s is obviously a symmetric function. The normalization is determined as

$$N = \begin{cases} (2\Delta)^{1-\lambda-\mu}(\Delta-a)^{\lambda} \; 2^{-1}(\Delta+a)^{\mu+1}B^{-1}(\lambda,-\lambda-\mu) \\ (2\Delta)^{1-\lambda-\mu}(a-\Delta)^{\lambda} \; a(\Delta+a)^{\mu}\, B^{-1}(\lambda,\mu) \end{cases} \tag{4.7}$$

for $|a| < \Delta$ and $\Delta < a$, respectively. B denotes the Beta function.

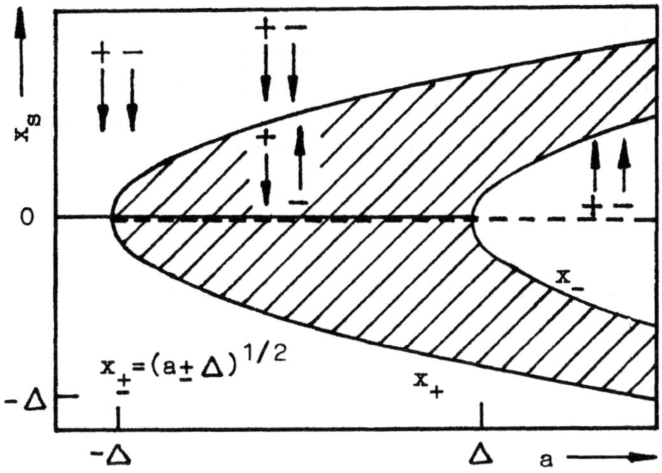

Fig. 4.
Linear stability analysis. Solid (broken) lines correspond to stable (unstable) states. The arrows indicate the flow for the different realizations of I_t.

The extreme values of P_s given by

$$x^2_{1/2} = (\alpha + 3a \pm ((\alpha - 2a)^2 + 5\Delta^2))^{1/2})/5 \quad , \tag{4.8}$$

can be inside or outside the support. Correspondingly $P_s(x)$ has the qualitative shape shown in Fig. 5.
For large α the system can hardly follow the noise. P_s changes from monomodal to bimodal shape crossing the line

$$a_c = (\alpha^2 + \Delta^2)^{1/2} - \alpha \approx \Delta^2/2\alpha - \Delta^4/8\alpha^3 \quad . \tag{4.9}$$

The shift of the bifurcation is smaller than for the corresponding GWN where $a_c = \Delta^2/2\alpha = \Gamma$. In the fast motion limit the system lives on the deterministic states, $x^2_2 \longrightarrow a$ as $\alpha \longrightarrow \infty$.
For small α the system has time enough to reach the stable stationary states the number of which may be larger than in the deterministic case (noise induced states). With increasing α the corresponding maxima of P_s are shifted and change their relative weights.
For $a/\Delta > 1$ the support consist of two unconnected branches. Only this regime has been recently discussed in an experimental context /16/.

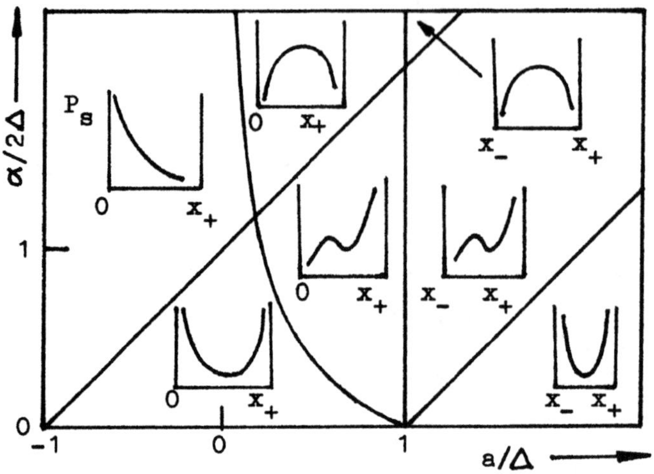

Fig. 5. Qualitative shape of P_s for $x > 0$ in the parameter plane (phase diagram).

Near the bifurcation line the maxima of P_s behave analogously to the order parameter in a continuous phase transition,

$$x_{max}=A(\Delta,\alpha)(a-a_c(\Delta,\alpha))^{1/2} \quad . \qquad (4.10)$$

Both the critical value a_c and the amplitude A depend on the noise parameters but not the critical exponent which is only determined by the choice of the functions f and g, as has been found also for the GWN /14/ and the OUP /15/.

Although for both OUP and DMP the autocorrelation is exponential the noise induced states are different due to the different nature of the driving processes. According to the appearance of noise induced states we find more complicated bifurcation patterns than in the deterministic case.

The long time behaviour of higher moments has been investigated for several nonlinear models by different authors /20/. It seems to be an open question how the transitions changing the long time behaviour are connected with the transitions of P_s.

The results can be applied to the problem of the influence of a stochastic voltage on the onset of an electrohydrodynamic instability in nematic liquid crystals. In /21/ a nonlinear equation was proposed which reads in reduced variables

$$\dot{x}(t)=ax(t)-x^3(t)+(x(t)-dx^3(t))I_t \quad . \qquad (4.11)$$

The corresponding P_s is /17/

$$P_s(x)= N(1-dx^2) \; |x|^{-2(\lambda+\mu)-1} \left(\frac{a+\Delta}{1+d\Delta}-x^2\right)^{\lambda-1} \left(x^2-\frac{a-\Delta}{1-d\Delta}\right)^{\mu-1} , \quad (4.12)$$

with λ, μ given by (4.6). Since the exponents are identical to those of the previous problem the boundaries in the parameter plane are the same as in Fig. 5. The instability occurs when nonzero values of x become probable, i.e. crossing the bifurcation line.

References

/1/ W.Horsthemke and M.Malek-Mansour, Z.Phys. B24, 307 (1976).

/2/ R.C.Bourret, U.Frisch, and A.Pouquet, Physica 65, 303 (1973).

/3/ V.E.Shapiro and V.M.Loginov, Physica 91A, 563 (1978).

/4/ P.Reinecker and H.Haken, Z.Phys. 250, 300 (1972); A.A.Ovchinnikov and N.S.Erichman, Zh.eksper.teor.Fiz. 67, 1474 (1974); A.Madhukar and W.Post, Phys.Rev.Lett. 39, 1424 (1977).

/5/ S.M.Girvin and G.D.Mahan, Phys.Rev. B20, 4896 (1979).

/6/ H.Sumi, J.chem.Phys. 67, 2943 (1977); K.Kitahara and J.W.Haus, Z.Phys. B32, 419 (1979); I.B.Rips and V.Capek, phys.stat.sol.(b) 100, 451 (1980).

/7/ U.Behn, phys.stat.sol.(b) 113, 219 (1982).

/8/ Y.Inaba, J.Phys.Soc.Japan 52, 3144 (1983).

/9/ R.Kubo, in: Fluctuation, Relaxation and Resonance in Magnetic Systems, Ed.D.terHaar, Oliver & Boyd, Edinburgh 1962.

/10/ U.Behn, phys.stat.sol.(b) 100, 129 (1980); ibid. 104, 203 (1981); ibid. 118, K151 (1983).

/11/ U.Behn and W.Weller, phys.stat.sol.(b) 99, 71 (1980).

/12/ D.Paquet and P.Leroux Hugon, Phys.Rev.Lett. 49, 1023 (1982).

/13/ R.L.Stratonovich, Topics in the Theory of Random Noise, Gordon and Breach, New York 1967, Vol. 2.

/14/ A.Schenzle and H.Brand, Phys.Rev. A20, 1628 (1979); R.Graham and A.Schenzle, Phys.Rev. A25, 1731 (1982); L.Brenig and N.Banai, Physica 5D, 208 (1982).

/15/ J.M.Sancho, M.SanMiguel, H.Yamazaki, and T.Kawakubo, Physica 116A, 560 (1982); J.M.Sancho, M.SanMiguel, S.L.Katz, and J.D. Gunton, Phys.Rev. A26, 1589 (1982).

/16/ S.Kabashima and T.Kawakubo, in: Systems far from Equilibrium, Lecture Notes in Physics Vol. 132, Ed. L.Garrido, Springer, Berlin, Heidelberg, New York 1980.

/17/ U.Behn and K.Schiele, Noise Induced Transition in a Simple Nonlinear Model Driven by Dichotomous Markovian Noise, to be published.

/18/ R.F.Pawula, IEEE Trans.Inf.Theory 13, 33 (1967); Intern.J.Control 25, 283 (1977); V.I.Klyatskin, Stokhasticheskie uravneniya i volny v sluchaino i neodnorodnykh sredakh, Izd.Nauka, Moskva 1980; Y.Pomeau, J.Stat.Phys. 24, 189 (1981).

/19/ K.Kitahara, W.Horsthemke, R.Lefever, and Y.Inaba, Progr.Theor. Phys. 64, 1233 (1980).

/20/ M.Suzuki, K.Kaneko, and F.Sasagawa, Progr.Theor.Phys. 65, 828 (1981); K.Ishii and K.Kitahara, ibid. 68, 665 (1982).

/21/ S.Kai, T.Kai, M.Takata, and K.Hirakawa, J.Phys.Soc.Japan 47, 1379 (1979).

THE INFLUENCE OF IMPURITIES ON THE SPIN-LATTICE RELAXATION IN SOME UNIAXIAL FERROELECTRICS

U. Geppert and A. Kühnel
Sektion Physik der Karl-Marx-Universität Leipzig
7010 Leipzig, DDR

1. Introduction

Recent electron spin echo experiments by Brunner, Völkel, Windsch et al. /1/ have shown that the longitudinal and the transverse relaxation times in some uniaxial ferroelectrics exhibit opposite anomalous behaviour at the phase transition: While the transverse relaxation time T_2 anomalously decreases at the Curie temperature T_c, the longitudinal relaxation rate T_1^{-1} falls off drasticly at T_c (Fig. 1 and 2). It is easily understood that T_2^{-1} increases at T_c because of the critical slowing down of the essential degrees of freedom at the phase transition. However, it is rather hard to understand, why some relaxation mechanisms are omitted at the phase transition. The described phenomenon has been found in crystals of TGS, TSCC, KDA, and Rochelle salt.

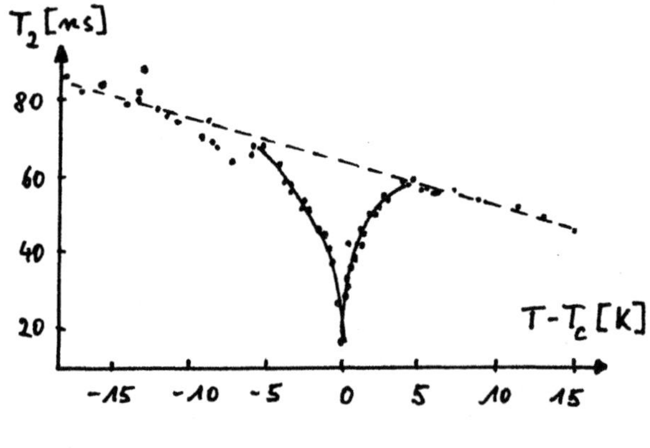

Fig. 1. Temperature dependence of the transverse relaxation time T_2 for TSCC (after /1/)

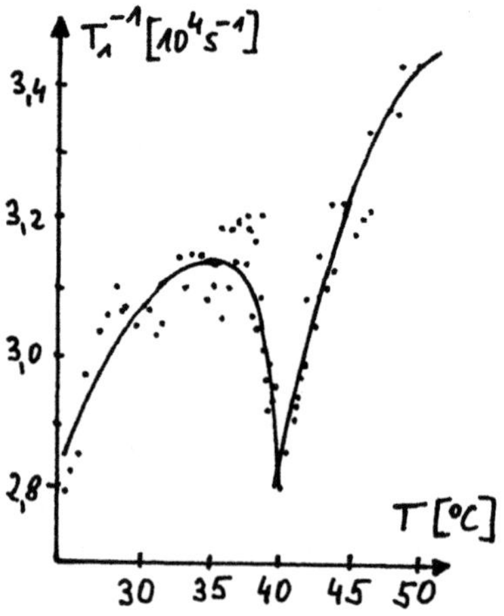

Fig. 2. Longitudinal relaxation rate versus temperature for TGS (after /1/)

There are several attempts to explain the opposite anomalies in the relaxation rates T_1^{-1} and T_2^{-1}. We applied first a mode-mode coupling theory /2/ and found that the coupling of an acoustic mode to a damped soft optical mode or to energy fluctuations yields an explanation of the opposite behaviour of T_1^{-1} and T_2^{-1}. However, the quantitative agreement with the experimental results is not sufficient. Blinc et al. /3/ and Blinc and Prelovšek /4/ assumed that the observed opposite anomalies are caused by local freezing out of impurity modes. These authors are able to explain the anomalies observed in KDA in a good quantitative agreement with the experiments. In their picture it is essential that a local mode is freezing out above T_c. This effect is well established in KDA, however, it is not found in the other materials under consideration. Völkel /5/ explained the opposite anomalies as resulting from a hard mode the damping of which becomes anomalously large at the phase transition. The microscopic origin of this increasing damping is still an open question. We have shown /6/ that the lattice anharmonicity leads to a dedamping of the relaxator mode at the EPR frequency in TGS providing us with an explanation of the opposite anomalies in TGS. In the present paper, we study the influence of defects coupled linearly to the order parameter (OP) on the spin-lattice relaxation times.

2. The Model

The experimentally investigated materials are uniaxial ferroelectrics, therefore, we need only a scalar OP. If we shall refer to a certain material, we shall have in mind TGS. So we assume a relaxator type OP mode. Völkel et al. /7/ found experimental evidence for a very slow motion of the paramagnetic centers having a characteristic frequency of about $10^7 \ldots 10^8$ s^{-1}, which is much less then the EPR frequency ($\omega_{EPR} \sim 10^{10}$ s^{-1}). We are going to treat this defect motion interacting with the order parameter relaxation. To this end we apply the model proposed by Sasvári and Schwabl /8/. In that model the spacial position of the defects is assumed to be quenched. The defects obey an internal degree of freedom which is coupled linearly to the OP. The defects cause a local T_c randomly distributed in the lattice and act as a random field right in the sense of the theory of relaxation.

The scalar OP φ and the variable α for the internal defect motion are assumed to be continuous variable corresponding to the usual mesoscopic point of view in the theory of phase transitions. Statics of the model are given by the following Ginzburg-Landau-Wilson functional /8/:

$$H = \int d^dx \left[\frac{1}{2}(r_o + \phi_R(x))\psi^2 + \frac{1}{2}\delta(\nabla\psi)^2 + \frac{1}{2}m(\frac{\partial\psi}{\partial z})^2 + \frac{1}{4}u\psi^4 - g\psi\tau + \frac{1}{2}\tau^2\right], \quad (1)$$

where we assume for the effect of the random defects on the Curie temperature

$$[\phi_R(x)] = 0, \quad [\phi_R(x)\,\phi_R(x')] = \Delta\,\delta(x-x'). \quad (2)$$

The square brackets mean the configurational average; Δ regulates the dependence of T_c on the defects. We took into account in the gradient terms in (1), that the ferroelectrics under consideration are uniaxial. The term containing ψ^4 describes the lattice anharmonicity.

Dynamics of the OP and the defect mode are given by two coupled Langevin equations:

$$\dot{\psi} = -\Gamma\frac{\delta H}{\delta\psi} + \zeta, \qquad \dot{\tau} = -\gamma\frac{\delta H}{\delta\tau} + \Theta. \quad (3)$$

The OP relaxes much faster ($\Gamma \sim 10^{12}\ s^{-1}$) then the defect mode ($\gamma \sim 10^8\ s^{-1}$), and the relation $\gamma \ll \Gamma$ holds. The noise is assumed to be a Gaussian white noise:

$$\langle\zeta(x,t)\zeta(x',t')\rangle = 2\Gamma\delta(x-x')\delta(t-t'),$$
$$\langle\Theta(x,t)\Theta(x',t')\rangle = 2\gamma\delta(x-x')\delta(t-t'). \quad (4)$$

Equations (1) to (4) fix the simple version we use here of the model by Sasvári and Schwabl /8/ which has been investigated in a different version earlier by Grinberg, Ma, and Mazenko /9/ and by Lubensky /10/.

3. The Correlation Functions

Response functions (RF) are defined in the usual way introducing the fields E and β conjugated to ψ and τ, respectively. The linear response to those fields defines the response functions /8/

$$[\langle\varphi(k,\omega)\rangle] = R_{\varphi\varphi}(k,\omega)E(k,\omega) + R_{\varphi\tau}(k,\omega)\beta(k,\omega),$$
$$[\langle\tau(k,\omega)\rangle] = R_{\tau\varphi}(k,\omega)E(k,\omega) + R_{\tau\tau}(k,\omega)\beta(k,\omega). \qquad (5)$$

Because of the linear coupling between φ and τ, the RF $R_{\varphi\tau}$, $R_{\tau\varphi}$, and $R_{\tau\tau}$ can be expressed by $R_{\varphi\varphi}$ /8/, which we denote by R in the following. The desired correlation function (CF) C results from the RF via the fluctuation-dissipation theorem. As another consequence of the linear coupling between φ and τ, we can eliminate τ in the equation of motion for φ, and we get after Fourier transformation the following equation of motion for φ /8/:

$$R_o^{-1}(k,\omega)\varphi(k,\omega) = \xi(k,\omega) + E(k,\omega) - \phi_R * \varphi - u(\varphi*\varphi)*\varphi, \qquad (6)$$

where ξ is another noise term

$$\xi = \frac{\xi}{\Gamma} + \frac{g\theta}{\gamma - i\omega} \qquad (7)$$

The last two terms in (6) are assumed to be small perturbations; the asterisk in those terms denotes the convolution of the factors connected by it. In the sense of perturbation theory, R_o is the unperturbed RF and is defined by

$$R_o^{-1} = r_o - g^2 + \delta k^2 - i\omega\tilde{\Gamma}_o^{-1} + m\left(\frac{k_z}{k}\right)^2. \qquad (8)$$

The CF corresponding to (8) is shown in Fig. 3; it increases for all frequencies ω, if $T \rightarrow T_c$.
According to the theory of relaxation (cf. e. g. /11/) the relaxation rates are given by

$$T_1^{-1} = J(\omega_{EPR}),\ T_2^{-1} = J(0), \qquad (9)$$

where $J(\omega)$ is the integral of $C(k,\omega)$ over all wave vectors k. In a first rough approximation one can substitute the integral over k by $C(0,\omega)$ near a ferroelectric phase transition. The considered opposite anomalies of T_1^{-1} and T_2^{-1} can be explained, if the inequalities

$$C(nr_o, \omega_{EPR}) > C(r_o, \omega_{EPR}),$$
$$C(nr_o, 0) < C(r_o, 0) \qquad (10)$$

are fulfilled for $n > 1$, i. e. approaching T_c. The unperturbed CF (8) does not obey (10).

We apply perturbation theory and calculate the contributions of the convolution terms in (6) in the lowest order linear in u and Δ. This procedure is a perturbation theory starting with mean-field approximation. The contributions to the self-energy are in the case of zero external momentum

$$\Sigma_u = -3u \int d^dq d\Omega\, C_o(q,\Omega), \quad \Sigma_\Delta = \Delta \int d^dq R_o(q, \omega). \tag{11}$$

The contributions (11) to the self-energy can be calculated in closed form /12/. We shall not write down these lengthy formulas, but show in Fig. 4 the numerical results for TGS.

Two cubic inequalities for Δ result from (10). These inequalities yield a lower limit for values of Δ, which are necessary to explain the opposite anomalies in question. The upper limit of Δ is given by the condition of applicability of perturbation theory, according to which the resulting logarithmic terms must not be too large /8/. Unfortunately, there are no experimental data available for Δ, and we use $\Delta K_4 \delta^{-3/2} = 6.6$ in the numerical calculations; this value fits into the admitted interval for Δ.

4. Discussion

The curves in Fig. 4 have been calculated for TGS using the parameters in the Ginzburg-Landau-Wilson functional as given in the paper /13/ by Nattermann. We have computed the one-particle CF $C(k = 0, \omega)$. However, as it is known from background relaxation apart from the phase transition, direct processes are not relevant in that temperature region, but Raman processes determine the relaxation in the temperature range around T_c. Therefore, we have to calculate the two-particle CF describing Raman processes, and we must find the opposite behaviour near T_c in that CF.

We know from former calculations for perfect crystals /6/ that the interesting behaviour of the CF (Fig. 4) at $\omega = 0$ and at $\omega \approx \omega_{EPR}$ is found in the integrals of the CF over wave vectors, too. This result was obtained approximately in closed form and by numerical calculations. Furtheron, the two-particle CF calculated from a CF similar to that in Fig. 4 increases at $\omega = 0$ and decreases at $\omega \approx \omega_{EPR}$

approaching T_c, and the considered anomalies can be found from the CF actually required /6/. However, as long as there are no experimental data available for Δ , more detailed numerical computation are of minor use.

The model investigated in this paper provides us with a possible explanation of the opposite anomalies in the longitudinal and in the transverse spin-lattice relaxation times in some uniaxial ferroelectris at T_c, but a quantitative comparison between theoretical and experimental results is impossible at present.

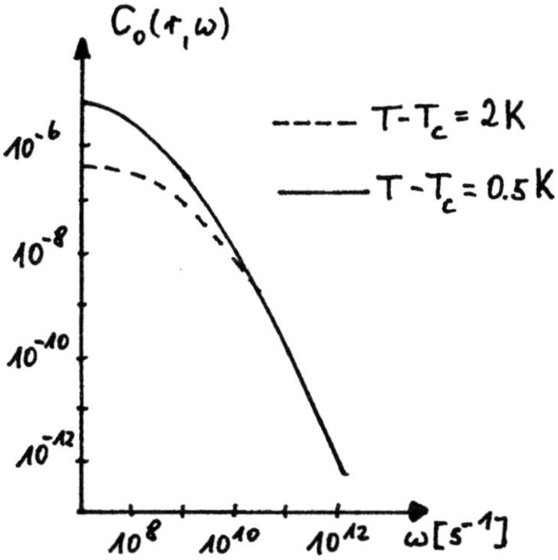

Fig. 3. Unperturbed correlation function for TGS

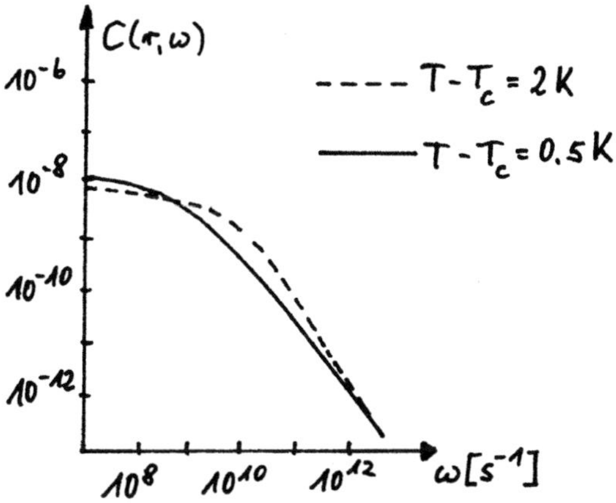

Fig. 4. Correlation function for TGS including the self-energy parts (11)

References

/1/ W. Brunner, G. Völkel, W. Windsch, I. N. Kurkin, and V. I. Shlenkin, Solid State Commun. 26, 853 (1978).

/2/ U. Geppert, T. Nattermann, and A. Kühnel, Ferroelectrics 29, 55 (1980).

/3/ R. Blinc, P. Cevc, P. Prelovšek, N. I. Silkin, G. M. Ribeiro, L. V. Gonzaga, A. S. Chaves, and R. Gazzinelli, Phys. Rev. B 25, 311 (1982).

/4/ R. Blinc and P. Prelovšek, Solid State Commun. 42, 893 (1982).

/5/ G. Völkel, Wiss. Z. Karl-Marx-Univ. Leipzig, Math.-Naturwiss. R. 32, 33 (1983).

/6/ U. Geppert and A. Kühnel, Wiss. Z. Karl-Marx-Univ. Leipzig, Math.-Naturwiss. R. 32, 16 (1983).

/7/ G. Völkel, W. Brunner, W. Windsch, and D. Steudel, phys. stat. sol. (b) 95, 99 (1979).
/8/ L. Sasvári and F. Schwabl, Z. Phys. B 46, 269 (1982).
/9/ G. Grinstein, S. Ma, and G. F. Mazenko, Phys. Rev. B 15, 258 (1977).
/10/ T. C. Lubensky, Phys. Rev. B 11, 3573 (1975).
/11/ I. V. Aleksandrov, Theory of Magnetic Relaxation, Moskow, Nauka 1975 (in Russian).
/12/ U. Geppert and A. Kühnel, submitted to phys. stat. sol.
/13/ T. Nattermann, phys. stat. sol. (b) 85, 291 (1978).

COMPUTER STUDIES OF MAGNETIC EXCITATIONS IN STRUCTURALLY DISORDERED ITINERANT-ELECTRON FERROMAGNETS

S. Krompiewski

Institute of Molecular Physics, Polish Academy of Sciences, Poznań
ul. Smoluchowskiego 17/19, Poland

Abstract

Structurally disordered ferromagnets are studied with use of the Hubbard hamiltonian. The emphasis is put on short-wavelength magnetic excitations which have been found in addition to usual long-wavelength ones. The computer results for one- and three-dimensional systems show that these excitations, centred around a wave vector corresponding to the maximum of the structure factor, are gapless in qualitative accordance with recent experimental data.

1. Introduction

The existence of spin-wave excitations in amorphous alloys has been established experimentally for many years now [1,2]. These excitations have been measured by various experimental techniques and show a usual quadratic long-wavelength magnon energy spectrum. However, there is still some controversy about the short-wavelength excitations found around the first peak of the structure factor. The initial measurements of these excitations [3,4] have suggested a considerable gap and relatively sharp spectrum while recent experiments [5,6] suggest no gap at all.

Many attempts have been made in order to adequatly describe magnetic excitations in structurally disordered ferromagnets. Most existing theories have described spins as localized and interacting through a Heisenberg interaction given by the hamiltonian

$$H = - \sum_{ij} J\left(|r_i - r_j|\right) S_i S_j \ . \qquad (1)$$

The simplest expression for the energy of spin-wave excitations, within this model, is that resulting from the so-called quasicrystalline approximation (QCA). The QCA consists in simply averaging, in a spherically symmetric way, the ideal crystal expression

$$\varepsilon_k = 2S \frac{1}{N} \sum_{ij} J\left(|r_i - r_j|\right)\left[1 - e^{ik\left(r_i - r_j\right)}\right] \qquad (2)$$

and reads

$$\mathcal{E}_k^{QCA} = 2S\,\frac{N}{V}\int_0^\infty 4\pi r^2\, g_2(r)\, J(r)\left(1 - \frac{\sin kr}{kr}\right) dr\,. \tag{3}$$

Incidentally, if the product $g_2(r)\,J(r)$ is sharply peaked at the atomic average distance r_o=2.5 Å the expression (3) can be shown to have a shallow dip at about $k = 3Å^{-1}$ which coincides with a typical position of a structure factor maximum in amorphous ferromagnets [7,8]. Other, more sophisticated, approximations lead to no dip at all [9] or to a dip more pronounced than that of the QCA [10].

2. Model and formalism

It seems quite obvious that a band (or itinerant-electron) model should be more appropriate for describing metallic ferromagnets than its localized spin counterpart - the Heisenberg model. A spin-wave in the band picture is a bound state of an electron with a given spin (minority spin) and a hole with an opposite spin. In spite of its formal complexity the band model has proved to be quite successful in describing neutron scattering processes in metals [11]. In this paper we adopt a simplest possible band model i.e. the single-band Hubbard hamiltonian

$$H = \sum_{ij\sigma} t_{ij}\, c^+_{i\sigma}\, c_{i\sigma} + U\sum_i c^+_{i\uparrow}\, c_{i\uparrow}\, c^+_{i\downarrow}\, c_{i\downarrow}\quad , \tag{4}$$

where t_{ij} is the hopping integral, U - the intraatomic Coulomb repulsion, $c^+_{i\sigma}(c_{i\sigma})$ - the creation (annihilation) operator of electrons at site i with spin σ (σ = $\uparrow$ or $\downarrow$), and the sums go over sites distributed at random.

In order to find a transverse spin magnetic susceptibility we first define the one-body (electron) Green functions

$$g_{ij\sigma}(t) = -\,i\;\Theta(t)\langle[\,c_{i\sigma}(t),\,c^+_{j\sigma}\,]_+\rangle\,. \tag{5}$$

These functions have to be determined selfconsistently from their equations of motion, within the Hartree-Fock approximation, and from the fluctuation-dissipation theorem:

$$(\omega - U\,n_{i-\sigma})\,g_{ij\sigma}(\omega) = \delta_{ij} + \sum_l t_{il}\,g_{lj\sigma}(\omega)\quad , \tag{6}$$

$$\langle c^+_{j\sigma}\,c_{i\sigma}(t)\rangle = -\frac{1}{\pi}\int_{-\infty}^{\infty} d\omega\; e^{-i\omega t}\; f(\omega)\;\mathrm{Im}\, g_{ij\sigma}(\omega)\quad , \tag{7}$$

where $$n_{i-\sigma} = \langle c^+_{i-\sigma}\,c_{i-\sigma}\rangle \tag{8}$$

and $$f(\omega) = \left[\exp \frac{1}{k_B T} (\omega - \mu) + 1 \right]^{-1} \quad . \tag{9}$$

Then a two-body (magnon) Green function is introduced

$$\chi_{ij}(t) = -i\,\Theta(t) \left\langle \left[c^+_{i\uparrow}(t)\, c_{i\downarrow}(t),\, c^+_{j\downarrow}\, c_{j\uparrow} \right]_- \right\rangle . \tag{10}$$

Within the random phase approximation it obeys the following matrix equation

$$\left[\underline{1} - U\, \underline{\chi}^{\circ}(E) \right] \underline{\chi}(E) = \underline{\chi}^{\circ}(E) \quad , \tag{11}$$

where $\chi^{\circ}(E)$ is the energy-Fourier component of

$$\chi^{\circ}_{ij}(t) = -\langle c_{i\uparrow}(t)\, c_{j\uparrow} \rangle\, g_{ij\downarrow}(t) - \langle c^+_{j\downarrow}\, c_{i\downarrow}(t) \rangle\, g_{ij\uparrow}(t) \tag{12}$$

and, in terms of the eigenvalues $\lambda^{\sigma}_{\alpha}$ and the eigenfunctions $u^{\sigma}_{\alpha i}$ of the hamiltonian (1), reads

$$\chi^{\circ}_{ij}(E) = \sum_{\alpha\beta} \left\{ \left[f(\lambda^{\uparrow}_{\alpha}) - f(\lambda^{\downarrow}_{\beta}) \right] / (\lambda^{\downarrow}_{\beta} - \lambda^{\uparrow}_{\alpha} - E) \right\} u^{\uparrow}_{\alpha i}\, u^{\uparrow}_{\alpha j}\, u^{\downarrow}_{\beta i}\, u^{\downarrow}_{\beta j} \,. \tag{13}$$

The zeros of the determinant of the expression in the bracket in (11) determine the energies of the magnetic excitations while the imaginary part of χ gives the scattering law (spectral density function):

$$S(q,E) = -\frac{1}{\pi}\frac{1}{N} \sum_{ln} \exp\left(iq\,(R_l - R_n) \right) \operatorname{Im} \chi_{ln}(E) \quad . \tag{14}$$

Another quantity of our interest is the structure factor $S(q)$ which is to be computed to check whether the position of its peak is correlated with the short-wavelength magnetic excitations. The calculations of all these quantities have to be repeated for various configurations and finally averaged.

3. Computations and results

The studied structures have been simulated by N hard spheres (or hard linear segments in the 1-d problem) packed in a random way for a given packing fraction η , and the periodic boundary conditions have been imposed. The strong ferromagnetism has been assumed with $N\uparrow = N$ spin-up electrons and $N\downarrow = 0.4\,N$ spin-down electrons. The hopping integral has been taken in an exponential form

$$t_{ij} = -|t| \exp\left(1 - |R_i - R_j| / R_{av} \right) \tag{15}$$

(with R_{av} being the average spacing) and assumed to vanish beyond the given interaction range R_o. The one-dimensional problem has been solved for N = 20 (see [12] for details). The results concerning the scattering law are presented in Fig.1 by rectangles which mark the loci of the S(q,E) maxima. To make a comparision with the ordered system dispersion energy curve (dots) easier the q-axis for the ordered system has been rescaled by a factor of η to bring its length to that of the disordered system $L = Na/\eta$ (where a = 1 is the length of the linear segment, and $\eta = 0.75$). It results from Fig.1 that the disorder produces no gap about $q_o = 1.6\pi$ where, as it is seen from Fig.2, the structure factor is a maximum. On the other hand, however, the spin-wave stiffness constant of the disordered system (steepness around q=0) seems to be lower than in a crystal whereas the spin-wave bandwidth is higher.

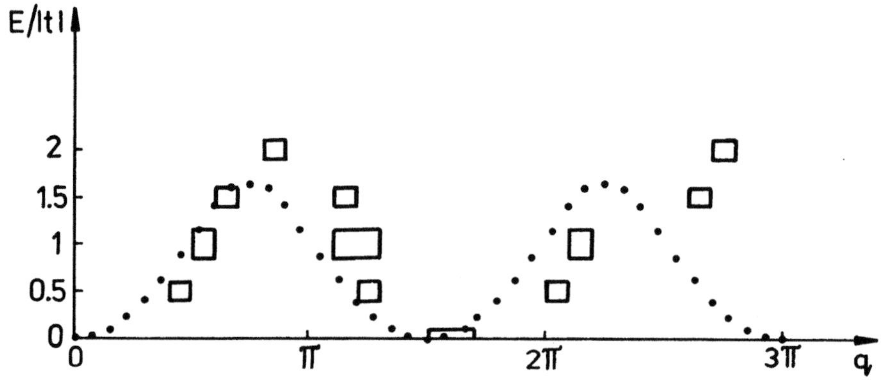

Fig.1 Ordered system magnon energy spectrum (dots) and the loci of the disordered system S(q,E) maxima (rectangles).

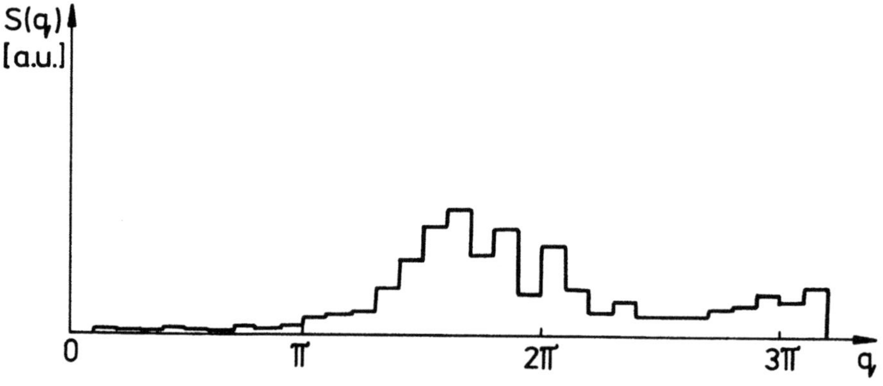

Fig.2 Configurationally averaged structure factor for the one-dimensional problem.

As regards the three-dimensional model a system of N=32 atoms has been considered. The starting configuration was the f.c.c. structure. Next the packing fraction has been reduced down to $\eta < \eta_{fcc}$ by extending the lattice with the sphere diameters kept fixed $(2r = 1)$. The spheres, no longer in contact, have been then displaced by the simple Monte-Carlo prescription. The computations have been performed for $\eta = 0.3$, $R_o = 1.75$, $T = 0$ and $U/|t| = 12$. The structure factor has been found to have a maximum at about $q_o = 2\pi$. The scattering law has been computed for a few energy values and its overall behaviour has turned out to be similar to that for the one-dimensional problem. Again, for E=0 a sharp magnon peak at q=0 has been found and a broad one about $q=q_o$. For increasing E the magnon peak shifts towards higher q values and the short-wavelength one gets split about q_o.

4. Conclusions

Both the calculations of the one-dimensional model and the preliminary results for the three-dimensional case lead to the following conclusions:

i) short-wavelength modes around q_o, where the structure factor is a maximum are gapless,

ii) numerical values of q_o in $Å^{-1}$, when compared with typical experimental values, are correct within 10%,

iii) the short-wavelengths excitations of the disordered system resemble the second Brillouin zone excitations of the ordered system with the scattering vector, q_o, acting as a smeared-out lattice-vector.

References

1 Y. Ajiro, K. Tamura and H. Ende, Phys. Lett. 35A, 275 (1971).

2 J.D. Axe, L. Passel and C.C. Tsuei, AIP Conf. Proc. 24, 119 (1975).

3 H.A. Mook, N. Wakabayashi and D. Pan, Phys. Rev. Lett. 34, 1029 (1975).

4 H.A. Mook and C.C. Tsuei, Phys. Rev. B 16, 2184 (1977).

5 G. Shirane, J.D. Axe, C.F. Majkrzak and T. Mizoguchi, Phys. Rev. B 26, 2575 (1982).

6 D. McK. Paul, R.A. Cowley, W.G. Stirling, N. Cowlam and H.A. Davies, J. Phys. F 12, 2687 (1982).

7 U. Krey, J. Magn. Magn. Mat. 6, 27 (1977).

8 T. Kaneyoshi, J. Phys. Soc. Jap. 45, 1835 (1978).
9 G.J. Yates and R.C. Jones, J. Phys. C 12, 1725 (1979).
10 V.A. Singh and L.M. Roth, J. Appl. Phys. 49, 1642 (1978).
11 W. Marshall and S.W. Lovesey, Theory of thermal neutron scattering, Oxford University Press,(1971) p. 274.
12 S. Krompiewski, Phys. Rev. Lett. 51, 1092 (1983).

SOME REMARKS ON THE SPIN FLUCTUATION THEORIES FOR ITINERANT ELECTRONS

B. Lorenz and H. Fehske
Sektion Physik der Karl-Marx-Universität
Karl-Marx-Platz
DDR-701 Leipzig

I. Introduction

For more than two decades the magnetism of transition metals and alloys has been the matter of much experimental and theoretical work. As is well known neither localized spin models nor the conventional Stoner theory yield a satisfactory description of the magnetic properties of transition metals. The non integral value (in units of μ_B) of the ground-state magnetization, the Curie-Weiss behaviour of the paramagnetic spin susceptibility, and the temperatur dependence of the magnetization observed in these materials show characteristic features of both localized and band magnetism. For an interpolation between the two limits the Hubbard model has been proposed /1/. The tight binding Hamiltonian is given by

$$H = -t \sum_{\langle ij\rangle,\sigma} (c^+_{i\sigma} c_{j\sigma} + h.c.) + U \sum_i n_{i\uparrow} n_{i\downarrow} - \mu \sum_{i,\sigma} n_{i\sigma} \quad . \qquad (1)$$

t and U are the nearest neighbour overlap and onsite Coulomb integrals, respectively. $c^+_{i\sigma}$ is the creation operator of an electron in the Wannier state i with spin projection σ and $n_{i\sigma} = c^+_{i\sigma} c_{i\sigma}$. The chemical potential μ is introduced in order to fix the mean electron number n.

For $U = 0$ the Hamiltonian (1) describes a single band of non-interacting electrons. In the atomic limit (t=0) the electrons and their magnetic moments are completely localized. Although a rigorous proof of the existence of long range magnetic ordering in the Hubbard model is missing the model (1) is applied in various approximations to the description of itinerant electron magnetism in transition metals.

In recent years spin fluctuation theories based on the functional integral representation of the partition function of (1) have attracted increasing attention. These theories provided an accurate description and a current understanding of itinerant magnetism /2 to 5/ (for a review see /6/). Methodically, the onsite interaction is linearized by the Hubbard-Stratonovich transformation /7/ and the thermodynamics of noninteracting particles in time dependent local fields has to be solved. Then the partition function of (1) is obtained as a functional integral over the fields. Obviously, the linearization of the operator $n_{i\uparrow}n_{i\downarrow}$ is not unique and depends on the representation of the interaction term, e.g.

$$n_{i\uparrow}n_{i\downarrow} = \frac{1}{2}(n_{i\uparrow}+n_{i\downarrow}) - \frac{1}{2}(n_{i\uparrow}-n_{i\downarrow})^2 \quad , \tag{2}$$

$$n_{i\uparrow}n_{i\downarrow} = \frac{1}{4}(n_{i\uparrow}+n_{i\downarrow})^2 - \frac{1}{4}(n_{i\uparrow}-n_{i\downarrow})^2 \quad . \tag{3}$$

In (2) the identity $n_{i\sigma}^2 = n_{i\sigma}$ has been used and one local spin field is introduced by the Hubbard-Stratonovich transformation. Using (3) charge and spin fields connected with the charge density operator $n_{i\uparrow}+n_{i\downarrow}$ and the spin density operator $n_{i\uparrow}-n_{i\downarrow}$, respectively, appear in the functional integral. Consequently, in dependence on the representation of the interaction term we distinguish one-field and two-field methods, both yielding an exact expression for the partition function of (1). However, evaluating the functional integral different approximations are applied. Firstly, the static approximation neglects the time dependence of the charge and spin fluctuation fields and the functional integral becomes an ordinary integral. Furthermore, the saddle point approximation, the self-consistent single site approximation etc. are used for the calculation of thermodynamic quantities. Within an approximate treatment the results depend strongly upon the applied methods. Until now the effects of the various approximations, especially that of the static one, have not been investigated very thoroughly.

To describe the magnetic properties we regard as a characteristic quantity the local magnetic moment m_{loc}, the square of which is given by

$$m_{loc}^2 = \langle (n_{i\uparrow}-n_{i\downarrow})^2 \rangle = n - 2\langle n_{i\uparrow}n_{i\downarrow} \rangle \quad . \tag{4}$$

$\langle \ldots \rangle$ denotes the thermodynamic average with respect to the Hamiltonian (1). The r.h.s. of (4) follows from the identity $n_{i\sigma}^2 = n_{i\sigma}$. Because of translational invariance m_{loc} does not depend on the site index i (this assumption is valid in the paramagnetic, ferromagnetic, and antiferromagnetic phases). The quantity (4) can be calculated from the partition function $Z = \mathrm{Tr}\exp(-\beta H)$ by

$$m_{loc}^2 = n + \frac{2}{N\beta}\frac{\partial \ln Z}{\partial U} \quad . \tag{5}$$

N denotes the number of sites and $\beta = 1/k_B T$ is the inverse temperature. Alternatively, adding a locally coupled magnetic field $h_i(n_{i\uparrow} - n_{i\downarrow})$ to H we get

$$m_{loc}^2 = \frac{1}{\beta^2 Z} \left. \frac{\partial^2 Z}{\partial h_i^2} \right|_{h_i = 0} . \tag{6}$$

Formulas (5) and (6) are equivalent definitions of m_{loc}. However, in an approximate treatment of Z this equivalence can be destroyed. Than the question arises whether (5) or (6) should be used to calculate the local moment.

In the spin fluctuation theories discussed above m_{loc} is calculated from (6) and in the paramagnetic phase it shows the following dependence on U and T /2/. In the ground state there is a region with zero local moment ($U < U_c$). For $U > U_c$ m_{loc} grows gradually to the atomic limit value with increasing U. At low temperatures m_{loc} is proportional to $T^{1/2}$ for $U < U_c$, i.e. the local magnetic moment increases with temperature /2,5,8/. This qualitative behaviour is in contradiction to exact results and other approximate solutions for the model (1), In the ground state solution of the Hubbard chain m_{loc} is always larger than zero and varies monotonically with U between band and atomic limit /9/. Furthermore, finite chain extrapolations show a gradual decrease of the local moment with increasing temperature if U is not too large /10/. For some limits of (1) the local moment and its expected dependence on U and T are discussed in /11/ and disagreement with the functional integral methods is found too.

In order to clarify the problems mentioned above we investigate critically the static approximation in the one- and two-field schemes, especially with respect to the behaviour of the local magnetic moment. In section II the basic formulas in the functional integral theory are derived and m_{loc} is expressed by functional integrals according to (5) and (6). In section III we consider a simple, exactly solvable two-site Hubbard model and compare the results of the static approximations with the exact solution. The main conclusions are summarized in section IV.

II. Functional Integral Representation

1) One-field method

With the identity (2) and the Hubbard-Stratonovich transformation the partition function is given by

$$Z = \int \prod_i \mathcal{D}\xi_i(\tau) \exp\left\{-\frac{\pi}{\beta}\int_0^\beta \sum_i \xi_i^2(\tau)\,d\tau\right\} \quad \tilde{Z}(\xi_i) \tag{7}$$

with $\tilde{Z}(\xi_i) = Tr\, T_\tau \, \exp\left\{-\int_0^\beta \tilde{H}(\tau)\,d\tau\right\}$,

$$\tilde{H}(\tau) = H^o(\tau) - \sum_{i\sigma}\left\{\sqrt{\frac{2\pi U}{\beta}}\,\sigma\,\xi_i(\tau) + \sigma h_i - \frac{U}{2}\right\} n_{i\sigma}(\tau) \quad , \tag{8}$$

$$H^o(\tau) = -t \sum_{\langle i,j\rangle,\sigma} (c_{i\sigma}^+(\tau)\, c_{j\sigma}(\tau) + h.c.) - \mu \sum_{i\sigma} n_{i\sigma}(\tau) \ . \tag{9}$$

The time dependent operators are taken in the interaction representation and T_τ denotes the time ordering operator. $\xi_i(\tau)$ is the time dependent spin field and h_i is the static magnetic field introduced in the previous section. The spin variables σ take the values ± 1 .

For the derivation of the local moment it is convenient to perform the substitution

$$\xi_i'(\tau) = \sqrt{\frac{2\pi U}{\beta}}\,\xi_i(\tau) + h_i \quad . \tag{10}$$

Then the Hamiltonian $\tilde{H}(\tau)$ does not depend on the h_i and the partition function takes the form

$$Z = \left(\frac{\beta}{2\pi U}\right)^{N/2} \int \prod_i \mathcal{D}\xi_i'(\tau) \exp\left\{-\frac{1}{2U}\int_0^\beta \sum_i (\xi_i'(\tau) - h_i)^2 d\tau\right\} \quad \tilde{Z}(\xi_i') \ . \tag{11}$$

The local magnetic moment is obtained according to (5) as

$$m_{loc}^2 = -\frac{1}{\beta U} + \frac{1}{\beta U^2}\left\langle \int_0^\beta \xi_i'^2(\tau)\,d\tau \right\rangle \tag{12}$$

and according to (6) as

$$m_{loc}^2 = -\frac{1}{\beta U} + \frac{1}{(\beta U)^2}\left\langle \left[\int_0^\beta \xi_i'(\tau)\,d\tau\right]^2 \right\rangle \ . \tag{13}$$

The angular brackets denote the functional average at zero static fields, $h_i = 0$.

Obviously, in the static approximation (12) and (13) yield identical expressions given by

$$m_{loc}^2 = -\frac{1}{\beta U} + \frac{1}{U^2}\langle \xi_i^2 \rangle_{st} \quad . \tag{14}$$

2) Two-field method

If we use the representation (3) the partition function is obtained as

$$Z = \int \prod_i \mathcal{D}\xi_i(\tau)\mathcal{D}\eta_i(\tau) \exp\left\{-\frac{\pi}{\beta}\int_0^\beta \sum_i (\xi_i^2(\tau)+\eta_i^2(\tau))\, d\tau\right\} \; \tilde{Z}(\xi_i,\eta_i) \tag{15}$$

with $\tilde{Z}(\xi_i,\eta_i) = Tr\, T_\tau \exp\left\{-\int_0^\beta \tilde{H}(\tau)\, d\tau\right\}$,

$$\tilde{H}(\tau) = H^o(\tau) - \sum_{j\sigma}\left\{\sqrt{\frac{\pi U}{\beta}}\,(\sigma\xi_j(\tau) + i\,\eta_j(\tau)) + \sigma h_j\right\} n_{j\sigma} \tag{16}$$

$H^o(\tau)$ is given in (9) and $\eta_j(\tau)$ is the local field coupled to the charge fluctuation at site j. With the substitution

$$\xi_i'(\tau) = \sqrt{\frac{\pi U}{\beta}}\;\xi_i(\tau) + h_i \quad , \quad \eta_i'(\tau) = \sqrt{\frac{\pi U}{\beta}}\;\eta_i(\tau) \tag{17}$$

we get

$$Z = \left(\frac{\beta}{\pi U}\right)^N \int \prod_i \mathcal{D}\xi_i(\tau)\mathcal{D}\eta_i(\tau) \exp\left\{-\frac{1}{U}\int_0^\beta \sum_i [(\xi_i'(\tau)-h_i)^2 + \eta_i'^2(\tau)]\, d\tau\right\} \; \tilde{Z}(\xi_i',\eta_i') \, . \tag{18}$$

Using the definition (5) the local moment is defined as

$$m_{loc}^2 = -\frac{2}{\beta U} + n + \frac{2}{\beta U^2}\left\langle \int_0^\beta [\xi_i'^2(\tau) + \eta_i'^2(\tau)]\, d\tau \right\rangle . \tag{19}$$

On the other hand, from (6) we obtain

$$m_{loc}^2 = -\frac{2}{\beta U} + \frac{4}{\beta^2 U^2}\left\langle \left[\int_0^\beta \xi_i'(\tau)\, d\tau\right]^2 \right\rangle \quad . \tag{20}$$

Note that in the static approximation from (19) and (20) follow the formulas

$$m_{loc}^2 = -\frac{2}{\beta U} + n + \frac{2}{U^2}\langle \xi_i^2 + \eta_i^2 \rangle_{st} \tag{21}$$

and

$$m_{loc}^2 = -\frac{2}{\beta U} + \frac{4}{U^2}\langle \xi_i^2 \rangle_{st} \quad , \tag{22}$$

respectively, which are not equivalent as in the one-field method. In the next section we will show that different results are obtained from (21) and (22).

III. Two-Site Hubbard Model

As an exactly solvable example we consider the model (1) for two sites only. Although there is no magnetic long range order in finite systems the qualitative behaviour of some thermodynamic functions e.g. the specific heat, the entropy, the local magnetic moment etc. is very similar to that of an infinite paramagnetic system /11,12/. The Hamiltonian is given by

$$H = -t\sum_{\sigma}(c^{+}_{1\sigma}c_{2\sigma} + h.c.) + U(n_{1\uparrow}n_{1\downarrow} + n_{2\uparrow}n_{2\downarrow}) - \mu\sum_{\sigma}(n_{1\sigma} + n_{2\sigma}) . \quad (23)$$

The model (23) is easily solved and the thermodynamic properties are discussed in /11,12/. Furthermore, it has the advantage that it can be treated within the functional integral formalism using only the static approximation. No further approximation has to be applied as e.g. for the infinite model (1). This allows us to investigate the effect of the static approach by comparison with the exact solution. In what follows, we shall consider the case $n = 1$ for which $\mu = U/2$. Since the results at $T = 0$ can be given analytically we restrict our discussion to the energy and the local moment in the ground state.

1) Exact solution

The partition function of (23) is given by /11,12/

$$Z = 2\exp\left\{\frac{\beta U}{2}\right\}\left[\operatorname{ch}\left\{\frac{\beta}{2}\sqrt{U^2 + 16t^2}\right\} + 3\exp\left\{\frac{\beta U}{2}\right\} + 4\operatorname{ch}\beta t\right] . \quad (24)$$

The energy and the local moment at $T = 0$ are obtained as

$$E = -\frac{1}{\beta}\ln Z + 2\mu = \frac{U}{2}\left[1 - \sqrt{1 + \left(\frac{4t}{U}\right)^2}\right] , \quad (25)$$

$$m^2_{loc} = 1 + \left[1 + \left(\frac{4t}{U}\right)^2\right]^{-1/2} . \quad (26)$$

2) One-field method

Using the static approximation and performing the trace in (11) we get

$$Z^{st} = \frac{\beta}{2\pi U} \int d\xi_1 \, d\xi_2 \, \exp\{-\beta F(\xi_1, \xi_2)\} \tag{27}$$

with the free energy density

$$F(\xi_1, \xi_2) = \frac{1}{2U}(\xi_1^2 + \xi_2^2) - \frac{2}{\beta} \ln\left(2\,\mathrm{ch}\left\{\frac{\beta}{2}(\xi_1 + \xi_2)\right\} + 2\,\mathrm{ch}\left\{\frac{\beta}{2}\left[(\xi_1 - \xi_2)^2 + \left(\frac{2t}{U}\right)^2\right]^{\frac{1}{2}}\right\}\right) \tag{28}$$

(We suppress the prime at the fields now).

At $T = 0$ the saddle points of (27) contribute only to the integral and the energy is obtained by minimizing (28) with respect to ξ_1, ξ_2. The extremal values, the energy, and m_{loc} are presented in table 1. The regions $U \leq t$ and $U > t$ are distinguished.

Table 1

	ξ_1 $(= -\xi_2)$	E	m_{loc}
$U \leq t$	0	$U - 2t$	0
$U > t$	$\pm \frac{1}{2}\sqrt{U^2 - t^2}$	$-t^2/U$	$1 - (t/U)^2$

3) Two-field method

In this case the partition function in the static approximation is written as

$$Z^{st} = \left(\frac{\beta}{\pi U}\right)^2 \int d\xi_1 \, d\xi_2 \, d\eta_1 \, d\eta_2 \, \exp\{-F(\xi_1, \xi_2, \eta_1, \eta_2)\} \,, \tag{29}$$

$$F(\xi_1, \xi_2, \eta_1, \eta_2) = \frac{1}{U}(\xi_1^2 + \xi_2^2 + \eta_1^2 + \eta_2^2) - \frac{1}{\beta} \ln \sum_{i=1}^{4} e^{-\beta\lambda_i} \tag{30}$$

$$\lambda_{1/2} = -z_1 \pm \sqrt{z_2^2 + t^2} \,, \qquad \lambda_{3/4} = -\lambda^*_{1/2} \,,$$

$$z_{1/2} = \frac{1}{2}(\xi_1 \pm \xi_2) + \frac{i}{2}(\eta_1 \pm \eta_2) \,.$$

(i is the imaginary unit).

For $T = 0$ the minimization of (30) yields the results shown in table 2 for $U \leq 2t$ and $U > 2t$. Both definitions (5) and (6) are used for the calculation of m_{loc}.

The energy and the local magnetic moment obtained in the static approximations as well as from the exact solution are shown in figures 1 and 2, respectively.

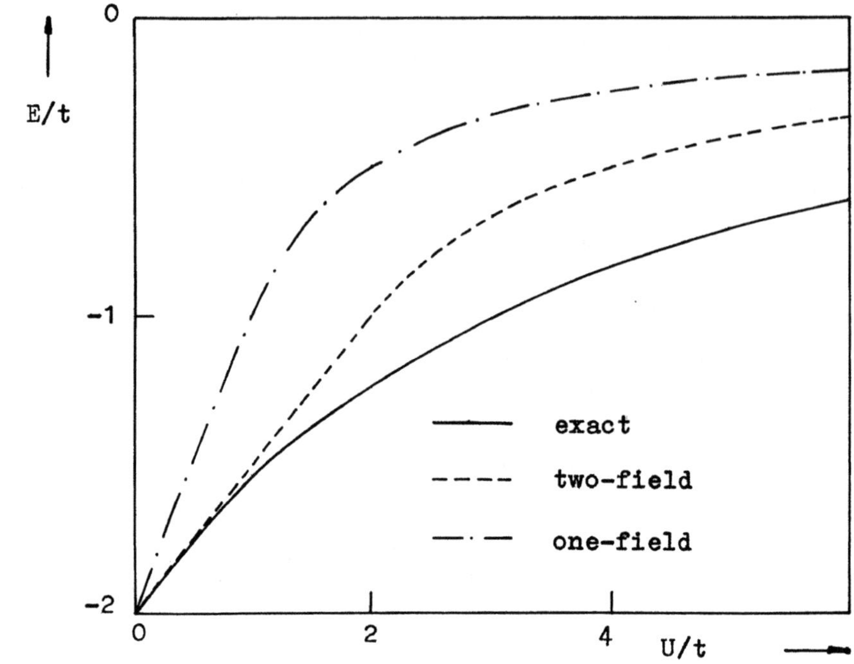

Fig. 1 Ground state energy of the two-site model

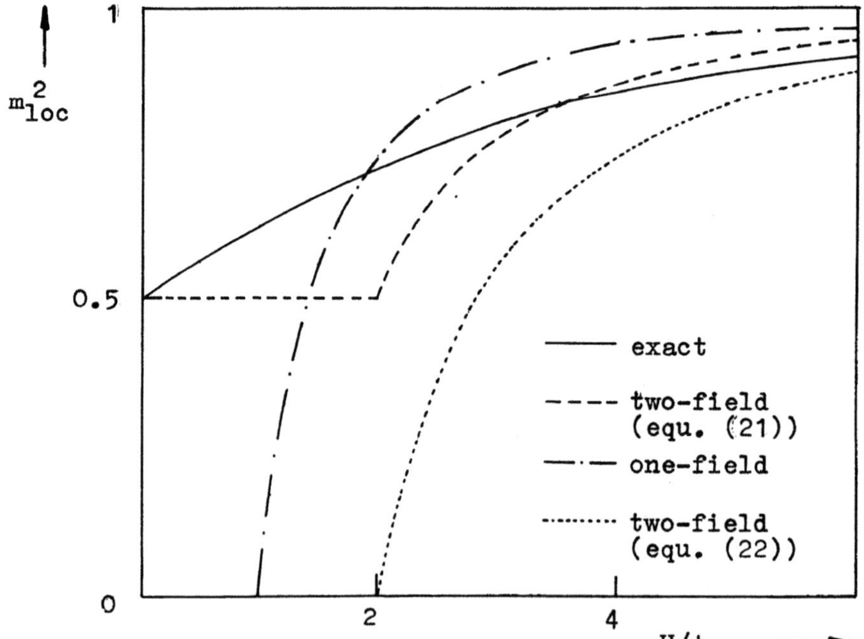

Fig. 2 Local magnetic moment in the ground state of the two-site model

Table 2

	ξ_1 $(=-\xi_2)$	η_1 $(=\eta_2)$	E	m_{loc}, according to (5)	(6)
$U \leqslant 2t$	0	$iU/2$	$U/2-2t$	$1/2$	0
$U > 2t$	$\pm\sqrt{U^2/4-t^2}$	$iU/2$	$-2t^2/U$	$1-2\,(t/U)^2$	$1-4\,(t/U)^2$

IV. Discussion and Conclusions

Comparing the ground state energies in figure 1 we conclude that the two-field description within the static approximation yields better results than the one-field approach. Especially, the correct slope of the energy at $U=0$ is obtained only by the two-field scheme. In the range of large U there is antiferromagnetic short range order since the extremal values of the spin fields have opposite signs ($\xi_1=-\xi_2$). This antiferromagnetic spin coupling is induced by the hopping term as can be seen in the perturbation expansion of (1) with respect to t. Note that in a single site approximation as applied to the infinite system this correlation is neglected and the results obtained at $T=0$ are equivalent to those of the Hartree-Fock approximation.

The ground-state local moments are shown in figure 2. With the definition (6) the one- and two-field methods yield similar results and fail in the region of small U where m_{loc} becomes zero contrary to the exact solution. It is interesting to note that in the infinite system m_{loc} shows strictly the same dependence on U /2,13/. Due to the wrong results at small U the temperature dependence of m_{loc} is given erroneously too (cf. the discussion in section I). This faulty behaviour is due to the fact that the operator identity $n_{i\sigma}^2=n_{i\sigma}$ is not preserved in the static approximation. However, a reasonable result is obtained in the two-field description calculating m_{loc} from (5). Both atomic and band limits are reproduced and the exact moment is considerably better approximated. The edge at $U=2t$ is clearly due to the static approximation and is expected to vanish if the time dependence of the fields is taken into account /14/. With the definition (5) one has to calculate the correlation function $\langle n_{i\uparrow}n_{i\downarrow}\rangle$ instead of $\langle (n_{i\uparrow}-n_{i\downarrow})^2\rangle$. Hence, the large error due to the treatment of $\langle n_{i\sigma}^2\rangle$ in the static approximation is avoided.

From our results we conclude

i) In the functional integral theory of the Hubbard Hamiltonian reasonable results can be obtained in the static approximation using the two-field method.

ii) In the calculation of correlation functions appreciable errors are obtained if operators like $n_{i\sigma}^2$ appear in the functional integral. The local magnetic moment, for example, should be calculated according to (5) rather than (6).

iii) The subdivision of the paramagnetic ground state into a phase without local moments (small U) and a phase with local moments (large U) and the qualitative temperature dependence of m_{loc} in the former phase as obtained in /2,3,5,8/ are given erroneously.

References

1. J. Hubbard, Proc. Roy. Soc. A 276, 238 (1963).
2. H. Hasegawa, J. Phys. Soc. Japan 49, 178 (1980).
3. H. Hasegawa, J. Phys. Soc. Japan 46, 1504 (1979); 49, 963 (1980); 50, 802 (1981).
4. J. Hubbard, Phys. Rev. B 19, 2626 (1979); B 20, 4584 (1979); B 23, 5974 (1981).
5. S. N. Evangelou, H. Hasegawa, and D. M. Edwards, J. Phys. F 12, 2035 (1982);
S. N. Evangelou and D. M. Edwards, J. Phys. C 16, 2121 (1983).
6. Electron Correlation and Magnetism in Narrow-Band Systems ed. T. Moriya, Solid-State Sciences, Vol. 29, Springer, Berlin/Heidelberg/New York 1981.
7. R. L. Stratonovich, Dokl. Akad. Nauk SSSR 115, 1097 (1957).
8. H. Fehske, phys. stat. sol. (b) 120, 611 (1983).
9. E. H. Lieb and F. Y. Wu, Phys. Rev. Letters 20, 1445 (1968).
10. H. Shiba, Prog. theor. Phys. (Kyoto) 48, 2171 (1972).
11. B. Lorenz, phys. stat. sol. (b) 119, 555 (1983).
12. U. Bernstein and P. Pincus, Phys. Rev. B 10, 3626 (1974).
13. M. Sakoh and M. Shimizu, J. Phys. Soc. Japan 40, 974 (1976).
14. D. Paquet and P. Leroux Hugon, Phys. Rev. Letters 49, 1023 (1982).

NECESSARY AND SUFFICIENT CONDITIONS FOR THE OVERBLOCKING EFFECT

H. Nencka-Ficek
Institute of Molecular Physics,
Polish Academy of Sciences, Poznań
Poland

Alexander and Pincus [1] revealed and explained a very interesting phenomenon during their investigations of a fully frustrated fcc Ising model. Namely, the ground state of the model is such that each elementary plaquette contains at most one frustrated bond when a lattice dimension is $d \leqslant 4$, and for $d > 4$ any ground state contains a finite density of triangles with three frustrated bonds. This effect was called the overblocking effect. Derrida et al [2] were able to show the generalization of this effect, taking into account simple hypercubic fully frustrated d-dimensional lattices. They have asked the following two questions: 1/ "whether this effect is lattice dependent and 2/ whether it is possible to obtain a simple geometric understanding of it". The partial answer to these questions can be found in [2].

We would like to give some contribution to understanding the overblocking effect. Therefore we present the necessary and sufficient condition for this effect [3]. However, in order to give such the conditions, we must use a trick. This trick consists in using the Ising model in a d-dimensional hypercubic lattice with multi-spin interactions /Fig 1/.

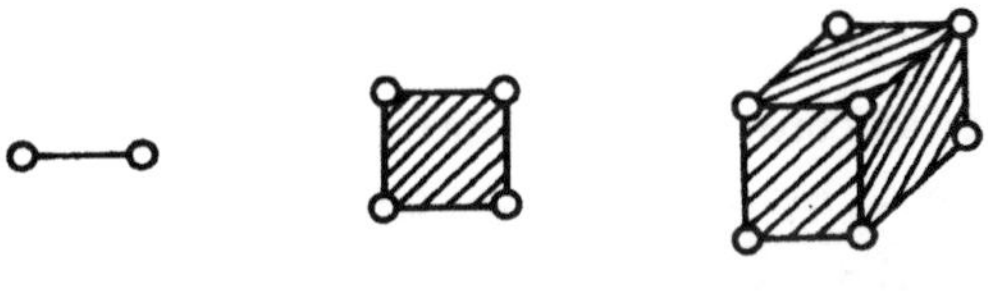

Fig 1 Multi-spin interactions: the case of 2-,4-,8-spin interactions.

We consider the model whose interactions are described by the Hamiltonian :

$$-\beta H = K(b_k) \sum_{\{b_k\}} A(b_k^j) \prod_{i \in b_k^j}^{w} \sigma_{i,j} \quad (1)$$

where $\beta = 1/k_BT$, $K(b_k)$ is a reduced coupling constant $(=\beta J(b_k))$ for multi-spin interactions placed on a b_k-dimensional elementary hypercube in the lattice, $A(b_k)$ is a random variable /±1/ for such the multi-spin interactions, $\sigma_{i,j}$ /±1/ are the Ising spins. The sum runs over all configurations of 2^k-spin interactions / k=1,2,...d/ placed on a b_k hypercube of the lattice, the product is over $w=2^k$ spins sitting at the sites of the hypercube b_k. Generalising the idea of the plaquette as a closed k-dimensional hypercontour $L_k = \bigcup_{i=1}^{n} b_k^i$ /i=1,... ..,n-1/ such that $b_k^i \cap b_k^{i+1} \neq \emptyset$, $b_k^1 \cap b_k^n \neq \emptyset$, we are able to consider the case of the frustration for the multi-spin interactions. /Fig 2/.

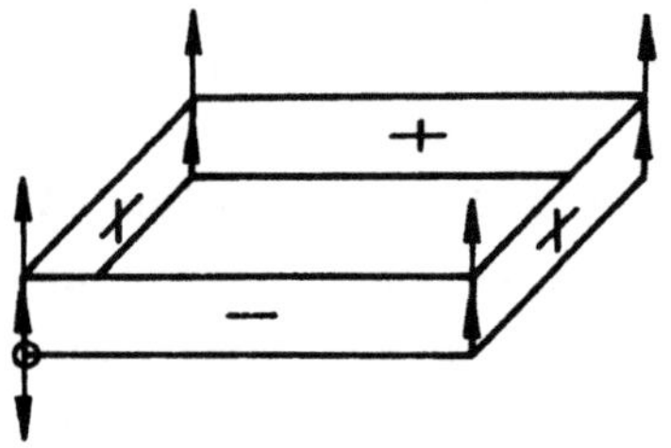

Fig 2 Frustration for 4-spin interactions on the 2-dimensional plaquette.

The frustration function for a k-dimensional plaquette is the following

$$\Phi(b_k) = A(b_k^1)A(b_k^2)A(b_k^3)A(b_k^4) \quad (2)$$

However, when we consider the frustration function for a hypercube then it turns out that the frustration depends on the configurations of the negative interactions, independent of the odd number of negative interactions in the hypercube. Although, taking into account the fact that each k-dimensional hypercube is built of $\mathfrak{M}$ $(= \sum_{i=1}^{k-1} i)$, (k-1)-dimensional plaquettes we are able to give the following definition of the frustration for the hypercube :

$$\Phi_h = (-1)^l \quad (3)$$

$$l = \begin{cases} 1 \;, & \text{if there is one at least k-dim frustrated plaquette} \\ 0 \;, & \text{otherwise.} \end{cases}$$

If we consider the simplest case of the multi-spin interactions, namely the 4-spin interactions /Fig 1/ in the 3-dimensional cubic lattice then we shall see that there are three different two-dimensional plaquettes /Fig 3/ and the frustration functions for them are the following :

$$\begin{aligned} \Phi_1 &= A_1A_2A_3A_4 \\ \Phi_2 &= A_1A_2A_5A_6 \\ \Phi_3 &= A_3A_4A_5A_6 \end{aligned} \tag{4}$$

If we demand the 2-dimensional plaquettes to be frustrated we shall find that it is impossible for all of them to be frustrated simultaneously. Only two of them can be frustrated. We call such an effect the overblocking effect. Let us introduce a formalism in order to derive the necessary and sufficient condition for the overblocking effect.

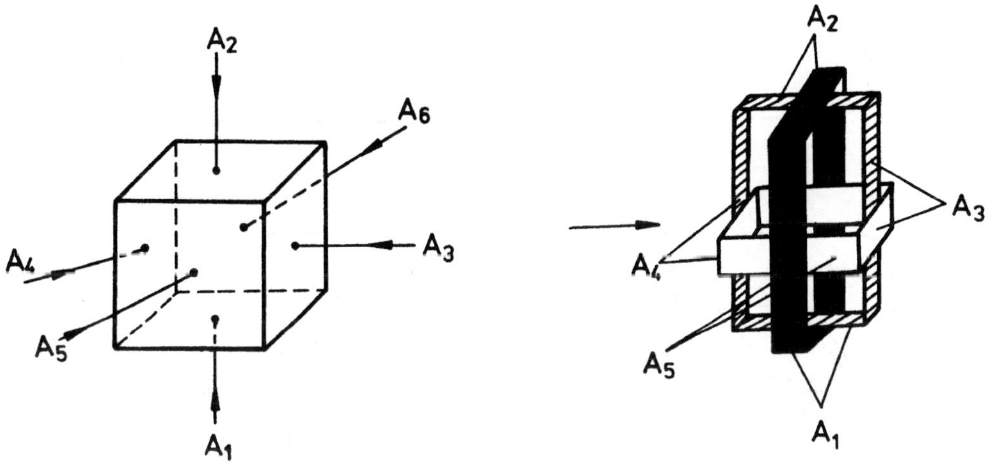

Fig 3 The configuration of the random variable $A(b_2) = A_i$ for the 4-spin interactions on the 2-dimensional plaquettes for a cube.

Let us define by

$$\mathcal{X}_k^i \equiv (A(b_k^i), A(b_k^{i+2})) \tag{5}$$

a pair of random variables $A(b_k)$ on the opposite k-dimensional walls in the same hypercube, and by

$$L_k^{i,i+1} \equiv [\mathcal{X}_k^i ; \mathcal{X}_k^{i+1}] \tag{6}$$

the k-dimensional plaquette which contains the walls b_k^i and b_k^{i+2}. Let us introduce the following mapping /Fig 4a/

$$f : \mathcal{X}_k^i \longleftrightarrow x_i \in T \tag{7}$$

where f maps each of the pair $\mathcal{X}_k^i$ of the physical space M^d to some point x_i of the abstract space T.

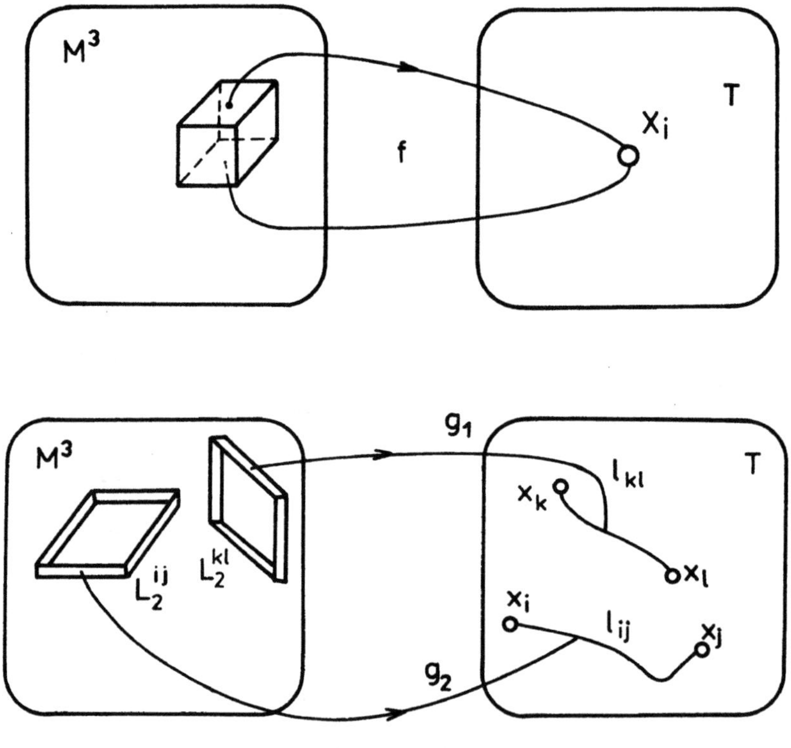

Fig 4 The mappings of: a/ the opposite walls onto the point x of the abstract space T, b/ 2-dimensional plaquettes into the curves of the space T.

There is another mapping /Fig 4b/

$$g: L_k^{ij} \longleftrightarrow l_{ij} \subset T \qquad (8)$$

which maps each of the k-dimensional plaquettes I_k of the physical space to a curve l_{ij} of the space T. The curve l_{ij} is ended by points x_i and x_j.

We can see /Fig 5/ that a loop α in T corresponds to three different 2-dimensional plaquettes in the same cube of the lattice.

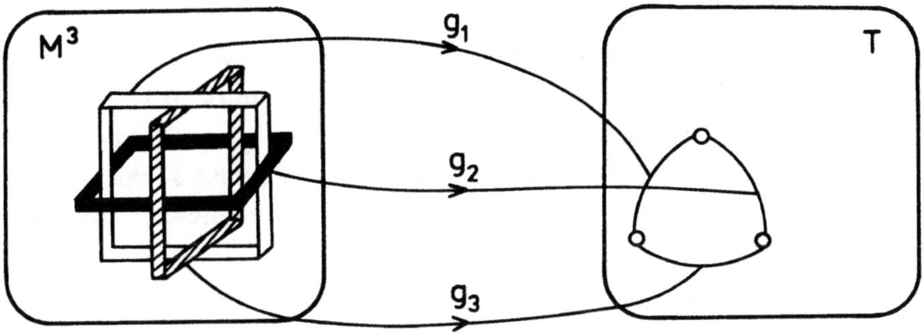

Fig 5 The mapping of a cube into a loop of the space T.

The mappings f and g map the topological properties of the lattice only. They are independent of the internal properties of the problem /the Ising model with the frustration/. Therefore we are forced to introduce a tool which will enable us to investigate the internal properties of the model. Let us introduce the following

Definition

The point $x_c = f(\mathcal{X}_k^i)$ is black, when the product

$$A(b_k^i) \cdot A(b_k^{i+2}) = -1 \ ,$$

the point $x_b = f(\mathcal{X}_k^j)$ is white, when the product

$$A(b_k^j)\ A(b_k^{j+2}) = +1 \ .$$

We can see that the curve corresponding to the frustrated plaquette which is independent of its dimension, has one black and one white vertices •——○ . Now, we can give the following two theorems.

Theorem I. The necessary condition for the overblocking effect.

The existence of at least one loop in the abstract space T is a necessary condition for the overblocking effect [3] .

Proof. Let us assume that there are graphs $G\{l^{ij}, l^{ir}, l^{ip}, .. l^{im}\}$ built of the curves $l^{ij}, l^{ir}, \ldots, l^{im}$.
The k-dimensional plaquettes corresponding to them are assumed to be frustrated. Then

$$l_{ij} = g\left[A^i, A^{i+2} ; A^j, A^{j+2} \right] = g\left[-1,+1\right],$$
$$l_{ir} = g\left[A^i, A^{i+2} ; A^r, A^{r+2} \right] = g\left[-1,+1\right],$$
$$l_{ip} = g\left[A^i, A^{i+2} ; A^p, A^{p+2} \right] = g\left[-1,+1\right],$$
$$\ldots$$
$$l_{im} = g\left[A^i, A^{i+2} ; A^m, A^{m+2} \right] = g\left[-1,+1\right],$$

and each of the k-dimensional plaquettes is frustrated. Therefore the overblocking effect cannot arise.

Now, let us assume that there are the closed graphs $G = \{l^{ij}, l^{jr}, l^{rp}, \ldots, l^{mi}\}$. When we assume that all of the k-dimensional plaquettes are frustrated, then we obtain

$$l_{ij} = g[-1,+1],\ l_{jr} = g[-1,+1],\ l_{rp} = g[-1,+1],\ \ldots,$$
$$\ldots,\ l_{mi} = g[\pm 1,-1] .$$

If the number of the curves is even then there is the function $g[+1,-1]$. If the number of the curves l is odd then there is the function $g[-1,-1]$. It means that there is no frustration and the overblocking effect arises.

q.e.d.

Theorem II. The sufficient condition for the overblocking effect.

The sufficient condition for the overblocking effect is the existence of one loop at least with odd number of vertices [3] .

Proof. Let the closed loop α be in the abstract space T. We demand that the colours of the nearest neighbour vertices are contrasted. Such a case corresponds to the frustration on each of the k-dimensional plaquettes. Let α_N be the N-vertices loop in space T. Each of the vertices $x \in \alpha_N$ has two nearest neighbours of the same colour in contrast to the colour of the vertex x. It should be N/2 vertices of the same colour. However, if N is odd it is impossible to create a loop with the vertices of the contrast colour. There must be two neighbouring vertices of the same colour. It means that there is such a k-dimensional plaquette in the space M^d, which is nonfrustrated. So the overblocking effect exists.

q.e.d.

Therefore, the necessary and the sufficient condition for the overblocking effect is the existence of a closed loop in some abstract space T.

Now, we are in the position to show that the overblocking effect can arise for the Ising model with two-spin interactions in some 3d-lattices. If we replace the random variables $A(b_k)$ /k arbitrary/ by the random variable $A(b_1)$ for two-spin interactions we can constract the lattice in the 3-dimensional space. The overblocking effect exists in the Ising model with 2-spin interactions in such the lattice.

References

1. S.Alexander and P.Pincus, J.Phys.A 13, 263 /1980/.
2. B.Derrida, Y.Pomeau, G.Toulouse, J.Vannimenus, J.Phys./Paris/ 40, 617 /1979/.
3. H.Nencka-Ficek, Thesis, Poznań 1983.

EXPERIMENTAL RESULTS FOR PHASE TRANSITION IN RANDOM FERROMAGNETS

K. Westerholt[+] and G. Sobotta[++]
[+]Institut für Experimentalphysik IV,[++]Institut für Theor . Physik III
Ruhr-Universität Bochum
D-4630 Bochum, Federal Republic of Germany

We are summarizing recently obtained experimental results on the critical behaviour of random ferromagnets. For the case of strong randomness in the vicinity of the critical concentration x_c of magnetic atoms, one observes the change of the critical exponents towards $\alpha \to -1$, $\beta \to 1/2$, $\gamma \to 2$, $\delta \to 5$.

This contribution is mainly based on a recent paper[1] by the authors, which reports on investigations of the solid solution $Eu_x\ Sr_{1-x}\ S_{0,50}\ Se_{0.50}$. In addition, some new results[2] from similar systems, namely $Eu_x\ Sr_{1-x}\ S$, $Eu_x\ Ba_{1-x}\ S$ and $Eu_x\ La_{1-x}\ S$ are discussed.

For different disordered magnetic systems, the critical exponent of the susceptibility γ is rather well known[1,3] and even the critical exponents of the magnetization β and δ are known for several systems[1,3]. These results seem to be in fairly good agreement with theoretical results[4], predicting for strong randomness a change of the critical exponents towards $\alpha \to -1$, $\beta \to 1/2$, $\gamma \to 2$, $\delta \to 5$, $\nu \to 1$ and $\eta \to o$. For a complete comparison between experiment and theory, it is desirable to measure several critical exponents independently. To approach this aim, the authors[1] tried to determine simultaneously α, β, γ and δ for the solid solution system $Eu_x\ Sr_{1-x}\ S_{0,50}\ Se_{0,50}$. To extract the critical exponents, the magnetic universal function and the magnetic equation of state in form of modified arrot plots have been applied. This analysis allowed to fix the temperature range, in which one may expect power law behaviour for the specific heat. In order to observe the complete change in the specific heat, caused by disorder the mentioned system has been investigated for the concentrations $x = 1.00$, $x = 0.90$, $x = 0.80$ and $x = 0.75$, the critical concentration for ferromagnetisme is at $x_c = 0.78$. In the following table (1), the critical exponents are listed up, for the non ferromagnetic sample with $x = 0.75$ only γ is given.

Table 1 Critical exponents of $Eu_x Sr_{1-x} S_{0.50} Se_{0.50}$

x	$T_c(x)$ /K	α	β	γ	δ
1.00	9.67 ± 0.02	- 0.15	0.33 ± 0.02	1.38 ± 0.05	4.80 ± 0.2
0.90	7.40 ± 0.02	- 0.31	0.37 ± 0.02	1.57 ± 0.05	5.00 ± 0.3
0.80	5.88 ± 0.02	- 0.72	0.44 ± 0.02	1.84 ± 0.05	5.00 ± 0.3
0.75				2.20 ± 0.07	

It is worth to notice, that the determination of α becomes very hard for lower concentrations and the missing error bars should indicate that the precision is not of the same quality as for the magnetic critical exponents. Direct fits of the specific heat with up to five free parameters give slightly lower values[2] of α, although one has to keep in mind that for a large negative values of α, correction to scaling may become important.

The large observed critical region with apparently well defined critical exponents for strongly random ferromagnets has been the subject of a further investigation. In table 2, these results are summarized, where t_{min} and respectively t_{max} are the minimal and maximal normalized temperature differences.

Table 2 Critical region for $Eu_x Sr_{1-x} S_{0.50} Se_{0.50}$

x	$T_c(x)$ /K	t_{min}	γ_{AC}	t_K	t_{max}	γ_{AC}
1.00	9.63	10^{-3}	1.37	5×10^{-1}	1.35×10^{0}	1.06
0.90	7.40	5×10^{-3}		5.5×10^{-1}	2×10^{0}	1.08
0.80	5.40	10^{-2}		7×10^{-1}	4×10^{0}	1.06

$t = (T-T_c)/T_c$

These data have been obtained by a AC-susceptibility measurement in zero field and shows small differences to the results in table 1, where the magnetic exponents have been obtained by extrapolation out of the range between $H_{min} \cong 100$ G ($H_{int\ min} \cong 30$ G) and $H_{max} \cong 3000$ G ($H_{int\ max} \cong 2500$ G), while the AC-experiment has been performed at $H_{min} \cong 1$ G ($H_{int\ min} \cong 0.001$ G) and a frequency of 87 Hz. It is worth noticing that for x = 0.80 the enhanced value of $\gamma_{AC}=2.10$ compared to $\gamma = 1.84$ leads via the Rushbrook relation $\alpha= 2(1-\beta)- \gamma$ to $\alpha\cong-1$ which is closer to the value of the specific heat exponent α, as obtained from the direct fit ($\alpha=-1$, T_c(x=0.80) = 5.46K). These small, but unambiguous differences in the critical exponents are subject of further investigations.

The results in the high temperature range show that the Landau value $\gamma_L=1$ is not yet reached at temperatures as high as 2...4 T_c. From the universal function $M/H^{-1/\delta} = M_{\lessgtr}(t/H^{1/\beta\delta})$, the scaling behaviour up to $H \cong 10^3-10^4$ G allows to estimate that for strong randomness (x=0.80), where $\beta\cong0.5$ and $\delta\cong5$, the relation of the critical ranges in temperature and magnetic field is given by

$$\frac{t_K}{H_K^{\;2/5}} \lesssim (1\ \mathrm{G})^{2/5}$$

for $t_K \cong 10^0$ and $H_K \cong 10^{2.5}$. In terms of the normalized magnetization $m=M/M_s$, where M_s is the saturation value, we find scaling up to m_K 0.7 for x=0.80, thus $m_K/t_K^{1/2} \cong 0.7$. These relations show, that for the Landau range $t_L > 4$ for the concentration x=0.80, there is no Landau range in the magnetic field to be expected, which shows that investigations of the critical behaviour of strongly random ferromagnets are very important for a large range in T and H and not only in a narrow vicinity of the critical point. In particular this result demonstrates that a molecular field analysis, for instance to obtain the magnetic moment, does not make any sense for strong random magnets. By AC-susceptibility technic as mentioned above γ_{AC} has been measured for similar systems. The result are given in table 3.

Table 3 The critical exponent γ_{AC}

$Eu_x S_{1-x} S$	x	$T_c(x)/K$	γ_{AC}
	1.00	16.52	1.38 ± 0.02
	0.70	8.45	1.71 ± 0.02
	0.55	$\cong 2.7$	-
$Eu_x Ba_{1-x} S$			
	1.00	16.52	1.38 ± 0.02
	0.92	13.19	1.40 ± 0.02
	0.74	7.19	1.74 ± 0.02
	0.65	$\cong 6.04$	-
$Eu_x La_{1-x} S$			
	1.00	16.52	1.38 ± 0.02
	0.85	23.80	2.16 ± 0.04
	0.70	24.70	2.30 ± 0.06
	0.60	$\cong 21.5$	-

In all these systems for the lowest concentrations, a diverging susceptibility is no longer observed and therefore no reasonable critical exponent γ_{AC} can be given. The most interesting result is found for $Eu_x La_{1-x} S$, which becomes a conductor above a concentration of 5% La atoms. Thus, instead of a Heisenberg magnet with well defined local exchange interactions, one gets RKKY-interactions via delocalized electrons. Even at high concentrations of Eu of about x = 0.85 the disorder seems to be fairly important, as the large value of $\gamma_{AC} = 2.16 \pm 0.04$.

In contrast to the $Eu_x Sr_{1-x} S_{0.50} Se_{0.50}$ solid solution, these systems a probably slightly inhomogeneous for higher dilution. This is hard to prove by direct methods, but even small macroscopic concentration modulations, leading to a position dependence in the critical temperature, seems to smear out the AC-susceptibility in a small range of about $t \cong 10^{-1} \ldots 10^{-2}$. Out of this range, one may obtain even reasonable scaling. This effect is detectable, because for example for $d\,T_c(x)\,/\,dx \cong 50\,K$ and for $T_c = 5$ K, one gets a change $\Delta T_c/T_c$ of about 0.01 for a concentration change of only 0.001. Such a small deviation requires an averaging volume with a linear lenght larger than $\sqrt[3]{N}$ a, a being the lattice constant, where $1/\sqrt{N} \cong 0.001$ leads to a length of 10 a, which has to be exceeded by the correlation length $\xi = 0.5a\ t^{-\nu}$ with $\nu \cong 1$ here. These rough estimates show that even small deviations from the statistical fluctuations are detectable, if one has an appropriate reference system to compare a smeared out susceptibility with a sharp one.

The reported experiments show, that the predicted new critical exponents for strongly random systems are qualitatively in good agreement with the obtained data. A more sophisticated experimental analysis shows, that for a complete understanding of the experiments one has to work out a crossover theory for the full range in the temperature, the concentration and even in the magnetic field.

References

1. K. Westerholt and G. Sobotta, J.Phys. F13 (1983) 2371

2. K. Westerholt, H. Bach and R. Römer, DPG-Frühjahrstagung, Münster 1984, to appear in J. Magn. Magn. Mat. (1984)

3. G. Sobotta, J. Magn. Magn. Mat. 28 (1982) 1

4. G. Sobotta and D. Wagner, J. Magn. Magn. Mat. (1977)
   G. Sobotta and D. Wagner, J. Phys. C11 (1978) 1467
   G. Sobotta and D. Wagner, J. Magn. Magn. Mat. 15-18 (1980) 257
   G. Sobotta and D. Wagner, DPG-Frühjahrstagung, Münster 1984, to appear in J. Magn. Magn. Mat. (1984) and ref. 3.

DOMAIN WALL-TYPE EXCITATIONS IN AN ANISOTROPIC FERROMAGNET

I.G. Gochev

Joint Institute for Nuclear Research, Dubna
Head Post Office P.O.Box 79
101 000 Moscow USSR

In the theory of domain wall-type excitations two approximations are usually made in ferromagnets, namely, a classical and a continuum approximation. I should like to discuss here some new results concerning such excitations in a ferromagnet with exchange anisotropy. Recently, the domain wall /DW/ problem in this case has been solved exactly without these approximations [1].

The Hamiltonian of the exchange anisotropy ferromagnet is the well known Heisenberg-Ising Hamiltonian:

$$\mathcal{H}=\frac{J}{2}\sum_{m,\delta}\left[\frac{1}{g}\left(S_m^x S_{m+\delta}^x+S_m^y S_{m+\delta}^y\right)+S_m^z S_{m+\delta}^z\right] \qquad /1/$$

Here J is the exchange integral, $J>0$; g is the anisotropy constant, $g>1$; the interactions between nearest neighbours are taken into account only.

In the classical approximation the spin operators $\hat{\vec{S}}_m$ are replaced by the classical vectors $\vec{S}_m$ of fixed length S:

$$S_m^x=s\sin\theta_m\cos\varphi_m\ , \quad S_m^y=s\sin\theta_m\sin\varphi_m\ , \quad S_m^z=s\cos\theta_m .$$

In the standard theory of DW [2] the second approximation, i.e. the continuum approximation, is usually made. The continuum approximation is valid only if the DW thickness is much greater than the lattice constant. Then it follows that the continuum approximation is valid in the case of weak anisotropy, $g-1\ll 1$. Thus, one can ask how to deal with the case of strong anisotropy. I shall solve this problem first of all in the frame of the classical picture.

1. Plane DW in a discrete system

The DW in the case under the consideration is of the Bloch DW type, i.e. the spins lie in a fixed plane. For this type of DW the angle φ_m is a constant and, according to /1/, the corresponding classical energy function takes the form:

$$W\{\theta_m\} = \frac{Js^2}{2}\sum_{m,\delta}\left[1 - \frac{1}{g}\sin\theta_m \sin\theta_{m+\delta} - \cos\theta_m \cos\theta_{m+\delta}\right] \qquad /2/$$

From the condition of minimum, $\frac{dW}{d\theta_m} = 0$, one obtains [3]:

$$g\sin\theta_m \sum_{\delta}\cos\theta_{m+\delta} = \cos\theta_m \sum_{\delta}\sin\theta_{m+\delta} \qquad /3/$$

In the one-dimensional case the exact DW solution of the nonlinear difference equation /3/ is known [1] :

$$\cos\theta_m = th(m-m_o)\sigma \quad , \quad \sigma = ln\left(g+\sqrt{g^2-1}\right),$$

$$S_m^z = s\,th(m-m_o)\sigma, \quad S_m^y = 0 \quad , \quad S_m^x = s\left[ch(m-m_o)\sigma\right]^{-1}, \qquad /4/$$

$$W = 2s^2 J\, th\,\sigma .$$

Since the position of the DW is arbitrary, m_o is a free parameter. In the case of weak anisotropy, i.e. $g = 1+2\eta^2$, $\eta \ll 1$, $\sigma \simeq 2\eta$, from Eq. /3/ the well known results follow for the DW in the classical continuum model [2]:

$$\cos\theta = th\,2\eta\xi \, , \quad W = 4Js^2\eta$$

Here ξ is a continuum variable.

In the opposite limit of a strong anisotropy, $\sigma \to \infty$ the saturation value of the DW energy, $W = 2s^2 J$, is exactly equal to the energy of DW in the Ising model. Thus, it is obvious that the exact DW energy coincides with the energy of DW for the continuum model only in the case of weak anisotropy.

Now, let us consider plane domain walls in two and three dimensional ferromagnets. For a plane DW the spin density depends on one variable only and, thus, the plane DW problem is effectively an one-dimensional problem. In the continuum approximation the problem of DW in two and three dimensional ferromagnets reduces to the already solved one-dimensional continuum problem. However, in the discrete case, there exist several differently oriented plane DW and the corresponding one-dimensional problems differ. For example, in the cubic structure, there exist two different DW with the normal vector $\vec{n}_A = (1,1,1)$ /A-type/ and $\vec{n}_B = (1,0,0)$ /B-type/, respectively. One can check that the one-dimensional problem for the A-type DW reduces to the problem solved before /cf. Eq. /3//. This means that we obtain for the cubic structure from Eq./4/ the exact description of the spin density and the energy density for the A-type DW. For the B-type DW the solution of the corresponding nonlinear difference equation is known only for two limiting cases $\sigma \ll 1$ or $\sigma \to \infty$. In the weak anisotropy limit we have $W_A = W_B$. This is an expected result since in the continuum model the DW energy is independent of the DW orientation. In the opposite case $\sigma \to \infty$ we get $W_B < W_A$. From these results we conclude that the DW energy in a discrete system depends on the orientation of the DW.

We have found excitations of a classical spin model. We believe that these excitations are related to the excitations of the quantum Heisenberg-Ising model. One can ask what is the relation of the DW excitations to the excitations on quantum picture? This question is a part of the more general problem, namely, what is the relation between classical solitons and quantum excitations? We shall discuss this correspondence in the one-dimensional case.

2. Quantum approach to the DW problem in a linear spin chain.

In the case of Heisenberg-Ising ferromagnet /1/ the operator $\hat{S}^z = \sum_m \hat{S}^z_m$ commutes with the Hamiltonian and the stationary states can be labelled by the eigenvalues of $\hat{S}^z$. In such a way one gets the well known classification scheme: a ground state without magnons, spin waves /or one-magnon states/, two-magnon states and so on. This classification scheme is valid for any value of spin magnitude S . Many authors have suggested that the magnon bound state /spin complex/ is a quantum analog of the classical soliton [4,5]. As it has been shown in the series of papers [4,5] after the quasiclassical quantization of the soliton solutions of nonlinear equations the soliton energy is equal to the spin complex energy in the quantum picture. The spin density /the expectation value of S^z_m in the spin complex state/ coincides with the classical result for the soliton in the case of weak nonlinearity [4]. All these results have been obtained recently in the continuum picture. Having the discrete DW solution /4/ we can check now the validity of this correspondence in the general case.

We expect that the heavy spin complex, localized near the end of semiinfinite chain, should be an analog of the classical DW. In the case of $S = \frac{1}{2}$ the exact results for the energy and the wave function of n -magnon bound state are known [6]:

$$E_n = \frac{J}{2}\, th\,\sigma \cdot th\, n\sigma, \quad |\Psi_n\rangle = \sum_{\{m_i\}} B_{m_1\dots m_n} S^-_{m_1}\cdots S^-_{m_n}|0\rangle,$$
$$B_{m_1\dots m_n} = \prod_{i=1}^{n}\left[\frac{ch(i-1)\sigma}{ch\,\sigma}\right]^{m_i}. \qquad /5/$$

First of all, it is seen that $E_n \xrightarrow[n\to\infty]{} W$. Next, the expectation value of the S^z_m can be written in the form [1]:

$$\langle\Psi_n|S^z_m|\Psi_n\rangle = \frac{1}{2} - \sum_{K=0}^{\infty}(-1)^K \exp\{K\sigma[2(n-m)-1-K]\}. \qquad /6/$$

In the general case the quantum spin density /6/ differs from the classical result /4/ for S^z_m . Both results coincide only in the case of weak anisotropy.

Up to now we have restricted ourselves, as well as all other authors [4,5], to two characteristics of DW only, namely, the energy W and the spin density S^z_m . The complete description of the DW includes two additional characteristics: x and y components of spins. One can show that $\langle\Psi_n|S^x_m|\Psi_n\rangle$ vanishes, whereas in the classical picture $S^x_m = s[ch(m-m_0)\sigma]^{-1}$. From this result we conclude that the main assumption in the classical theory, $\vec{S}_m$ is a vector of fixed length, is violated in magnon bound states.

For the purpose of finding a quantum analog of DW it will be usefull to recall the correspondence between the classical spin wave and quantum excitations. Recently, it has been shown [7] that in the spin wave approximation the Glauber coherent state is a complete quantum analog of the classical spin wave. This coherent state is a linear combination of states with different number of magnons, whereas the quantum spin wave is a state with fixed number /$n=1$/ of magnons. Having in mind this correspondence in the linear theory let us try to find the quantum analog of the classical DW /nonlinear excitation!/ as a superposition of heavy spin complexes: $|\phi\rangle = \sum_n C_n|\Psi_n\rangle$. It turns out that this is possible in our case by mapping this problem to the problem of calculating the equilibrium distribution function of an ideal Fermi gas with equidistant spectrum. The final answer for $|\phi\rangle$ can be written in a simple form [1]:

$$|\phi_{m_0}\rangle = A\sum_{K=-\infty}^{\infty}\exp\left(-\frac{K\sigma^2}{2}\right)\cdot|\Psi_{N_0+K}\rangle,$$
$$m_0 = N_0 + \frac{1}{2}, \quad N_0 \to \infty. \qquad /7/$$

In the limit $N_0 \to \infty$ this state is complete quantum analog of the classical DW:

$$\langle \phi_{m_0} | S_m^\alpha | \phi_{m_0} \rangle = S^\alpha_{m,\, class.} \; ; \quad \mathcal{H} | \phi_{m_0} \rangle = W | \phi_{m_0} \rangle .$$

The state $|\phi_{m_0}\rangle$ can be rewritten in a different form /see also [3,8] /:

$$|\phi_{m_0}\rangle = \prod_{m=1}^{\infty} \begin{pmatrix} \cos \frac{\theta_m}{2} \\ \sin \frac{\theta_m}{2} \end{pmatrix} \qquad /8/$$

Here θ_m are classical angles /4/. The $|\phi\rangle$ state is a direct product of the Bloch one-side coherent states [9], so $|\phi\rangle$ is a minimum uncertainty packet and the correlation functions in this state factorize:

$$\langle \phi | S^\alpha_{m_1} S^\beta_{m_2} | \phi \rangle = \langle \phi | S_{m_1} | \phi \rangle \langle \phi | S_{m_2} | \phi \rangle , \quad m_1 \neq m_2$$

The properties of the $|\phi\rangle$ state indicate that this is a coherent state for the considered interacting spin system. Up to now the coherent states have been found for noninteracting systems only.

References

1. I.Gochev, Zh.Eksp.Teor.Fis. 85, 199 /1983/.
2. L.Landau, E.Lifshitz, Electrodynamics of continous media, "Nauka", Moscow, 1982 /in Russian/.
3. A.Bishop, E.Domany and J.Krumhansl, Ferroelectrics, 16, 183 /1977/.
4. A.Kosevich and B.Ivanov, Zh.Eksp.Teor.Fis. 72, 2000 /1977/.
5. J.Tjon and J.Wright, Phys.Rev. B15, 391 /1977/; A.Bishop, J.Phys. C13, L67 /1980/; H.Fogedby, J.Phys. C13, L195 /1980/; T.Schneider, Phys.Rev. B24, 5327 /1981/; I.Gochev, Phys.Lett. A89, 31 /1982/; A.Kosevich, B.Ivanov, A.Kovalev, "Nonlinear magnetization waves: Dynamical and topological solitons", "Naukova dumka", Kiev, 1983 /in Russian/.
6. I.Gochev, Zh.Eksp.Teor.Fis. Pis'ma Red. 26, 136 /1977/.
7. S.Rezende and N.Zagury, Phys.Rev. B4, 201 /1971/.
8. V.Pokrovskii and S.Khokhlachev, Zh.Eksp.Teor.Fis. Pis'ma Red. 22, 371 /1975/.
9. F.Arecchi, E.Courtens, R.Gilmore and H.Thomas. Phys.Rev. A6, 2211 /1972/.

LIST OF PARTICIPANTS

1.	M.Ausloos	Institut de Physique B5, Universite de Liege, B-4000 Liege 1, Belgium
2.	J.Barnaś	Instytut Fizyki, Politechnika Poznańska, ul.Piotrowo 3, PL-60-965 Poznań, Poland
3.	K.Bartkowiak	Instytut Fizyki, Zakład Magnetyków, Uniwersytet im.A.Mickiewicza, ul.Grunwaldzka 6, PL-60-780 Poznań, Poland
4.	U.Behm	Sèktion Physik, Karl-Marx-Universität, Karl-Marx-Platz 10-11, DDR-701 Leipzig, DDR
5.	L.Biegała	Instytut Niskich Temperatur i Badań Strukturalnych Polskiej Adademii Nauk, Pl.Katedralny 1, Wrocław, Poland
6.	M.Błaszczyk	Instytut Fizyki, Zakład Teorii Ciała Stałego, Uniwersytet im.A.Mickiewicza, ul.Matejki 48/49 PL-60-769 Poznań, Poland
7.	W.Borgieł	Instytut Fizyki, Uniwersytet Sląski, ul.Uniwersytecka 4, 40-007 Katowice, Poland
8.	A.Borowiec	Instytut Fizyki Teoretycznej, Uniwersytet Wrocławski, ul.Cybulskiego 36, PL-50-205 Wrocław, Poland
9.	T.Burkhardt	Dep.of Physics, Temple University of the Commonwealth System of Higher Education, College of Liberal Arts Philadelphia, Pensylvania 19122, USA
10.	M.Cieplak	Instytut Fizyki Teoretycznej, Uniwersytet Warszawski ul.Hoża 69, PL-00-681 Warszawa, Poland
11.	C. di Castro	Istituto di Fisica, Università degli Studi di Roma "La Sapienza", Piazzale Aldo Moro 2, I-00185 Roma, Italy
12.	Z.Domański	Instytut Niskich Temperatur i Badań Strukturalnych PAN pl.Katedralny 1, PL-Wrocław, Poland
13.	M.Droz	Department de Physique Theorique, Universite de Genève Bld. d'Yvoy 32, CH-1211 Genève 4, Switzerland
14.	A.Drzewiński	Instytut Niskich Temperatur i Badań Strukturalnych PAN pl.Katedralny 1, PL-Wrocław, Poland
15.	M.Dudek	Instytut Fizyki Teoretycznej, Uniwersytet Wrocławski ul.Cybulskiego 36, PL-50-205 Wrocław, Poland
16.	K.Durczewski	Instytut Niskich Temperatur i Badań Strukturalnych PAN pl.Katedralny 1, PL-Wrocław, Poland
17.	Z.Dmitruk	Instytut Fizyki, Uniwersytet im.Marii Curie-Skłodowskiej, pl.M.Curie-Skłodowskiej 1, PL-20-031 Lublin, Poland
18.	Z.Galasiewicz	Instytut Fizyki Teoretycznej, Uniwersytet Wrocławski, ul.Cybulskiego 36, PL-50-205 Wrocław, Poland
19.	H.Geistlinger	Sektion Physik, Karl-Marx-Universität, Karl-Marx-Platz,10-11, DDR-701 Leipzig, DDR
20.	R.Gielerak	Instytut Fizyki Teoretycznej, Uniwersytet Wrocławski, ul.Cybulskiego 36, PL-50-205 Wrocław, Poland

21.	I.Gotchev	Joint Institute Nucl.Research, Dubna, Head Post Office, 101 000 Moscow, USSR
22.	Z.Haba	Instytut Fizyki Teoretycznej, Uniwersytet Wrocławski ul.Cybulskiego 36, PL-50-205 Wrocław, Poland
23.	J.Hańćkowiak	Instytut Fizyki Teoretycznej, Uniwersytet Wrocławski ul.Cybulskiego 36, PL-50-205 Wrocław, Poland
24.	D.Huckaby	Texas Christian University, Fort Worth, USA
25.	F.Igloi	Central Research, Institute for Physics, Hungarian Academy of Science, P.O Box 49, H-1525 Budapest 114, Hungary
26.	S.Jankowski	Instytut Podstaw Elektroniki, Politechnika Warszawska, PL-00-665 Warszawa, ul.Nowowiejska 15/19, Poland
27.	H.Jarzęcka	Instytut Fizyki, Zakład Teorii Ciała Stałego, Uniwersytet im.A.Mickiewicza, ul.Matejki 48/49 PL-60-769 Poznań, Poland
28.	K.Jezuita	Wyższa Szkoła Inżynierska, Instytut Fizyki PL-26-600 Radom ul.Malczewskiego 29, Poland
29.	J.Jędrzejewski	Instytut Fizyki Teoretycznej, Uniwersytet Wrocławski ul.Cybulskiego 36 , PL-50-205 Wrocław, Poland
30.	G.Kamieniarz	Instytut Fizyki, Uniwersytet im.A.Mickiewicza, u.Grunwaldzka 6, PL-60-780 Poznań, Poland
31.	W.Kinzel	Institut für Kernphysik, Kernferschungslange (KFA) Postfach 1913, D-5170 Jülich 1, FRG
32.	A.Komoda	Instytut Fizyki Teoretycznej, Uniwersytet Wrocławski, ul.Cybulskiego 36, PL-50-205 Wrocław, Poland
33.	J.Kowalska	Instytut Fizyki, Uniwersytet Sląski, ul.Uniwersytecka 4, 40-007 Katowice, Poland
34.	S.Krompiewski	Instytut Fizyki Molekularnej, Polskiej Akademii Nauk, ul.Smoluchowskiego 17/19, 60-179 Poznań, Poland
35.	S.Krukowski	Zakład Fizyki Teoretycznej , Polska Akademia Nauk al.Lotników 32/46, PL-02-668 Warszawa, Poland
36.	S.Krzemiński	Instytut Fizyki, Politechnika Wrocławska, ul.Wybrzeże Wyspiańskiego 27, PL-Wrocław, Poland
37.	A.Kühnel	Sektion Physik, Karl-Marx-Universität, Karl-Marx-Platz, 10-11, DDR-701 Leipzig, DDR
38.	J.M.J. Van Leeuwen	Laboratorium voor Technische Natuurkunde, Technische Hogeschool, Delft, Postbus 5046, 2600 GA Delft, Netherlands
39.	S.Lipiński	Instytut Fizyki Molekularnej Polskiej Akademii Nauk, ul.Smoluchowskiego 17/19, PL-60-179 Poznań, Poland
40.	J.Lorenc	Instytut Fizyki Teoretycznej, Uniwersytet Wrocławski ul.Cybulskiego 36, PL-50-205 Wrocław, Poland

41. B.Lorenz — Karl-Marx-Universität, Sektion Physik, Karl-Marx-Platz 10-11, DDR-701 Leipzig, DDR

42. J.Lukierski — Instytut Fizyki Teoretycznej, Uniwersytet Wrocławski ul.Cybulskiego 36, PL-50-205 Wrocław, Poland

43. J.Łuczka — Instytut Fizyki, Uniwersytet Sląski, ul.Uniwersytecka 4, PL-40-007 Katowice, Poland

44. R.Łyżwa — Instytut Niskich Temperatur i Badań Strukturalnych PAN, pl.Katedralny 1, PL-Wrocław, Poland

45. K.Mikke — Instytut Energii Atomowej, 05-400 Otwock-Swierk, Poland

46. J.Milczarek — Instytut Energii Atomowej, 05-400 Otwock-Swierk, Poland

47. T.Natterman — Humboldt Universität, Sektion Physik, Postfach 1297, DDR-1086 Berlin, DDR

48. W.Nawrocka — Instytut Fizyki Teoretycznej, Uniwersytet Wrocławski, ul.Cybulskiego 36, PL 50-205 Wrocław, Poland

49. H.Nencka-Ficek — Instytut Fizyki Molekularnej, Polska Akademia Nauk ul.Smoluchowskiego 17/19 PL-60-179 Poznań, Poland

50. Cz.Oleksy — Instytut Fizyki Teoretycznej, Uniwersytet Wrocławski ul.Cybulskiego 36, PL-50-205 Wrocław, Poland

51. Z.Onyszkiewicz — Instytut Fizyki, Zakład Teorii Ciała Stałego, Uniwersytet im.A.Mickiewicza, ul.Matejki 48/49, PL-60-769 Poznań, Poland

52. T.Paszkiewicz — Joint Institute Nuclear Research, Dubna, Head Post Office 101 000 Moscow, USSR

53. F.Pelzer — Institut für Theoretische Kernphysik, (TKP), Universität Karlsruhe, Kaiserstrasse 12, Postfach 6380, D-7500 Karlsruhe 1, FRG

54. Z.Petru — Instytut Fizyki Teoretycznej, Uniwersytet Wrocławski, ul.Cybulskiego 36, PL-50-205 Wrocław, Poland

55. Z.Prętczyński — Instytut Podstawowych Problemów Techniki, PAN PL-02-668 Warszawa, al.Lotników 32/46, Poland

56. A.Pękalski — Instytut Fizyki Teoretycznej, Uniwersytet Wrocławski ul.Cybulskiego 36, PL-50-205 Wrocław, Poland

57. R.Piasecki — Wyższa Szkoła Pedagogiczna, Instytut Fizyki, PL-45-052 Opole, ul.Oleska 48, Poland

58. M.Pietruszka — Instytut Fizyki, Uniwersytet Sląski, ul.Uniwersytecka, PL-40-007 Katowice, Poland

59. A.Podolska-Strycharska — Instytut Fizyki Teoretycznej, Uniwersytet Wrocławski, ul.Cybulskiego 36, PL-50-205 Wrocław, Poland

60. W.Prystasz — Instytut Niskich Temperatur i Badań Strukturalnych PAN, pl.Katedralny 1, PL-Wrocław, Poland

61. Z.Przełoźny — Akademia Górniczo-Hutnicza im.St.Staszica, PL-30-059 Kraków, ul.A.Mickiewicza 30, Poland

62. A.Ratuszna — Instytut Fizyki, Uniwersytet Sląski, PL-40-007 Katowice, ul.Uniwersytecka 4, Poland

63. W.Rochowska — Instytut Fizyki Teoretycznej, Uniwersytet Wrocławski ul.Cybulskiego 36, PL-50-205 Wrocław, Poland

64. P.Rusek — Instytut Fizyki, Politechnika Wrocławska,ul.Wybrzeże Wyspiańskiego 27, PL-Wrocław, Poland

65. L.Sasvari — Eötvös Lorand University Institute of Theoretical Physics, VIII-Puskin u.5-7, H-1088, Budapest 8, Hungary

66. W.Selke — Universität Saarbrucken, Institut für Theoretische Physik, Saarbrucken, FRG

67. J.Sivardière — Centre d'Etudes Nucl.(CEA) Av.des.Martyrs 53, B.P.85, Centre de Tri, F-38041 Grenoble Cedex, France

68. D.Skrajna — Instytut Fizyki Teoretycznej, Uniwersytet Wrocławski ul.Cybulskiego 36, PL-50-205 Wrocław, Poland

69. J.Sitarz — Instytut Fizyki Molekularnej, Polska Akademia Nauk, ul.Smoluchowskiego 17/19, PL-60-179 Poznań, Poland

70. G.Sobotta — Ruhr Universität Bochum, Theoretische Physik III, Postfach 102148, Universitätstr. 150, D-4630 Bochum, FRG

71. H.Stachowiak — Instytut Niskich Temperatur i Badań Strukturalnych PAN pl.Katedralny 1, PL-Wrocław, Poland

72. J.Stelmach — Instytut Fizyki Teoretycznej, Uniwersytet Wrocławski ul.Cybulskiego 36, PL-50-205 Wrocław, Poland

73. M.Stęślicka — Instytut Fizyki Doświadczalnej, Uniwersytet Wrocławski ul.Cybulskiego 36, PL-50-205 Wrocław, Poland

74. R.B.Stinchcombe — Department of Theoretical Physics, University of Oxford, 1 Keble Road, Oxford OX1 3NP, England

75. Z.Strycharski — Instytut Fizyki Teoretycznej, Uniwersytet Wrocławski ul.Cybulskiego 36, PL-50-205 Wrocław, Poland

76. A.Sütő — Central Research, Institute for Physics, P.O.Box 49, Hungarian Academy of Science, H-1525 Budapest 114, Hungary

77. J.Szlanta — Wyższa Szkoła Inżynierska, Instytut Fizyki PL-26-600 Radom ul.Malczewskiego 29, Poland

78. J.Sznajd — Instytut Niskich Temperatur i Badań Strukturalnych PAN p.Katedralny 1, PL-Wrocław, Poland

79. J.Tarski — ICTP, Trieste, Strada Costiera 11, I-34100 Trieste, Italy

80. W.Trimper — Martin Luther Universität, Halle, DDR-4010 Halle, Sektion Physik, Friedemann - Bach Platz.6, DDR

81. J.Ulner — Instytut Niskich Temperatur i Badań Strukturalnych PAN, pl.Katedralny 1, PL-Wrocław, Poland

82. K.Walasek — Politechnika Wrocławska, Instytut Fizyki, PL-Wrocław, ul.Wybrzeże Wyspiańskiego 27, Poland

83. K.Walasek — Politechnika Wrocławska, Instytut Fizyki, PL-Wrocław, ul.Wybrzeże Wyspiańskiego 27 , Poland

84. W.Wasilewski — Wyższa Szkoła Inżynierska, Instytut Fizyki, PL-26-600 Radom ul.Malczewskiego 29, Poland

85, W.Weller — Sektion Physik, Karl-Marx-Universität, Karl-Marx-Platz 10-11 DDR-701 Leipzig, DDR

86. K.Wentowska — Instytut Fizyki, Politechnika Warszawska, PL-00-662 Warszawa ul.Koszykowa 25, Poland

87. B.Westwański — Instytut Fizyki, Uniwersytet Sląski, PL-40-007 Katowice, ul.Uniwersytecka 4, Poland

88. G.Wietrowski — Instytut Fizyki, Uniwersytet Łódzki, PL-90-136 Łódź, ul.G.Narutowicza 68, Poland

89. K.Witański — Instytut Fizyki Teoretycznej, Uniwersytet Sląski, PL-40-007 Katowice, ul.Uniwersytecka 4, Poland

90. R.Wojciechowski — Instytut Fizyki, Uniwersytet im.A.Mickiewicza, PL-60-780 Poznań, ul.Grunwaldzka, Poland

91. W.Zarek — Instytut Fizyki Teoretycznej, Uniwersytet Sląski, PL-40-007 Katowice, ul.Uniwersytecka 4, Poland

92. P.Zieliński — Instytut Fizyki Jądrowej, Uniwersytet Jagielloński, PL-30-059 Kraków, ul.Reymonta 4, Poland

93. B.Zięba — Akademia Górniczo-Hutnicza, Instytut Fizyki, PL-30-059 Kraków, ul.A.Mickiewicza 30, Poland

94. E.Zipper — Instytut Fizyki, Uniwersytet Sląski, PL-40-007 Katowice ul.Uniwersytecka 4, Poland

95. J.Zittartz — Institut für Theoretische Physik, Universität zu Köln, Zülpicher Str.77, D-5000 Köln 41, FRG

96. D.Żymierska — Instytut Fizyki, Polska Akademia Nauk, PL-02-668 Warszawa, al.Lotników 32/46 , Poland

INTERFACE ROUGHENING IN RANDOM FIELD MODELS

Thomas Nattermann
Sektion Physik der Humboldt-Universität, Bereich 04, 1040 Berlin, DDR

1. Introduction

Interfaces appear in condensed matter physics in various contexts (e.g. liquid-vapour transitions, magnetism, ferroelectricity, incommensurate systems etc.) and under different names (e.g. surfaces, domain walls, discommensurations, solitons), for recent reviews see [1,2,3] and references therein. Strictly speaking, the interface is the boundary between two phases, embedded in an infinite system, which can be enforced by appropriate boundary conditions or external fields.

For the sake of definiteness let us consider an Ising system. Refering to Fig. 1, the four solid lines on the surface of the cube of volume L^d are supposed to be the borders of a reference plane perpendicular to the z-axis. On the faces of the cube, the spins right / left the dashed line are $\downarrow$ / $\uparrow$. In the state of perfect order, the border between the up and down spin phase is planar.

If the interface is exposed to disorder (e.g. quantum, thermal, quenched randomness), it will make distortions $f(\underset{\sim}{r})$ parallel to the z-direction ($\underset{\sim}{r}$ denotes the postion vector on the reference plane, see Fig. 2). In general, the notion of the interface as a well defined geometrical object becomes vague, overhang and bubble configurations make $f(\underset{\sim}{r})$ to a non-unique function of $\underset{\sim}{r}$. However, it is still possible to introduce an interface free energy F_I as the difference of two bulk energies: If we denote the free energy of the system with the above boundary conditions by $F_{\downarrow\uparrow}$ and that with spin $\uparrow$ on all faces of the cube by $F_{\uparrow\uparrow}$

$$F_I(L) = F_{\downarrow\uparrow}(L) - F_{\uparrow\uparrow}(L) = \sigma(L)L^{d-1} \sim L^{x(d)} \qquad (1)$$

Here $\sigma(L)$ denotes the interface free energy density (or interface tension). If $x(d) < 0$ for arbitrary weak disorder, the system does not

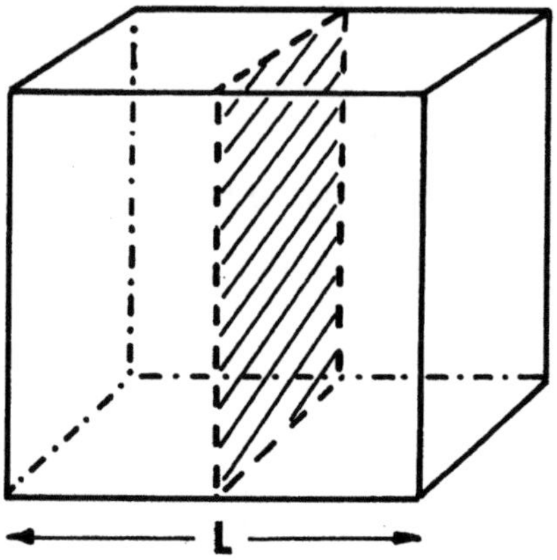

Fig. 1: Planar interface in a perfectly ordered system

'feel' the different boundary conditions from which one concludes the absence of long range order. Since $x(d)$ is expected to increase with the dimensionality d, $x(d_l)=0$ defines the lower critical dimensionality (lcd) d_l . This prescription for determing F_I and d_l can be extended also to other types of order [4]. However, in the further discussion we will restrict ourselves to systems with a discrete (Ising) symmetry of the order parameter.

From the profile structure, one associates an intrinsic width ξ with the interface which is expected to behave like the bulk correlation length. But the interface as whole wanders so that the statistical width $w = [\langle f^2 \rangle]^{1/2}$ may diverge in the thermodynamic limit

$$w(L) = \lambda P\left(\frac{L}{\lambda}, \frac{L}{\xi}\right) \sim \lambda\left[(L/\lambda)^{y(d)} - (\xi/\lambda)^{y(d)}\right], \quad L \gg \xi \quad (2)$$

λ denotes the scale on which the interface is locally fuzzy. In this sense λ is a measure of the strength of the disorder. Since in the critical region the bulk correlation length is the only length scale of the system, one should expect $\lambda \sim \xi$ there.

For $y(d) < (>)\ 0$, the interface is smooth (rough), Since $y(d)$ decreases with increasing d, $y(d) \to 0$ for $d \to d_R$ defines an upper critical dimension for interface roughening d_R. For $0 < y(d) < 1$ the interface is rough, but effectively smooth since $w/L \to 0$ for $L \to \infty$. On the contrary, for $y(d) > 1$, $w/L \to \infty$ for $L \to \infty$, i.e. the interface is not longer well defined on a macroscopic scale even for small disorder. Thus, alternatively to (1) one may determine the lcd for an Ising system from $y(d_l)=1$. If $y(d)$ is linear in d

$$y(d) = \frac{d_R - d}{d_R - d_l} \quad . \tag{3}$$

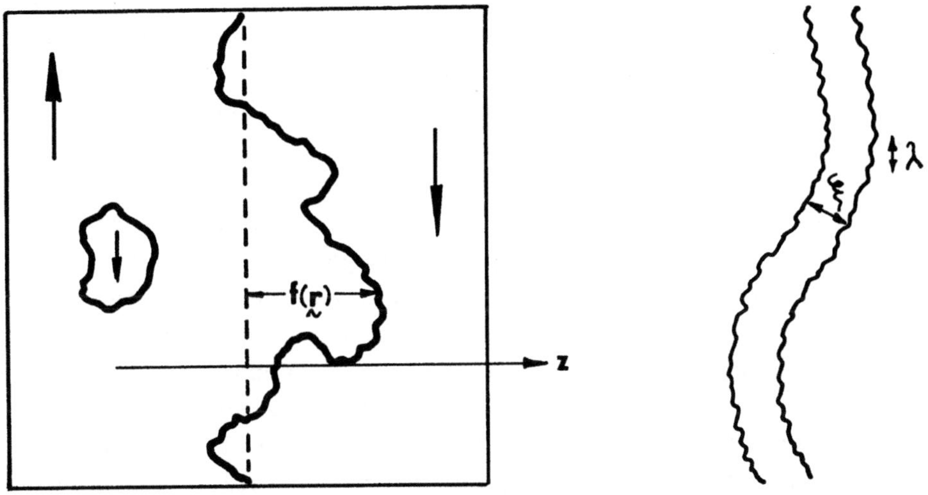

Fig. 2: (a) Cross-section of a distorted interface with a droplet configuration. (b) Part of the interface enlarged, showing its intrinsic width ξ and the fuzzyness scale λ.

If there is an influence of an underlying lattice structure, the divergence of w with L may occur only above a certain disorder strength $\lambda_R(d)$. At λ_R the system undergoes a roughening transition from a smooth to a rough interface. From $\lambda_R(d) \longrightarrow 0$ for $d \longrightarrow \hat{d}_R$ a lcd for the occurence of a roughening transition is defined. If λ is increased further, σ may vanish as $\xi^{-\bar{\mu}}$ which determines the transition to the disordered phase.

There is a number of reasons to study interface properties: The determination of the lcd is a non-trivial task for disordered and frustrated systems [4]. The interface profile and the correlator $\langle f_k f_{-k} \rangle$ of the Fourier transformed f_k of $f(\underset{\sim}{r})$ are directly accessible in experiments. y(d) is related to the domain wall density exponent $\bar{\beta}=y/2(1-y)$ of the incommensurate (I) - commensurate (C) - transition.[5] The roughening transition is important in connection with crystal growth [6] . Moreover, since $\lambda \sim \xi$ in the scaling region, one can expect to be able to calculate bulk critical properties from interface models, which means an enormous reduction of the degrees of freedom.

Despite of its general formulation, this program has been worked off in detail so far only for thermally disordered systems with short range interaction [1,2] . Using for such systems the subscript $_T$, $d_{l,T}=1$, $\hat{d}_{R,T}=2$, $d_{R,T}=3$. For $2<d \leq 3$ there is a roughening transition T_R below the bulk critical temperature T_c. Bulk critical exponents have been calculated as a power series in (d-1).

In this lecture I am going to attack the much less investigated question for the influence of quenched randomness of strength h on the interface properties. This problem is of practical importance, since any real system includes a certain amount of imperfections. One could expect - and the hitherto existing results support this - that the phenomenological picture drawn above describes also interfaces in random systems. Our main concern will be the determination of $d_{l,h}$, $d_{R,h}$ and $\hat{d}_{R,h}$ (we use the subscript $_h$ in order to distinguish quantities from the corresponding purely thermal ones). Apart from Section 6 we restrict ourselves to weak disorder, i.e. to small T and h.

2. From Bulk to Interface Models

Usual starting point for the field-theoretical description of interfaces is the Ginzburg-Landau (GL) model for a single component order parameter $\phi(\vec{r})$. For free boundary conditions and $T < T_c$ there are two degenerated energy minima $\phi = \pm \phi_o$. Using anti-periodic boundary conditions we enforce a kink-solution (planar interface)

$$\phi_1(\vec{r}) = \phi_o \, \varkappa\left(\frac{z-f}{\xi_o}\right) \quad , \quad \vec{r}=(\underset{\sim}{r},z) \tag{4}$$

with $\varkappa(u) = \text{th}(u/2)$. ξ_o is the correlation length of the GL-theory. Inhomogeneous distortions $f(\underset{\sim}{r})$ can be incorporated into (4), if we substitute ξ_o by the local width in the z-direction $\xi_o \rightarrow \xi_o\sqrt{1+(\nabla f)^2}$. Plugging this solution into the GL-Hamiltonian, we end up with the capillary wave Hamiltonian [1,2]

$$\mathcal{H}_{cw} = \sigma_o \int d^{d-1}r \sqrt{1 + (\underset{\sim}{\nabla} f)^2} + \text{const.} \tag{5}$$

with $\sigma_o = \int dz\,(d\,\phi_1/dz)^2$. In deriving (5) droplets and overhangs have been neglected. Moreover, we assumed $K\,\xi_o \ll 1$ where K denotes the (extrinsic) curvature of the interface.

Statistical mechanics introduces a second length scale λ_T via $\sigma_o/T = \lambda_T^{-(d-1)}$, which measures the strength of the thermal disorder (λ_T is in some sense comparable to the Compton wave-length of an electron in 1+1 dimensions).

When discussing interfaces in lattice systems, we restrict ourselves here to anisotropic Ising models with coupling J in z-direction much weaker than perpendicular to it. The corresponding interface is described by the SOS-Hamiltonian [2,6] .

$$\hat{\mathcal{H}}_{SOS} = J \sum_{i,j} |f_i - f_j| + \sigma_o L^{d-1} \tag{6}$$

If disorder leads to a roughening of the interface one should expect, that the long wave-length fluctuations are capillary waves. The corresponding Hamiltonian is then

$$\mathcal{H}_o = \sigma_o L^{d-1} + \frac{1}{2}\hat{\sigma} \int d^{d-1}r\,(\underset{\sim}{\nabla} f)^2 + O(\,(\underset{\sim}{\nabla} f)^4\,) \tag{7}$$

$\hat{\sigma}$ denotes the effective stiffness constant of the interface which is a function of the disorder strength T and h [6,8] . In all expressions below we may change back to the isotropic continuum system by replacing $\hat{\sigma}$ by σ_o.

In nature, Ising interfaces can be realized not only by external conditions, but also by competing interactions . An old and well-known example are ferroelectric or magnetic domain walls. A more recent one are incommensurate systems where competing microscopic interactions lead to a formation of a regular domain wall lattice [3]. Physical examples can be found among adsorbed monolayers, ferroelectrics and -magnets, superconductors with modulated thickness in an external magnetic field, quasi-1- and 2-dimensional conductors and other systems. For their theoretical description several models like the ANNNI, chiral clock and Pott's model have been invented[7]. We consider here the Frenkel-Kontorova (FK) model which describes the behaviour of an elastic chain (adatoms) in an external potential (substrate). Denoting the position of the adatoms by Z_n, the lattice spacing of the free chain by a_o and that of the substrate by a, the FK-Hamiltonian is

$$\hat{\mathcal{H}}_{FK} = \sum_{n=1}^{N} \left[\frac{1}{2} (Z_{n+1}-Z_n-a_o)^2 + a^2 \gamma \cos \frac{2\pi Z_n}{a} \right] \tag{8}$$

It is convenient to introduce the distortions from the n-th potential minimum $a\varphi_n/2\pi$ and to change subsequently to the continuum des - cription

$$\varphi_n = \frac{2\pi}{a} Z_n - 2\pi n =: \varphi(na) \longrightarrow \varphi(z)$$

Coupling finally different chains, we end up with (p=1)

$$\mathcal{H}_{FK} = \int d^{d-1}r dz \left[\frac{1}{2} (\vec{\nabla}\varphi - \mu_o \vec{n}_z)^2 + \frac{\gamma}{a^2} \cos p\varphi \right] \tag{9}$$

$\mu_o = \frac{2\pi}{a}(\frac{a_o}{a} -1)$ characterizes the misfit between both lattices. (9) can be also considered as the truncated form of a GL-Hamiltonian for a complex order parameter $\Psi = A \exp i\varphi$.

The states of extremal energy of $\mathcal{H}_{FK}$ are n-soliton solutions $\varphi_n(z)$ [3]. For large soliton distance $l = L/n$, $\varphi_n(z)$ can be written as a superposition of 1-soliton solutions (phase walls) $\varphi_1(z)$. The latter has the form (4) with $\phi_o = \pi/p$ and $\mathcal{X}(u) = (4/\pi)\tan^{-1}e^u$, $\xi_o = a/p\,\gamma^{-1/2}$. Plugging $\varphi_n(z)$ into $\mathcal{H}_{FK}$ delivers for the energy density $\mathcal{E}(l)$

$$\mathcal{E}(l) = (\sigma_o - \mu) \frac{1}{l} + 4\sigma_o e^{-l/\xi} \cdot \frac{1}{l}$$

$\mu = 2\pi\mu_o/p$ plays the role of a chemical potential for domain walls. Minimization of $\mathcal{E}(l)$ yields for the wall density

$$\xi_o/l \sim (\ln(\mu - \sigma_o)^{-1})^{-1} \qquad \text{if } \mu > \mu_c = \sigma_o$$

For $\mu < \mu_c$ the commensurate state is stable, where $\varphi = 2\pi \cdot m/p$, m integer. Experimentally, one observes frequently a power law behaviour

$$l^{-1} \sim (\mu - \mu_c)^{\bar{\beta}} \tag{10}$$

with $\bar{\beta} \simeq 0 \ldots 1/2$, depending on the system under consideration [3,10,11]. Interface roughening may lead to a non-zero exponent [5,8] as we will indicate now. Indeed, the energy of a single wall is given by (5) (or (7)) with const = $-\mu \int d^{d-1}r$. If disorder roughens the walls, neighbouring solitons will collide (Fig. 3). The mean distance $l_{\parallel}$ between two wall collisions can be estimated by $w(l_{\parallel}) \simeq l$, or with (2) $l_{\parallel}/\lambda \sim (l/\lambda)^{1/y}$. Since the walls are free only on a scale $L \lesssim l_{\parallel}$, we have to replace σ_0 by the renormalized interface tension $\sigma(l_{\parallel})$ in $\mathcal{E}(l)$. As we will show below (eq. (28)) $\sigma(L) - \sigma \sim \hat{\sigma}\cdot(L/\lambda)^{2(y-1)}$ where $\sigma =: \sigma(\infty)$. Thus

$$\mathcal{E}(l) = (\sigma-\mu)\frac{1}{l} + 4\sigma_0 e^{-l/\xi_0}\cdot\frac{1}{l} + c\,\hat{\sigma}\,\frac{1}{l}\left(\frac{\lambda}{l}\right)^{2(1-y)/y}$$

which delivers $\bar{\beta} = y/2(1-y)$ or with (3) $\bar{\beta} = (d_R - d)/2(d-d_l)$[5].

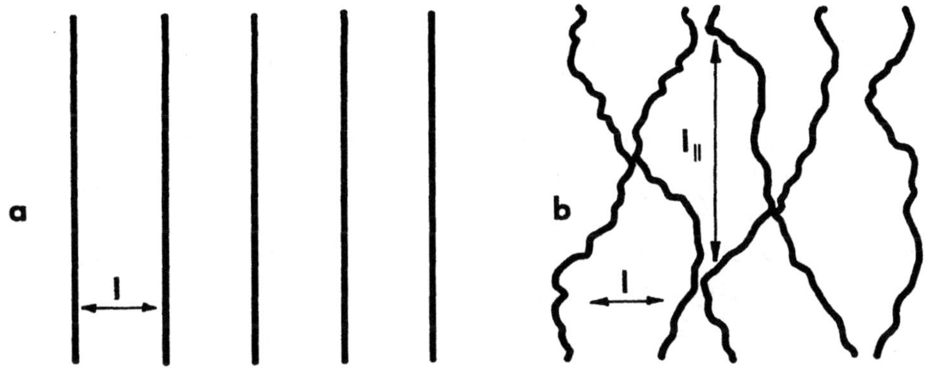

Fig. 3: Soliton lattice in a completely ordered (a) and a weakly disordered (b) incommensurate system.

We note however, that these considerations do not imply the persistance of true long range order in the I-phase for $d > d_l$. Strictly speaking, the interface, as defined in (1), of the incommensurate state is smeared over the whole cube. Indeed, the application of (1) to the soliton lattice yields

$$F_I \simeq |\mu-\mu_c| L^{d-2}\, pl \,/\, (1 + pl/L) \tag{11}$$

Thus, whereas for $l \gg L$ (i.e. for a single wall as in the C-phase) $F_I \sim L^{d-1}$ like for an Ising-interface, we get for $l \ll L$ (i.e. for the soliton lattice of the I-phase) $F_I \sim L^{d-2}$ like for a Bloch-wall. This has consequences for the long range order in the I-phase[9,12,13]. Thus, in connection with IC-transitions, d_l denotes the lcd of the C-phase. Moreover, we disregard wetting transitions of phase walls[7,9]

3. Random Fields (RF)

Randomness is a further source for the creation of interfaces. As a paradigmatic example for quenched randomness we consider the Ising model in a RF. Off-center impurities in systems exhibiting displacive transitions and randomly diluted antiferromagnets in a uniform field are physical realizations of this model. Postponing the discussion of lattice effects to Section 5, we have to add to the GL- Hamiltonian a term

$$\mathcal{H}_h = -\int d^d r \quad H(\vec{r})\ \phi(\vec{r}) \tag{12}$$

The RF is assumed to be Gaussian correlated with

$$\overline{H(\vec{r})} = 0, \qquad \overline{H(\vec{r})\ H(\vec{r}\,')} = h_o^2\ \delta(\vec{r} - \vec{r}\,') \tag{13}$$

where the bare denotes the configurational average. If the RF originates from impurities, h_o^2 is proportional to their concentration. Below we assume $\sigma_o, \hat{\sigma} \gg h_o \phi_o =: h$.

One of the basic questions is that for the lcd of RF Ising model. Imry and Ma [14] considered the stability of the completely ordered state $\phi = \pm \phi_o$ against a decay into domains of size L. By appropriate choice of the sign of ϕ inside a given domain, one can gain energy of the order of the mean fluctuation of the field energy $\Delta E_h \sim -hL^{d/2}$. This has to be compared with the domain wall energy loss $\Delta E_\sigma \sim \sigma(L)L^{d-1}$. Minimizing the total energy density $\varepsilon(L) = L^{-d}(\Delta E_\sigma + \Delta E_h)$ with respect to L yields a domain size

$$L \sim \xi_h = \left(\frac{\sigma}{h} \right)^{\frac{2}{d-2}} \quad , d<2 \quad . \tag{14}$$

According to this estimate, for $d \geq 2$ the ferromagnetic state is stable. In deriving (14), the L-dependence of σ has been assumed to be weak. An lcd $d_l=2$ is however contradictory to a result from perturbation theory [15], which, if extrapolated from its validity region $4<d \leq 6$ to lower d, suggests $d_{l,h}=d_{l,T}+2=3$.

For systems with continuous order parameter symmetry, where $d_{l,T}=2$, the domain wall is smeared over a region L which implies $\sigma \sim 1/L$. The Imry-Ma argument yields now $d_l=4$, in agreement with the perturbation theory. It has been therefore suggested, that a <u>rough</u>

interface (which eventually has $\sigma \sim L^{-1/2}$) could resolve this discrepancy about the lcd of the RF Ising model [17].

In order to find the relevant interface model, we have to insert (4) into (12):

$$\mathcal{H}_h = \phi_o \int d^{d-1}r \; dz \; H(\underset{\sim}{r},z) \; \varkappa\left(\frac{z - f(\underset{\sim}{r})}{\xi_o\sqrt{1+(\underset{\sim}{\nabla} f)^2}}\right) \qquad (15)$$

The full interface Hamiltonian is therefore

$$\mathcal{H}_I = \mathcal{H}_o + \mathcal{H}_h \qquad (16)$$

with $\mathcal{H}_o$ given in (5) or (7). For the sake of simplicity, we treat below the case of a sharp interface, i.e. $\xi_o \rightarrow 0$. Then

$$\mathcal{H}_h = -2\,\phi_o \int d^{d-1}r \int_0^{f(\underset{\sim}{r})} dz \; H(\underset{\sim}{r},z) \quad . \qquad (15')$$

$\mathcal{H}_I$ has been first considered by Grinstein and Ma [18] and in a slightly different form by Villain [19].

Further motivation for the investigation of interface roughening in the RF Ising model comes from the unexplained $\bar{\beta}$ ($\sim 1/2$)-values in 3D incommensurate ferroelectrics of the K_2SeO_4-family [10]. These are described by the FK-Hamiltonian (9) with p=6. The polarisation P is coupled to the phase φ via an additional term $P\cos(p\varphi/2)$ [20]. Introducing $\phi(\vec{r}) = \cos(p\varphi/2)$, random dipole moments act on $\phi(\vec{r})$ as a random field. The system can be therefore mapped on (15), (16) with $\varkappa(u)$ = th u. A 2D-analogon are Xe-monolayers adsorbed on a Cu (110) surface. Randomly distributed impurities act as a RF, breaking locally the Ising symmetry of the system [19].

4. Interface Roughening

A crude estimate of the interface roughening in the RF Ising model can be obtained from the phenomenological expression for the free energy $F_I(L,w) = \sigma_o L^{d-1} + F_1(L,w)$ of a wall of extension L and characteristic distortion $f(\underset{\sim}{r}) \simeq w$ (k_B=1) [21]:

$$F_1(L,w) = \frac{1}{2}\hat{\sigma} L^{d-1} \left(\frac{w}{L}\right)^2 - 2ch(|w| L^{d-1})^{1/2} - T \ln N(w) \quad (17)$$

The three terms on the r.h.s. of (17) correspond to contributions from the elastic energy, the field energy gain in the region bounded by the interface and the z=0 plane (see Fig. 2), and the entropy, respectively. N(w) is the number of different wall configurations with approximately the same distortion w, $N(w) \sim (w/\xi)^{d-1}$. c is a positive random number of order unity. $F_1(L,w)$ is minimized with

$$w(L) = \lambda_h (L/\lambda_h)^{\frac{5-d}{3}} Q(x(L)), \qquad d>2 \quad (18)$$

The fuzzyness scale is now related to the random field strength h

$$\lambda_h = \left(\frac{ch}{\hat{\sigma}}\right)^{\frac{2}{d-2}}, \qquad Q(x) = (1 + Q^{-1/2}x^{-1})^{2/3} \quad (19)$$

$$x(L) = \frac{1}{d-1} \frac{(ch)^{4/3}}{T\hat{\sigma}^{1/3}} L^{(d+1)/3} = \frac{1}{d-1} \left(\frac{\lambda_h}{L}\right)^{2(d-2)/3} \left(\frac{L}{\lambda_T}\right)^{d-1}$$

For $x(L) \gg 1$, $Q(x) \approx 1$ and (18) has the anticipated form (2), (3) with $\lambda = \lambda_h$, $d_{R,h}=5$ and $d_{l,h}=2$. In the same approximation, the bare interface tension σ_0 is corrected by a term

$$\Delta\sigma(L) = -\frac{3}{2}\hat{\sigma}(ch/\hat{\sigma})^{4/3} L^{2(2-d)/3} \quad (20)$$

The interface is indeed rough for $d<5$. However, for $d>2$ the L-dependent correction to σ_0 becomes small for large L, from which we conclude the stability of the ordered phase for small T and h. On the contrary, for $d<2$, $\sigma(L) \rightarrow 0$ for

$$L \sim \xi_h (\hat{\sigma}/\sigma_0)^{1/2(2-d)} \quad (21)$$

i.e. the interface tension vanishes on a scale of the size of the order of the Imry-Ma domains. It is remarkable, that the thermodynamic quantities like w^2 or $\Delta\sigma$ have a non-analytic h_0^2-dependence.

In the opposite limit $x(L) \ll 1$, $Q(x \rightarrow 0) \sim x^{-1/2}$ and one obtains conventional thermal roughening as discussed in the Introduction. Increasing L, for non-zero h there is always a cross-over to the region $x(L) \gg 1$. The characteristic length scale L_c for the cross--over is given by $x(L_c) \simeq 1$.

These considerations can be refined by an approximate renormalization group treatment as first performed by Grinstein and Ma [18] and Villain [19]. This is of particular importance at the critical dimensionalities $d_{R,h}=5$ and $d_{1,h}=2$ where distortions on <u>all</u> scales contribute to $w(L)$ and $\sigma(L)$, respectively. The approach consists in a repeated decimation and rescalation of variables.

(i) <u>Decimation</u>: The short wave-length fluctuations $f_1(\underset{\sim}{r})$ (varying on a scale a...ba, a=1, b>1) are integrated out of the partition function $Z=Z(h,T;L,a)$. To this aim, one divides $f = f_1+f_r$ and $H=H_1+H_r$ where f_r, H_r are almost constant on the scale a...ba. f_1 and H_1 are further decomposed into wave-packets $f_{1,\lambda}(\underset{\sim}{r})$ and $H_{1,\lambda}(\underset{\sim}{r},z)$ of extension b. This leads to an approximate separation of variables such that $\mathcal{H} = \mathcal{H}_r+\sum_\lambda \mathcal{H}_{1,\lambda}$ [18] with $\mathcal{H}_r= \mathcal{H}\{f_r,H_r\}$ and

$$\mathcal{H}_{1,\lambda} \simeq \int_b d^{d-1}r \left[\frac{1}{2}\hat{\sigma}\,(\underset{\sim}{\nabla} f_{1,\lambda})^2+ \int_{f_r}^{f_r+f_{1,\lambda}} dz\ H_{1,\lambda}(\underset{\sim}{r},z)\right]. \tag{22}$$

We may now perform the integration over $f_{1,\lambda}(\underset{\sim}{r})$

$$\begin{aligned} Z(h,T;L,a) &= \int Df(\underset{\sim}{r})\, e^{-\mathcal{H}/T} = \int\!\!\int_{\dots} Df_r(\underset{\sim}{r})\, e^{-\mathcal{H}_r/T}\prod_\lambda \left[Df_{1,\lambda} e^{-\mathcal{H}_{1,\lambda}/T}\right] \\ &\simeq Z(h,T;L,ba) \prod_\lambda \int df_{1,\lambda}\, e^{-F_1(b,f_{1,\lambda})/T} \\ &\simeq Z(h,T;L,ba)\, e^{-b^{d-1}\sigma_1(b)/T} . \end{aligned} \tag{23}$$

For $x(b)\gg 1$ a saddle point approximation yields

$$\sigma_1 = \Delta\sigma(b) - \text{const.}\ b^{1-d}T + O(T^2) \tag{24}$$

where $\Delta\sigma(b)$ is given in (20) with $c=c_1$. c_1 denotes the average over the random numbers $c_{1,\lambda}$ corresponding to the individual wave packets.

(ii) <u>Rescaling</u>: We apply now the scale transformation

$$f=bf' \ ,\ \vec{r}=b\vec{r}\,' \ ,\ T=b^{d-1}T' \ ,\ H(b\vec{r})=b^{-1}H'(\vec{r}\,') \ ,\ h=b^{(d-2)/2}h' \tag{25}$$

This leaves the form of $\mathcal{H}_r/T$ invariant, but changes the minimal and maximal length scale $ba \to a'=a$ and $L \to L'=L/b$. Thus $Z(h,T;L,ba) \longrightarrow Z(h',T';L/b,a)$. The procedure is now repeated until we reach the scale $L^{(l^*)}=L/b^{l^*}=a$. In this way a sequence of contributions $\Delta\sigma(h^{(l)},T^{(l)})$

$l=0,1,\ldots,l^*$, to the interface tension is generated. For $T \to 0$ (or all $x^{(l)}(b) \gg 1$) this yields

$$\sigma(h,0;L) = \sigma_0 - \sum_{l=0}^{l} c_l \, \hat{\sigma} \, (h/\hat{\sigma})^{4/3} \, b^{2(2-d)l/3} . \qquad (26)$$

For $d<2$, the large scales $l \lesssim l^*$ give the main contribution to σ such that the conclusions drawn from the simple estimate (17) remain essentially unchanged. However, for d=2, σ vanishes on a scale

$$L \sim \xi_h = \exp c \, \frac{\sigma_0 \hat{\sigma}^{1/3}}{h^{4/3}} \quad , \; d=2 \quad . \qquad (27)$$

An Imry-Ma argument suggests now also an instability with respect to domain formation of size ξ_h. For $d>2$, the sum (26) converges and

$$\sigma(h,0;L) \cong \sigma(h,0) + c \, h^{4/3} \hat{\sigma}^{-1/3} \, L^{2(2-d)/3} \quad . \qquad (28)$$

$\sigma(h,0)$ denotes the interface tension of the infinite wall at T=0.

If our initial parameters are such that $0 < x(b) \ll 1$, $\Delta\sigma \sim -Tb^{1-d}$ for the first iterations. However, since $x^{(l)}(b)=x(b^l)$, for $l \gtrsim l_c = \ln L_c / \ln b$, a cross-over to the strong field region $x^{(l)}(b) \gg 1$ occurs and the above discussion applies.

In the same way one finds, that w(L) is given by (17) for $2<d<5$, and $w(L) \sim (h/\hat{\sigma})^{2/3} \ln L$ for d=5. For $d>5$ the interface is smooth. The linear L-dependence of w for d=2 has been confirmed numerically [28]. An lcd $d_{l,h}=2$ seems to be also in agreement with the experimental findings, as discussed in a recent review by Grinstein [29] (see also [30]). Concerning the applicability of these results to incommensurate systems we note, that $\bar{\beta}=(5-d)/2(d-2)$ for $2<d\leqslant 5$. However, since the macroscopic symmetry of the order parameter in the I-phase is a continuous one, the lcd for the long range in the I-phase is now $d_l=4$ (compare (11) and the discussion below eq. (14)).

5. Lattice Effects

The influence of an underlying lattice can be studied by considering the discrete version $\hat{\mathcal{H}}_I$ of our interface model [22,23]

$$\hat{\mathcal{H}}_I = \hat{\mathcal{H}}_{SOS} + \sum_i \sum_{z=o}^{f_i} H_{iz} \text{ sign } f_i \tag{29}$$

Here $H_{iz} = H(\underset{\sim}{r}_i, z)$, i,z and f_i are integer numbers. We discuss now the question, whether the implicite assumption made in (7), namely that the low-lying excitations are capillary waves, is valid for (29). If this is the case, we have to find the dependence of $\hat{\sigma}$ on h, T and J.

Denoting by $\sigma(h,T;L,\theta)$ the interface tension of a plane wall whose normal makes an angle θ with the z-axis, $\hat{\sigma}(h,T)$ is given by

$$\hat{\sigma}(h,T) = \lim_{L\to\infty} \sigma(h,T;L,0) + \frac{\partial^2}{\partial\theta^2} \sigma(h,T;L,\theta)\bigg|_{\theta=0} \tag{30}$$

Actually, one has to expect, that this relation is fulfilled already above a critical length scale $\hat{L}$.

For purely thermally disordered systems, relation (30) can be exploited by various techniques, in particular for d=2 [6,8]. We are not aware of a corresponding calculation for quenched disorder.

However, an expression for $\hat{\sigma}(h,T)$ can be obtained, if one matches a scaling relation for the interface width basing on (29), with the continuum theory result (18) [23]. For h=0 this yields for d=2 $\sigma \sim T \exp cJ/T$ at low T, in agreement with the exact result [8]. For d=3 the matching procedure breaks down, which may signal the occurence of a roughening transition at finite T. For h > 0 one obtains in this way

$$\hat{\sigma}(h,0) \sim h\left[c(3-d)\left(\frac{J}{h}\right)^2 + 1\right]^{\frac{5-d}{2(3-d)}}, \quad d \lesseqgtr 3 \tag{31}$$

For d=3 this gives an exponential $(J/h)^2$ dependence of $\hat{\sigma}$. For d > 3 the matching breaks down from which one could expect the existence of a roughening transition at some critical RF strength h_R.

Qualitatively the same results are obtained from phenomenological considerations [22,23]. For d=2, the interface can be considered as a random walk on a square lattice which takes steps only in the positive x-direction. After each x-step, the interface rests or takes r steps parallel to the z-axis. Since $h,T \ll J$ the step density $\hat{L}^{-1}$ will be low, most likely $r=\pm 1$. Then the following phenomenological expression for the interface tension can be written down (a=1)

$$\sigma_{ph}(h,T;L,\hat{L}) = J(1+\hat{L}^{-1}) - T\hat{L}^{-1}(\ln 2\hat{L} + 1) - ch\,\hat{L}^{-1/2} \tag{32}$$

σ_{ph} is minimized with $\hat{L} = e^{J/T}G(y)/2$ where $y = \frac{ch}{T}e^{J/2T}$ and G(y) fulfilles the equation $\ln G(y) = - y\, G^{1/2}(y)$. In order to give the interface a mean tilt θ, θL z-steps of equal sign have to be added. Thus, with (30), (32) and $\sigma(h,T;L,\theta) \simeq \sigma_{ph}(h,T;L,\hat{L}/1+\theta\hat{L})/\cos\theta$

$$\hat{\sigma}(h,T) = \frac{1}{2}\, T\, e^{J/T}\, G(y)(1 + \tfrac{1}{2} y G^{1/2}(y)) \qquad (33)$$

For $y \gtrless 1$, (33) agrees with the result from the matching procedure. It follows from the derivation, that the capillary wave approach applies on a length scale $L \gg \hat{L} \simeq \hat{\sigma}/J$.

For d=3 a similar consideration yields for T=0, $\hat{\sigma} \sim \exp c(\frac{J}{h})^2$. Thus, the interface is rough even at low temperatures in a non-zero RF. Actually, one can gain free energy by moving parts of the interface upwards or downwards, since the arising step-lines are interfaces in a 2D system. The latter has a negative free energy on a sufficiently large scale $L > \xi_h$ (eq. (21)).

For $d > 3$ steps are interfaces in $d > 2$ dimensions, which have a positive free energy for $h, T \ll J$. Thus, interface roughening is unfavourable in this regime. From this discussion one is led to the conjecture $\tilde{d}_R = 1 + d_l$ for the lcd of the roughening transition.

6. Scaling Region

For purely thermal disorder, Wallace and Zia [1,2] have shown, that in the critical region, λ scales like the bulk correlation length $\xi \sim t^{-\nu}$, $t=(T-T_c)/T_c$. We will argue, that this has to be expected also for the RF results. To this end we repeat the considerations at the beginning of Section 2 for a renormalized GL-Hamiltonian, which includes only fluctuations with wave-length larger than ξ. With the scaling behaviour for the free energy density $\varepsilon \sim t^{2-\alpha}$ and $\sigma \sim \varepsilon\xi$ (or $\sigma \sim \phi^2/\xi^{1-\eta}$, $\phi \sim t^{\beta}$), we get $\sigma \sim t^{\mu}$, $\mu = 2-\alpha-\nu$.

For purely thermal disorder we use the scaling relation

$$2 - \alpha_{(T)} = \nu_{(T)} d \qquad (34)$$

i.e. $\mu_{(T)} = \nu_{(T)}(d-1)$ and $\lambda_T \sim (T/\sigma)^{1/(d-1)} \sim t^{-\nu_{(T)}} \sim \xi$. For the RF case a new relation [24, 31]

$$2 - \alpha_{(h)} = \nu_{(h)}(d - 2 + \eta_{(h)}) \qquad (34')$$

has been proposed. (34') has been derived from a variation of a

Pippard argument [24, 29] . An alternative derivation (to our knowledge not published so far) goes as follows. In a homogeneous external field H, $\xi \sim H^{-\nu/(\beta+\gamma)}$ for t=0. For a correlated region ξ^d there is an average excess field $H \sim h_o \xi^{d/2}/\xi^d$ or $\xi \sim (H/h_o)^{-2/d}$. Thus $\beta_{(h)}+\gamma_{(h)}=\nu_{(h)}d/2$, which yields with the d-independent scaling law $2(\beta+\gamma)=2-\alpha+\nu(2-\eta)$ (34') . With the same considerations one obtains $\xi \sim h_o^{-2\nu_{(T)}/\gamma_{(T)}}$ for $h_o \to 0$ and hence $(T_c(0)-T_c(h))/T_c(0) \sim$ $\sim h_o^{2\nu_{(T)}/\gamma_{(T)}\nu_{(h)}}$ (see also [32]).

Taking (34') for granted, we get $\mu_{(h)}=\bar{\mu}_{(h)}\nu_{(h)}=\nu_{(h)}(d-3+\eta_{(h)})$ and hence

$$\lambda_h \sim (h_o \phi/\sigma)^{2/(d-2)} \sim t^{-\nu_{(h)}} \sim \xi$$

as expected. The comparison of (34) and (34') suggests the relation $\varsigma_{(T)}(d') = \varsigma_{(h)}(d)$ between the critical exponents of the pure and the RF Ising model where $d' = d-2+\eta_{(T)}(d')$ [31]. Then $\bar{\mu}_{(h)}$ vanishes for $d \to 2$.

7. Ramifications

The results of the previous Sections can be extended in various ways. Here we consider three : the extension to long-range RF correlations, to dipolar interaction and to a different coupling between the RF and the interface coordinates. It turns out, that in all three cases the results can be brought into the form (2), (3), but with modified expressions for d_1, d_R and λ .

(i) Long-range RF correlations

We assume for the RF correlations

$$\overline{H(\vec{r})\,H(\vec{r}\,')} = h_o^2\ |\vec{r}-\vec{r}\,'|^{-(d-d')} \tag{35}$$

where d'=1,2 model randomly orientated symmetry breaking line or plane defects, respectively. Repeating the calculation with (13) replaced by (35) delivers $d_R=5$, $d_1=2+d'$ for $0<d'\leq 1$ and $d_R=4+d'$, $d_1=2+d'$ for $1 \leq d'<d-2$. For $d'\leq 0$ $d_R=5$ and $d_1=2$ as above.
The random rod model $\overline{H(\underset{\sim}{r},z)H(\underset{\sim}{r}',z')} = h_o^2\ \delta(\underset{\sim}{r}-\underset{\sim}{r}')$ [25] corresponds to d' =1 in this scheme.

(ii) Dipolar interaction

If there is an additional dipolar interaction between the spins, which are assumed to be parallel to the x-direction, the Fourier transformed quadratic part of the interface Hamiltonian reads [5]:

$$\mathcal{H}_{o,dip} = \frac{1}{2} \sum_{\underset{\sim}{k}} \left[\sigma \underset{\sim}{k}^2 + \frac{g}{d-2} k_x^2 \left(|\underset{\sim}{k}|^{2-d} - 1 \right) \right] f_{\underset{\sim}{k}} f_{-\underset{\sim}{k}} \quad (36)$$

Here g measures the strength of the dipolar interaction, $d \geqslant 2$. For $k_x \neq 0$ the dipolar interaction increases the stiffness of the wall which results in $d_R = 4$ and $d_l = 2$ for short range RF correlations. For d=3 this yields the IC-critical exponent $\bar{\beta} = 1/2$, as observed in some systems [10].

(iii) Modified coupling

In the preceeding calculations we considered a linear coupling between the RF and the order parameter. However, there are further possibilities. If there is e.g. a coupling between the RF and ϕ^p ($p > 0$, even) or $|\partial\phi/\partial z|^p$ ($p > 0$, integer), the corresponding part of the interface Hamiltonian has the form (15), but with a function $\mathcal{H}(u)$ essentially non-zero only for $|u| \lesssim 1$.

Physical examples for such a coupling can be found among systems with weakly random exchange constant or incommensurate systems, where random elastic deformations couple to $\sin(p\varphi/2)$. $\mathcal{H}(u) \sim (\cosh u)^{-n}$, $n > 0$ can be considered as a typical example for $\mathcal{H}(u)$ [26,27,28]. Effectively, the RF couples now only within the interface to the order parameter.

An estimate similar to (17) yields for the free energy increase due to an distortion w [26]

$$F_1(h,T;L,w) = \frac{1}{2} \hat{\sigma} L^{d-1} \left(\frac{w}{L}\right)^2 - 2ch\left(\xi_o L^{d-1}\right)^{1/2} \ln^{1/2} N(w) - T \ln N(w) \quad (37)$$

Here we have exploited the tails of the field energy distribution. Minimization yields

$$w(L) = \lambda_h' \left(L/\lambda_h' \right)^{\frac{5-d}{4}} \left[\ln^{-1/2} N(w) + \left(\lambda_T^2 / \lambda_h' L \right)^{\frac{d-1}{2}} \right]^{1/2} \quad (38)$$

(38) has for large L the form (2) with $d_l = 1$, $d_R = 5$ and $\lambda = \lambda_h' = \left[(d-1) ch\, \xi_o^{1/2} / \hat{\sigma} \right]^{2/(d-1)}$. For d=3 we get again $\bar{\beta} = 1/2$. Note, that there is again a non-analytic dependence of the thermodynamic quantities on h_o^2. The true lcd for long range order in the I-phase is $d_l = 4$.

A more refined treatment, including the discussion of possible depinning effects is still missing for this case (but see [27]). Moreover, it seems to be conceivable, that λ_h' scales differently from ξ.

8. Conclusions

In the present paper a first attempt to a statistical mechanics of interface roughening in random systems has been outlined. It proved to be difficult to obtain exact results even for weakly disordered systems such that crude approximations had to be used. The origin of these difficulties is that the interface is a non-local object: its shape and energy depends on the precise arrangement of the RFs (or the impurities) even if they are very distant. Therefore, the interface adopts shapes, that are unfavourable locally in order to minimize its total (free) energy. This has its mathematical correpondence in the non-analytic dependence of the thermodynamic quantities on h_o^2 (or the impurity concentration), which seems to rule out straightforward perturbation expansions (and eventually also the replica method). The approach in its present form is still to crude to determine bulk critical properties or the nature of the roughening transition. Thus, both from a physical and a methodological point of view, further attempts in this field seem to be of interest.

I wish to thank Professor W. Ebeling for providing the possibility to perform this work.

References

The present list is far from being complete. Many further references can be found in these papers.

1. Wallace, D.J.: Proceedings of Ecole d' Eté de Physique Théorique, Les Houches 1982, North Holland
2. Zia, R.K.P.: Poceedings of the Scottish Universities Summer School in Physics 1983, ed. K. Bowler and A.Mc Kane
3. Villain, J: Ordering in Strongly Fluctuating Condensed Matter Systems, ed. T. Riste, N.Y. Plenum Publ. Corp. 1980
   Bak, P. : Rep. Progr. Phys. 45, 587 (1982)
4. Anderson P.W. and Pond, C.M., Phys.Rev.Lett. 40, 903 (1978)
   Banavar, J.R. and Cieplak, M., Phys.Rev.Lett. 48, 832 (1982)
5. Nattermann,T., J.Phys. C 16, 4125 (1983)
6. Weeks, J.D.: Ordering in Strongly Fluctuating Condensed Matter Systems, ed. T. Riste, N.Y., Plenum Publ. Corp. 1980;
   Fisher, M.P.A., Fisher, D.S., and Weeks, J.D., Phys.Rev.Lett. 48, 368 (1982)
7. Selke, W., this conference
8. Fisher, M.E. and Fisher, D.S., Phys.Rev. B25, 3192 (1982)
9. Huse, D.A. and Fisher, M.E., Phys.Rev. B29, 239 (1984)

10. Blinc, R. et al., Phys. Rev.Lett. 46, 1406 (1981);
Moudden, A.H. and Als-Nielsen, J., unpublished
11. Pokrovskii, V.L. and Talapov, A.L., Phys.Rev.Lett. 42, 65 (1979)
Pokrovskii, V.L., Talapov, A.L. and Bak, P., BNL-preprint 1983
12. Coppersmith, S. et al., Phys.Rev.Lett. 46, 549 (1981)
13. Lyuksutov, I.F. and Feigel'man, M.V., ZhETF 85, 774 (1983)
14. Imry, Y. and Ma, S.-K., Phys.Rev.Lett. 35, 1399 (1975)
15. Grinstein, G., Phys.Rev.Lett. 37, 944 (1976);
16. Young, A.P., J.Phys.C 10, L257 (1977)
17. Binder, K., Imry, Y., and Pytte, E., Phys.Rev.B 24, 6736 (1981)
18. Grinstein, G. and Ma, S.-K., Phys.Rev.Lett. 49, 685 (1982), Phys. Rev. B 28, 2588 (1983)
19. Villain, J., J. de Physique 43, L551 (1982);
Villain, J., Semeria, B., Lancon, F., and Billard, L., J.Phys. C 16, 6153 (1983)
20. Nattermann, T., Solid St. Comm. 44, 869 (1982)
21. Nattermann, T., J.Phys. C 16, 6407 (1983)
22. Binder, K., Z. Physik B 50, 343 (1983)
23. Nattermann, T., Z. Physik B 54, 247 (1984)
24. Aharony, A., Imry, Y., and Ma, S.-K., Phys.Rev.Lett. 37, 1367 (1976)
25. Nattermann, T., J.Phys. C 16, 4116 (1983)
26. Nattermann, T., phys.stat.sol. (b) 119, 209 (1983)
27. Feigel'man, M.V. and Lyuksutov, I.F., Solid St. Comm. 48, 397 (1983);
Lyuksutov, I.F., Pis'ma ZhETF 38, 165 (1983)
28. Fernandez, J. et al., Phys.Rev.Lett. 51, 203 (1983)
29. Grinstein, G., Proceedings of the 29th Conference on Magnetism and Magnetic Materials, Pittsburgh 1983
30. Villain, J., preprint 1983
31. Krey, U., Z. Physik, 1984 (in press)
32. Aharony, A. and Pytte, E., Phys. Rev. B 27, 5872 (1983)

Springer Series in Synergetics

Editor: **H. Haken**

Volume 23
Synergetics of the Brain
Proceedings of the International Symposium on Synergetics at Schloß Elmau, Bavaria, May 2–7, 1983
Editors: **E. Başar, H. Flohr, H. Haken, A. J. Mandell**
1983. 192 figures. VIII, 377 pages. ISBN 3-540-12960-X

Volume 20
H. Haken
Advanced Synergetics
Instability Hierarchies of Self-Organizing Systems and Devices
1983. 105 figures. XV, 356 pages. ISBN 3-540-12162-5

Volume 17
Evolution of Order and Chaos
in Physics, Chemistry, and Biology
Proceedings of the International Symposium on Synergetics at Schloß Elmau, Bavaria, April 26–May 1, 1982
Editor: **H. Haken**
1982. 189 figures. VIII, 287 pages. ISBN 3-540-11904-3

Volume 16
L. A. Blumenfeld
Physics of Bioenergetic Processes
Volume Editor: H. Haken
1983. 30 figures. X, 132 pages. ISBN 3 540-11417-3

Volume 15
W. Horsthemke, R. Lefever
Noise-Induced Transitions
Theory and Applications in Physics, Chemistry, and Biology
1984. 56 figures. XV, 318 pages. ISBN 3-540-11359-2

Volume 14
W. Weidlich, G. Haag
Concepts and Models of a Quantitative Sociology
The Dynamics of Interacting Populations
1983. 71 figures. XII, 217 pages. ISBN 3-540-11358-4

Volume 13
C. W. Gardiner
Handbook of Stochastic Methods
for Physics, Chemistry and the Natural Sciences
1983. 29 figures. XIX, 442 pages. ISBN 3-540-11357-6

Volume 12
Nonlinear Phenomena in Chemical Dynamics
Proceedings of an International Conference, Bordeaux, France, September 7–11, 1981
Editors: **C. Vidal, A. Pacault**
1981. 124 figures. X, 280 pages. ISBN 3-540-11294-4

Volume 11
Chaos and Order in Nature
Proceedings of the International Symposium on Synergetics at Schloß Elmau, Bavaria, April 27–May 2, 1981
Editor: **H. Haken**
1981. 134 figures. VIII, 275 pages. ISBN 3-540-11101-8

Volume 10
Y. L. Klimontovich
The Kinetic Theory of Electromagnetic Processes
Translated from the Russian by A. Dobroslavsky
1983. XI, 364 pages. ISBN 3-540-11458-0

Volume 9
Numerical Methods in the Study of Critical Phenomena
Proceedings of a Colloquium, Carry-le-Rouet, France, June 2–4, 1980
Editors: **J. Della Dora, J. Demongeot, B. Lacolle**
1981. 83 figures. IX, 269 pages. ISBN 3-540-11009-7

Volume 8
Stochastic Nonlinear Systems
in Physics, Chemistry, and Biology
Proceedings of the Workshop, Bielefeld, Federal Republic of Germany, October 5–11, 1980
Editors: **L. Arnold, R. Lefever**
1981. 48 figures. VIII, 237 pages. ISBN 3-540-10713-4

Volume 7
L. A. Blumenfeld
Problems of Biological Physics
1981. 38 figures. IX, 224 pages. ISBN 3-540-10401-1

Springer-Verlag
Berlin
Heidelberg
New York
Tokyo

Lecture Notes in Physics

Vol. 173: Stochastic Processes in Quantum Theory and Statistical Physics. Proceedings, 1981. Edited by S. Albeverio, Ph. Combe, and M. Sirugue-Collin. VIII, 337 pages. 1982.

Vol. 174: A. Kadić, D.G.B. Edelen, A Gauge Theory of Dislocations and Disclinations. VII, 290 pages. 1983.

Vol. 175: Defect Complexes in Semiconductor Structures. Proceedings, 1982. Edited by J. Giber, F. Beleznay, J. C. Szép, and J. László. VI, 308 pages. 1983.

Vol. 176: Gauge Theory and Gravitation. Proceedings, 1982. Edited by K. Kikkawa, N. Nakanishi, and H. Nariai. X, 316 pages. 1983.

Vol. 177: Application of High Magnetic Fields in Semiconductor Physics. Proceedings, 1982. Edited by G. Landwehr. XII, 552 pages. 1983.

Vol. 178: Detectors in Heavy-Ion Reactions. Proceedings, 1982. Edited by W. von Oertzen. VIII, 258 pages. 1983.

Vol. 179: Dynamical Systems and Chaos. Proceedings, 1982. Edited by L. Garrido. XIV, 298 pages. 1983.

Vol. 180: Group Theoretical Methods in Physics. Proceedings, 1982. Edited by M. Serdaroğlu and E. Ínönü. XI, 569 pages. 1983.

Vol. 181: Gauge Theories of the Eighties. Proceedings, 1982. Edited by R. Raitio and J. Lindfors. V, 644 pages. 1983.

Vol. 182: Laser Physics. Proceedings, 1983. Edited by J.D. Harvey and D.F. Walls. V, 263 pages. 1983.

Vol. 183: J.D. Gunton, M. Droz, Introduction to the Theory of Metastable and Unstable States. VI, 140 pages. 1983.

Vol. 184: Stochastic Processes – Formalism and Applications. Proceedings, 1982. Edited by G.S. Agarwal and S. Dattagupta. VI, 324 pages. 1983.

Vol. 185: H.N. Shirer, R. Wells, Mathematical Structure of the Singularities at the Transitions between Steady States in Hydrodynamic Systems. XI, 276 pages. 1983.

Vol. 186: Critical Phenomena. Proceedings, 1982. Edited by F.J.W. Hahne. VII, 353 pages. 1983.

Vol. 187: Density Functional Theory. Edited by J. Keller and J.L. Gázquez. V, 301 pages. 1983.

Vol. 188: A.P. Balachandran, G. Marmo, B.-S. Skagerstam, A. Stern, Gauge Symmetries and Fibre Bundles. IV, 140 pages. 1983.

Vol. 189: Nonlinear Phenomena. Proceedings, 1982. Edited by K.B. Wolf. XII, 453 pages. 1983.

Vol. 190: K. Kraus, States, Effects, and Operations. Edited by A. Böhm, J.W. Dollard and W.H. Wootters. IX, 151 pages. 1983.

Vol. 191: Photon Photon Collisions. Proceedings, 1983. Edited by Ch. Berger. V, 417 pages. 1983.

Vol. 192: Heidelberg Colloquium on Spin Glasses. Proceedings, 1983. Edited by J.L. van Hemmen and I. Morgenstern. VII, 356 pages. 1983.

Vol. 193: Cool Stars, Stellar Systems, and the Sun. Proceedings, 1983. Edited by S.L. Balliunas and L. Hartmann. VII, 364 pages. 1984.

Vol. 194: P. Pascual, R. Tarrach, QCD: Renormalization for the Practitioner. V, 277 pages. 1984.

Vol. 195: Trends and Applications of Pure Mathematics to Mechanics. Proceedings, 1983. Edited by P.G. Ciarlet and M. Roseau. V, 422 pages. 1984.

Vol. 196: WOPPLOT 83. Parallel Processing: Logic, Organization and Technology. Proceedings, 1983. Edited by J. Becker and I. Eisele. V, 189 pages. 1984.

Vol. 197: Quarks and Nuclear Structure. Proceedings, 1983. Edited by K. Bleuler. VIII, 414 pages. 1984.

Vol. 198: Recent Progress in Many-Body Theories. Proceedings, 1983. Edited by H. Kümmel and M. L. Ristig. IX, 422 pages. 1984.

Vol. 199: Recent Developments in Nonequilibrium Thermodynamics. Proceedings, 1983. Edited by J. Casas-Vázquez, D. Jou and G. Lebon. XIII, 485 pages. 1984.

Vol. 200: H.D. Zeh, Die Physik der Zeitrichtung. V, 86 Seiten. 1984.

Vol. 201: Group Theoretical Methods in Physics. Proceedings, 1983. Edited by G. Denardo, G. Ghirardi and T. Weber. XXXVII, 518 pages. 1984.

Vol. 203: C. Marchioro, M. Pulvirenti, Vortex Methods in Two-Dimensional Fluid Dynamics. III, 137 pages. 1984.

Vol. 204: Y. Waseda, Novel Application of Anomalous (Resonance) X-Ray Scattering for Structural Characterization of Disordered Materials. VI, 183 pages. 1984.

Vol. 205: Solutions of Einstein's Equations: Techniques and Results. Proceedings, 1983. Edited by C. Hoenselaers and W. Dietz. VI, 439 pages. 1984.

Vol. 206: Static Critical Phenomena in Inhomogeneous Systems. Edited by A. Pękalski and J. Sznajd. Proceedings, 1984. VIII, 358 pages. 1984.

Selected Issues from

Lecture Notes in Mathematics

Vol. 909: Numerical Analysis. Proceedings, 1981. Edited by J.P. Hennart. VII, 247 pages. 1982.

Vol. 912: Numerical Analysis. Proceedings, 1981. Edited by G. A. Watson. XIII, 245 pages. 1982.

Vol. 920: Séminaire de Probabilités XVI, 1980/81. Proceedings. Edité par J. Azéma et M. Yor. V, 622 pages. 1982.

Vol. 921: Séminaire de Probabilités XVI, 1980–81 Supplément: Géométrie Différentielle Stochastique. Proceedings. Edité par J. Azéma et M. Yor. III, 285 pages. 1982.

Vol. 922: B. Dacorogna, Weak Continuity and Weak Lower Semicontinuity of Non-Linear Functionals. V, 120 pages. 1982.

Vol. 923: Functional Analysis in Markov Processes. Proceedings, 1981. Edited by M. Fukushima. V, 307 pages. 1982.

Vol. 926: Geometric Techniques in Gauge Theories. Proceedings, 1981. Edited by R. Martini and E.M.de Jager. IX 219 pages. 1982.

Vol. 927: Y. Z. Flicker, The Trace Formula and Base Change for GL (3). XII, 204 pages. 1982.

Vol. 928: Probability Measures on Groups. Proceedings 1981. Edited by H. Heyer. X, 477 pages. 1982.

Vol. 929: Ecole d'Eté de Probabilités de Saint-Flour X – 1980. Proceedings, 1980. Edited by P.L. Hennequin. X, 313 pages. 1982.

Vol. 930: P. Berthelot, L. Breen, et W. Messing, Théorie de Dieudonné Cristalline II. XI, 261 pages. 1982.

Vol. 931: D.M. Arnold, Finite Rank Torsion Free Abelian Groups and Rings. VII, 191 pages. 1982.

Vol. 932: Analytic Theory of Continued Fractions. Proceedings, 1981. Edited by W.B. Jones, W.J. Thron, and H. Waadeland. VI, 240 pages. 1982.

Vol. 934: M. Sakai, Quadrature Domains. IV, 133 pages. 1982.

Vol. 935: R. Sot, Simple Morphisms in Algebraic Geometry. IV, 146 pages. 1982.

Vol. 936: S.M. Khaleelulla, Counterexamples in Topological Vector Spaces. XXI, 179 pages. 1982.

Vol. 937: E. Combet, Intégrales Exponentielles. VIII, 114 pages. 1982.

Vol. 938: Number Theory. Proceedings, 1981. Edited by K. Alladi. IX, 177 pages. 1982.

Vol. 942: Theory and Applications of Singular Perturbations. Proceedings, 1981. Edited by W. Eckhaus and E.M. de Jager. V, 363 pages. 1982.

Vol. 953: Iterative Solution of Nonlinear Systems of Equations. Proceedings, 1982. Edited by R. Ansorge, Th. Meis, and W. Törnig. VII, 202 pages. 1982.

Vol. 956: Group Actions and Vector Fields. Proceedings, 1981. Edited by J.B. Carrell. V, 144 pages. 1982.

Vol. 957: Differential Equations. Proceedings, 1981. Edited by D.G. de Figueiredo. VIII, 301 pages. 1982.

Vol. 963: R. Nottrot, Optimal Processes on Manifolds. VI, 124 pages. 1982.

Vol. 964: Ordinary and Partial Differential Equations. Proceedings, 1982. Edited by W.N. Everitt and B.D. Sleeman. XVIII, 726 pages. 1982.

Vol. 968: Numerical Integration of Differential Equations and Large Linear Systems. Proceedings, 1980. Edited by J. Hinze. VI, 412 pages. 1982.

Vol. 970: Twistor Geometry and Non-Linear Systems. Proceedings, 1980. Edited by H.-D. Doebner and T.D. Palev. V, 216 pages. 1982.

Vol. 972: Nonlinear Filtering and Stochastic Control. Proceedings, 1981. Edited by S.K. Mitter and A. Moro. VIII, 297 pages. 1983.

Vol. 978: J. Ławrynowicz, J. Krzyż, Quasiconformal Mappings in the Plane. VI, 177 pages. 1983.

Vol. 979: Mathematical Theories of Optimization. Proceedings, 1981. Edited by J.P. Cecconi and T. Zolezzi. V, 268 pages. 1983.

Vol. 982: Stability Problems for Stochastic Models. Proceedings, 1982. Edited by V.V. Kalashnikov and V.M. Zolotarev. XVII, 295 pages. 1983.

Vol. 989: A.B. Mingarelli, Volterra-Stieltjes Integral Equations and Generalized Ordinary Differential Expressions. XIV, 318 pages. 1983.

Vol. 994: J.-L. Journé, Calderón-Zygmund Operators, Pseudo-Differential Operators and the Cauchy Integral of Calderón. VI, 129 pages. 1983.

Vol. 999: C. Preston, Iterates of Maps on an Interval. VII, 205 pages. 1983.

Vol. 1000: H. Hopf, Differential Geometry in the Large, VII, 184 pages. 1983.

Vol. 1003: J. Schmets, Spaces of Vector-Valued Continuous Functions. VI, 117 pages. 1983.

Vol. 1005: Numerical Methods. Proceedings, 1982. Edited by V. Pereyra and A. Reinoza. V, 296 pages. 1983.

Vol. 1007: Geometric Dynamics. Proceedings, 1981. Edited by J. Palis Jr. IX, 827 pages. 1983.

Vol. 1015: Equations différentielles et systèmes de Pfaff dans le champ complexe – II. Seminar. Edited by R. Gérard et J.P. Ramis. V, 411 pages. 1983.

Vol. 1021: Probability Theory and Mathematical Statistics. Proceedings, 1982. Edited by K. Itô and J.V. Prokhorov. VIII, 747 pages. 1983.

Vol. 1031: Dynamics and Processes. Proceedings, 1981. Edited by Ph. Blanchard and L. Streit. IX, 213 pages. 1983.

Vol. 1032: Ordinary Differential Equations and Operators. Proceedings, 1982. Edited by W. N. Everitt and R. T. Lewis. XV, 521 pages. 1983.

Vol. 1035: The Mathematics and Physics of Disordered Media. Proceedings, 1983. Edited by B.D. Hughes and B.W. Ninham. VII, 432 pages. 1983.

Vol. 1037: Non-linear Partial Differential Operators and Quantization Procedures. Proceedings, 1981. Edited by S.I. Andersson and H.-D. Doebner. VII, 334 pages. 1983.

Vol. 1041: Lie Group Representations II. Proceedings 1982–1983. Edited by R. Herb, S. Kudla, R. Lipsman and J. Rosenberg. IX, 340 pages. 1984.

Vol. 1045: Differential Geometry. Proceedings, 1982. Edited by A.M. Naveira. VIII, 194 pages. 1984.

Vol. 1047: Fluid Dynamics. Seminar, 1982. Edited by H. Beirão da Veiga. VII, 193 pages. 1984.

Vol. 1048: Kinetic Theories and the Boltzmann Equation. Seminar, 1981. Edited by C. Cercignani. VII, 248 pages. 1984.

Vol. 1049: B. Iochum, Cônes autopolaires et algèbres de Jordan. VI, 247 pages. 1984.

Vol. 1054: V. Thomée, Galerkin Finite Element Methods for Parabolic Problems. VII, 237 pages. 1984.

Vol. 1055: Quantum Probability and Applications to the Quantum Theory of Irreversible Processes. Proceedings, 1982. Edited by L. Accardi, A. Frigerio and V. Gorini. VI, 411 pages. 1984.

GPSR Compliance

The European Union's (EU) General Product Safety Regulation (GPSR) is a set of rules that requires consumer products to be safe and our obligations to ensure this.

If you have any concerns about our products, you can contact us on ProductSafety@springernature.com

In case Publisher is established outside the EU, the EU authorized representative is:

Springer Nature Customer Service Center GmbH
Europaplatz 3
69115 Heidelberg, Germany

Batch number: 10838062

Printed by Printforce, the Netherlands